Praise for *The Independent Fa*

"I love the way the Doughertys capture both the soaring majesty and the down-and-dirty reality of farm life. . . . This book encourages and inspires all of us, no matter if we're just wannabes or if we're old hands. Thoughtfully portrayed experience never goes out of style, and this book is one you'll want to refer to for years to come."

—From the foreword by
Joel Salatin, Polyface, Inc.

"In *The Independent Farmstead*, Shawn and Beth Dougherty have rooted a wealth of practical and useful farming information in the fertile soil of social and economic reality and timeless ecological wisdom. Their farm is a grass-based homestead, and their personal story is compelling, but their insights are important for beginning or experienced farmers of any type or scale who don't yet know—or have forgotten—what real farming is about."

—John Ikerd, professor emeritus
of agricultural economics, University of Missouri

"Playing off its title, this informative, companionable book could be called *The* Interdependent *Farmstead*: It notes how a successful operation relies on interactions among animals, soil, grass, sunlight, and community as well as human ingenuity and, invariably, humor. The book's wisdom is that building on these synergies helps one realize the potential of any given piece of land."

—Judith D. Schwartz, author of
Cows Save the Planet and *Water In Plain Sight*

"Shawn and Beth cover a broad range of topics in this readable and user-friendly book. They manage to touch on most of the essential information a small-scale farmer needs to graze a cow and make cheese, feed the waste milk to the pigs and make bacon, and practice sustainable land use and animal husbandry along the way."

—Sarah Flack, author of
The Art and Science of Grazing

"With grace and grit, Shawn and Beth show you how to cultivate and care for an often overlooked but integrally important part of our food chain—grass—as well as the diverse livestock that transform pasture into the most wholesome foods on earth."

—David Asher, author of
The Art of Natural Cheesemaking

The Independent Farmstead

The Independent Farmstead

Growing Soil, Biodiversity, and Nutrient-Dense Food with Grassfed Animals and Intensive Pasture Management

SHAWN and BETH DOUGHERTY

Foreword by JOEL SALATIN

Chelsea Green Publishing
White River Junction, Vermont

Project Manager: Alexander Bullett
Developmental Editor: Michael Metivier
Copy Editor: Eric Raetz
Proofreader: Eileen M. Clawson
Indexer: Linda Hallinger
Designer: Melissa Jacobson
Page Composition: Abrah Griggs

Printed in the United States of America.
First printing August, 2016
10 9 8 7 6 5 23 24 25

Our Commitment to Green Publishing
Chelsea Green sees publishing as a tool for cultural change and ecological stewardship. We strive to align our book manufacturing practices with our editorial mission and to reduce the impact of our business enterprise in the environment. We print our books using vegetable-based inks whenever possible. This book may cost slightly more because it was printed on paper from responsibly managed forests, and we hope you'll agree that it's worth it. *The Independent Farmstead* was printed on paper supplied by Versa that is certified by the Forest Stewardship Council.

Library of Congress Cataloging-in-Publication Data
Names: Dougherty, Shawn, 1960- author. | Dougherty, Beth, 1963- author.
Title: The independent farmstead : growing soil, biodiversity, and nutrient-dense food with grassfed animals and intensive pasture management / Shawn and Beth Dougherty.
Other titles: Growing soil, biodiversity, and nutrient-dense food with grassfed animals and intensive pasture management
Description: White River Junction, Vermont : Chelsea Green Publishing, 2016. | Includes index.
Identifiers: LCCN 2016017978| ISBN 9781603586221 (pbk.) | ISBN 9781603586238 (ebook)
Subjects: LCSH: Rotational grazing. | Permaculture. | Organic farming.
Classification: LCC SB199 .D68 2016 | DDC 631.5/8--dc23
LC record available at https://lccn.loc.gov/2016017978

Chelsea Green Publishing
White River Junction, Vermont, USA
London, UK
www.chelseagreen.com

For our parents, who showed us the path,
and our children, who travel it with us.

Contents

Foreword

On the way to speak at a conference recently, my host asked, "How much of your self-esteem comes from speaking and traveling?" In a discreet way, he was probing how much satisfaction I still derived from the farm after a half century. Without any hesitation or batting an eye, I shot back, "None."

My earliest childhood dreams and vocational aspirations centered on farming. Nothing gives me more contentment than to rise at daybreak, step out on the back porch, inhale deeply, survey the homestead surroundings, and know deep within my soul that I am nestled—cradled even—in a womb of abundance. And it's a womb I can caress, can encourage, can bring out more abundance and strength from than if it were left in a static state.

In a world where most people feel disunion with nature's pulse, on our farm I enjoy a daily embrace from this visceral representation of God's provision and sufficiency. In turn, I have the honor and privilege of embracing back, like "I love you back," in a participatory environmentalism wherein I am partner, not pillager. I am daily friend, not occasional visitor. Good farming and good food systems should synergize nature's desires: clean and abundant soil, air, and water. Most modern farming systems compete with these interests; they don't complement nature's desires.

Ultimately, on our farm, we don't grow food to feed people; we grow food as part of our land-healing ministry. We're glad that it also tastes good and feeds us, but ultimately our farming is about restorative capacity, balanced eco-health. Shawn and Beth Dougherty share a similar vision on their farm, as they explain in a beautiful story of family, community, partnering with nature, and life in *The Independent Farmstead*.

The local food tsunami sweeping American culture includes a hefty homestead movement. Homesteading is about as primal as it gets. Driven by preppers, self-reliance, the call to commune with nature, Dilbert cubicle burnout, and a host of other yearnings, the desire to own a safe haven property to feed, clothe, and shelter ourselves with is a pilgrimage to "life, liberty, and the pursuit of happiness." That the average urban westerner spends much of his spare time playing fantasy video games, living in a world of entertainment, or getting away from it all speaks to the soul-yearning pulse of the human heart to connect with life. Something beyond the commute, the office, the packaged groceries, and the credit card.

The leap to a homestead dances in the minds of many folks as the ultimate antidote for the rat race, the treadmill. I remember well the day I made that leap—although it was a much smaller leap than most. September 24, 1982, the day I left my town job, left the steady paycheck, left company health insurance, and returned to the family farmstead. With a wife and infant son, the leap appeared foolhardy to everyone—and they told me so in no uncertain terms.

But with a wife more frugal than I, a piece of raw land my parents had worked off-farm to procure, my dad's vision, a well-read stack of *Organic Gardening and Farming* and *Mother Earth News* magazines plus iconic chemical-free agriculture books, our little family settled into a rich but cash-poor life. We never looked back and never regretted the decision. Only in very recent years have we ever driven a vehicle manufactured in the same decade. I think I've been to a movie theater about six times in my life. We still don't have a TV. Who needs to eat out when the best food on the planet is in the freezer and cellar?

People addicted to celebrity culture, consumerism, processed food, fashion, and recreation have no idea what they're missing. My grandchildren's friends think our farmstead is the coolest place in the world. Our ponds brim with bass and bluegills, not to mention egret dives and barn swallow theatrics. Spiders, butterflies, and toads occupy little ones for hours. A well-run homestead is a place of awe, wonder, and mystery.

And yes, what makes it so is the farmer. This is not a park or wilderness area. It's actually teaming with more life, building soil more rapidly, sequestering carbon more efficiently due to the day-to-day, systematic, mindful injection of human innovation, work, and mechanical leverage. The notion that wild places are more natural, or more ecological, than a well-managed farmstead is poppycock. Since I came alongside our ecological womb as farmer many years ago, it supports far more wildlife, holds more water, has deeper and more fertile soil, and grows far more vegetation than it ever did—even before Europeans arrived.

A working, productive farm can be ecologically balanced and beneficial. But such a farm never strays far from ancient patterns. These are fairly simple. First, all functional ecosystems have animals—lots of animals. Second, grass is more efficient at converting sunbeams into biomass than trees. Third, animals move. Fourth, perennials occupy far more land than annuals, which are almost an after-thought. Fifth, carbon builds soil; chemical fertilizer does not. Sixth, food systems are local-centric primarily, with global commerce relegated to the exotic or luxurious. Seventh, feeding ourselves is the cornerstone of civilization.

Using these basic ideas as templates for their own homestead, Shawn and Beth Dougherty take us by the hand and lead us down their discovery path transforming a piece of land deemed unfit for farming by the state of Ohio, to one brimming with abundance by the hard work of their hands. I can vouch for its truth because every day is not perfect. I've slept with chickens, pulled calves from troubled cows in the middle of the night, waded through flood waters to rescue laying hens.

Who needs Disneyworld when you have these things to pump your adrenaline? Who needs fantasy games when you're matching wits with a sly fox or recalcitrant pig? I love the way the Doughertys capture both the soaring majesty and the down-and-dirty reality of farm life. One afternoon's heart-stopping gorgeous setting sun casting vibrant oranges and yellows off verdant pastures is replaced the next with torrential rain and devastating winds that rip flapping pieces of barn roof off and soak the nicest second-cutting hay stashed underneath.

I'm impressed by the shear breadth and eclecticism of Shawn and Beth's farmstead, which of course is why people come to them from all over to learn, see, and participate. That they can now reach more people with this book is tribute both to their teachers' hearts and to the functionality of their operation that they can devote this kind of time and attention to documenting their journey.

This book encourages and inspires all of us, no matter if we're just wannabes or if we're old hands. Thoughtfully portrayed experience never goes out of style, and this book is one you'll want to refer to for years to come. Thank you, Shawn and Beth, for letting us peek into your life window—it's a wonderful view.

—Joel Salatin
Polyface Farm, Swoope, Virginia
April 2016

INTRODUCTION

The One Cow Revolution

For some of us it is the fast pace of modern life, the peripatetic dash from place to place; for others, it is the invasiveness of the ever-present digital age, the perpetual noise and demand of incoming information. Some feel a concern for our planet, its soils, oceans, forests, climate—for the future of this Earth that we hold in stewardship. We long for simplicity, beauty, community, harmony. And of course there is the desire for clean, affordable food—unmodified, unprocessed, and unmedicated—and the security of local food sourcing, for ourselves and our children. It sparks in us an urgent desire.

To move to the country. Go back to the Simple Life. To raise our own food.

Many of us know people who have tried it. Lots of us work or go to church with someone who, a few years ago, got on an idealistic bandwagon, jacked up his or her family, sold the city house, and moved to the country. They were going to grow a big garden, keep chickens and maybe rabbits, raise lots of healthy food, and be a sort of modern Walton family. It never lasts. The end is always the same: after a year or two, they sell the country place and move back to town. It was fun, kind of, they say; it was fun, at first, they qualify. But it was a long drive to work and to the good city schools (of course we weren't going to lose the advantages of the good city schools); and after a hot afternoon at the soccer field it was tiring to come home to chickens that had to be fed and a garden badly in need of weeding. And the fuel necessary for the long commute blew the gas budget.

And frankly, when you came right down to it, raising your own food was just too expensive.

Yes, expensive! Take gardening. Just a few bucks' worth of seeds and some weeding, you think, and there it will be, your bounteous vegetable garden, overflowing with tomatoes ripe on the vine, cucumbers dying to jump into your salad, fresh sweet corn, watermelon, bok choy, parsnips. Step outside with your plate and fork and your free dinner is waiting. Only, it isn't like that. First of all, to break ground for a garden you have to have a tractor, plow, and rototiller and the fuel to run them. Seeds are expensive, and anyway unless you remember to start them indoors in March you have to buy plants, not seeds, for, say, fifty cents apiece. Before you can harvest even one tomato you may already have hundreds of dollars in your garden.

And the work is incredible. I mean, what would you rather do on a Saturday morning, hit the river with your Jet Ski, or go out in the garden and hoe weeds? Huge weeds, like something out of *Little Shop of Horrors*. You pull weeds for hours, and then when you want to fix lunch you have to go back out and pick stuff and bring it in and wash it off. People forget that gardens mean dirt, and I mean *dirt*. Dirt on your potatoes when you harvest them. Bugs on beans, and who knows if their feet were clean? And then those vegetables have to be peeled, seeded, and stemmed, and by the time you've done all that who wants to wait for them to cook? They're delicious, I don't deny that, the best you ever tasted, but who has time for million-dollar, three-months-in-the-growing, sweat-dripping-down-your-face delicious vegs?

And what about the bugs? Big fat green caterpillars with spikes on their tails that 'doze through your tomato patch stripping the leaves from the branches and leaving a trail of little green droppings like mini alfalfa bales. Armies of gray squash bugs marching on your zucchini and making a bad smell when disturbed. Swarms of tiny whiteflies that suck the juices from your spinach leaves, and platoons of striped Mexican bean beetles and their fat, fuzzy larvae like tiny yellow hedgehogs, turning a lush green bean patch into a forest of brown leaves and bare stems. How is anyone supposed to compete for food with such a creepy bunch? Not to mention the depredations of nematodes or fungi, or the plant illnesses resultant from less-than-perfect soil conditions. Tomatoes standing tall and loaded with fruit turn black and rot in three days with late blight; lush bean vines refuse to produce pods due to inadequate lime in the soil. Crowded cornstalks bear only a few small ears, and without good air flow cucumbers are stricken with mildew. After all our hard work, what have we to show for it?

Raising animals is lots harder than you think, too. Take chickens: two dozen hens should have provided our family with more eggs than we could ever eat, bursting with omega-3s or what have you, enough for us and some left over to sell. First we put three hundred dollars into materials for a henhouse because the cute one we saw online that we really wanted was over two thousand bucks. The one we built was pretty rustic, but it might have been okay. Then two dozen chicks cost seventy-five dollars at the feed store, and how were we to know that half of them would be roosters? By the time the little guys started crowing they were five weeks old and there were no more to be had at the store, so we ordered some female chicks to be sent from a hatchery. That was fun because Jenny and the girls got to choose lots of fancy varieties, but by the time they came the first lot was half-grown and they pecked the little ones, not that it mattered that much because the chicken wire we put in the poultry house windows wasn't raccoon-proof, and the third time one got in it killed all but four. After that we put in stronger wire, and the second batch of chicks did all right, but fancy hens don't lay that many eggs, and anyway winter came and they quit altogether. The next year they picked up again, but they went through sacks and sacks of feed, especially since we were also feeding hordes of rats and mice that were attracted by the chickens' grain. Then our oldest daughter made the varsity team in soccer, and she was the one who really liked the chickens, so since no one else wanted to do the chores we sold them on craigslist. Later we figured out we had spent about fifteen dollars for each dozen eggs we collected. The thing was a bust.

It just costs too much to raise your own food.

And so it goes. The end is always the same: our friends sell the place in the country and move back to town, a little proud of their experience and the hardships they have endured, a little sorry to say goodbye to Poco the pet pony (but not sorry they'll never clean her stall again), and ready with the benefit of their experience to save the next guy from making the same mistake: "Forget it. It costs too much. You just can't farm anymore."

Now this is an odd statement, when you come to think about it, because as few as seventy-five years ago a quarter of the population of the United States lived on, and made its living from, a small farm. Our grandparents were among these, people who grew their own food and a few cash crops and did other jobs—plowed the county roads, helped neighbors butcher—to make a little cash so they could buy what they couldn't grow. They didn't get rich, but they lived a long time, enjoyed hearty good health, and they ate really well.

Their children, however, took the fast track to the city and a university education. We, their grandchildren, knew the old homestead as a mecca visited all too seldom and all too briefly, a place where bobcats haunted dusty pine woods strung with spiderwebs, and white-faced cows stood chewing cuds and swishing tails beside ponds the color of clay tile. The small, thin-floored houses smelled magically of sulphur matches and stove gas, sweaty water pipes, dust, talcum powder, and divine cooking. A dappled pony was kept especially for grandchildren, on which we were set three at a time to hold on as well as we could while being led tamely around the yard, at which we shivered with the visceral terror of the city person encountering a Large Animal. To leave the farm at the end of a visit and go back to the city was to mourn with prematurely mature mourning, the soul-wrenching sorrow of mortals evicted from Paradise.

But eventually the Old Folks got older. With all the kids in the city, there was no one left at home to help with the work, and in time the demands of the farm just got to be too much. "We can't do this anymore," they said to one another. "You can't do this anymore," their children assured them. And we, their grandchildren, heard, "It can't be done."

Perhaps it was in this way that the myth first arose, the myth that says *you can't farm anymore*. It's an interesting myth, as myths go, because it is one that is brand-new with the present age. It has never been told before, could never have been told because until this very moment in time, it would have been preposterous. People would have had simply to look around them to see that it was not true. Man's existence has depended on the small farm for thousands of years, years in which we have not only survived, but developed civilization, mechanization, industrialization, digitization. Unlike other classic myths—that of the Great Flood, or the God-King, or the Virgin Birth, all of which persist in practically every culture in one form or another, myths about discrete events in a long-ago past—this myth is about Now. Unlike those other myths, in which Nature opens a window for a once-in-all-time abrogation of one of Her laws, this myth tells us that all Her laws are abrogated for all time, starting Yesterday. This myth says that *the land will no longer yield food to the laborer*.

And our modern experience, as far as it goes, confirms this. No one we know farms; at most, a few people keep pots of chemically boosted tomatoes on the patio. Food, our experience tells us, grows in factories, or is mined with tractors as big as McMansions, or extruded in polystyrene packaging from machines the size of football fields. We of the modern age, surrounded by stores filled with food that appears never to have grown anywhere, cut off from even a single unadulterated contact with this Earth that teems with plant and animal life in intimate, balanced, resonant relationship, experiencing reality only in its most limited and man-made forms—we accept the Myth because nothing in our experience contradicts it.

Sure, we know vaguely that somewhere there are big places where lots of hamburgers walk around and eat corn—or is it straw?—before they are sent to McDonald's. We have seen pictures of pristine warehouses where jacketed attendants hover over long lines of sparkling cages filled with gleaming

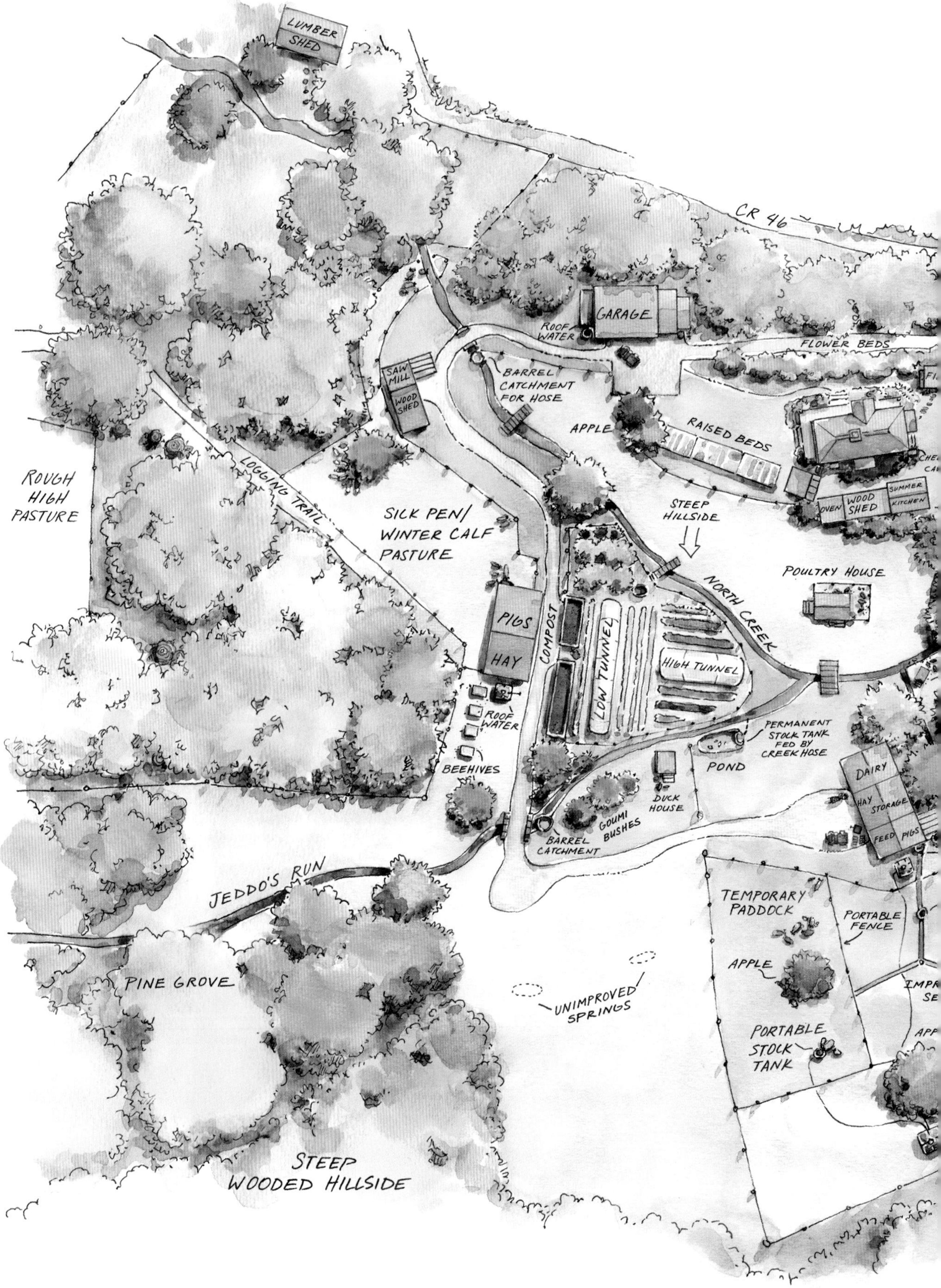

LUMBER SHED
CR 46
GARAGE
ROOF WATER
FLOWER BEDS
SAW MILL
WOOD SHED
BARREL CATCHMENT FOR HOSE
APPLE
RAISED BEDS
ROUGH HIGH PASTURE
LOGGING TRAIL
SICK PEN/ WINTER CALF PASTURE
OVEN
WOOD SHED
SUMMER KITCHEN
STEEP HILLSIDE
NORTH CREEK
POULTRY HOUSE
PIGS
HAY
COMPOST
LOW TUNNEL
HIGH TUNNEL
ROOF WATER
BEEHIVES
PERMANENT STOCK TANK FED BY CREEK HOSE
POND
DUCK HOUSE
GOUMI BUSHES
BARREL CATCHMENT
DAIRY
HAY STORAGE
FEED
PIGS
JEDDO'S RUN
TEMPORARY PADDOCK
PORTABLE FENCE
APPLE
PINE GROVE
UNIMPROVED SPRINGS
PORTABLE STOCK TANK
STEEP WOODED HILLSIDE

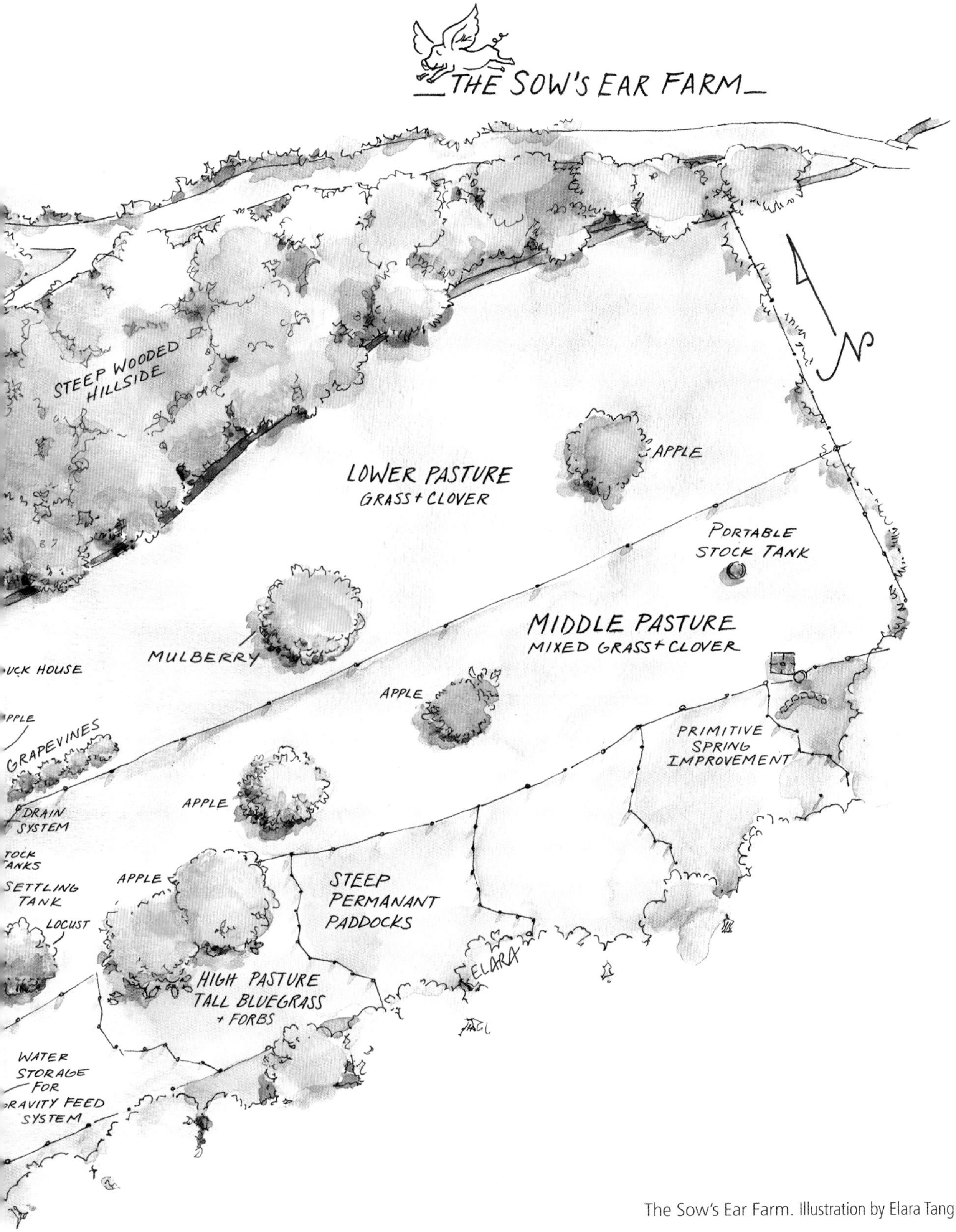

The Sow's Ear Farm. Illustration by Elara Tanguy.

white chickens laying laboratory-clean eggs, probably right into the cartons. Bell peppers materialize in cellophane bags on pretty bushes over thousands of acres in California. We don't know anyone whose day includes getting dirt under his fingernails, carrying feed to lots of animals, watching the weather with anxious attention, or getting up at night to check on a sick animal. None of our friends has ever milked a cow; many have never even seen a cow, not up close enough to be sure it really *was* a cow. Food doesn't come from farms anymore; food, as long as it is in the stores whenever we go Hunting and Gathering, is a fact of life, like air or television; a thing to be accepted, not questioned. The cute little farm with dairy cows and red barns is a thing of the past.

Or is it?

A Different Model

When we bought the Sow's Ear in 1996 and began the process of turning it into a family smallholding, we followed the usual path of neophyte homesteaders: we put in a garden (several gardens), bought chickens, acquired goats. We picked up how-to books on animal husbandry and organic vegetable growing; we ate lots of tomatoes, collected eggs from our flock of brown leghorns, drank goat's milk. It was fun, and our diets underwent a significant improvement; but we began to be conscious of a vague unease. Was what we were doing really farming? Something told us, as we lugged sack after sack of laying mash and sweet feed from the station wagon to the barn, that this importation of concentrated nutrients—many of them genetically modified—was *not* farming, not as we remembered our grandparents doing it.

We looked at the farms around us, larger hobby or commercial spreads, and wondered some more. The scale on which they operated was considerable, their "product"—as the farmer calls it—turning over with regularity, their cash flow presumably healthy. But were they really the islands of security they appeared to be? We considered one 450-cow dairy farm from which we bought baby bulls to raise for beef. What would happen to this place if its inputs were interrupted, even for one day? Milking machines and refrigerator tanks would cease to operate, silage augers and conveyor belts would stop running. Without petroleum fuels to power the equipment, the barns could not be cleaned, feed could not be trucked in. Even water might be cut off. The whole place would come grinding to a halt, and if something didn't happen quickly to restore its inputs, in a short time the animals would sicken and die. Was *this* really farming? We had our doubts.

And as our own husbandry projects broadened, we saw more ways in which our efforts, and those of the larger farms around us, differed from farms of the past. We began raising pigs, which meant more sacks of petroleum-produced concentrated feeds. The pork was delicious, but the price was disheartening. Yet we were following the instructions in our how-to books to the letter! Dissatisfaction led to research. We extended our reading beyond the many how-to-do-it books that stocked the shelves of the farm store—books that gave instructions for a scaled-down version of the commercial methods, replete with charts and diagrams on the nutritional needs of the bovine, breeding and feeding schedules for pigs, and lists of inoculations to improve the performance of hens in confinement—and we began reading books by Joel Salatin, Eliot Coleman, Allan Savory, Greg Judy. And a light went on.

These books, and the people who wrote them, were making some fundamental assertions about farming, all of which were derivative of one basic assumption: that the sun is the source of bioenergy for this planet. Animals and plants—including humans—live because the sun shines. The success—commercial and ecological—of these writers' farming endeavors was attributable, directly or indirectly, to their recognition of this fact. Pastured meat chickens, all-grass beef, and four-season vegetable harvest are all constructions resting on one constant: the successful farmer is the one who makes the best use of his sunlight. Reading these authors, we knew for certain that the redistribution of petro-produced, livestock-converted nutrients into which we had entered with such gusto was not, as we had suspected, real farming. The sun, rather than our expensive feeds of questionable provenance, should be the power source for our farm. But how?

The answer was *grass*: the permaculture, present in enormous communities over 40 percent of the planet's land mass, which can collect our quotidian solar energy—the sunlight falling on our farm every day—so that cows, goats, sheep, pigs, and poultry can harvest their dinners. Even more: in the digestive system of a dairy animal, that sunlight can become protein, fat, and lactose, nutrients supremely available not only for human food, but in such quantities as to make it possible to supplement practically every animal and operation on the farm. The puzzle was coming together: grass, the solar collector; ruminants, the converters; joined by chickens and pigs as batteries, self-reproducing storage units of surplus solar energy. Here at last was the secret of Grandfather's farm, that Mecca of good food, strong, hard-working men and women, and unquestioned food security: grass management. Otherwise known as *intensive rotational grazing*.

We took our lesson out in the fields and put it to work, and we watched our little farm come to life. As our ruminants moved over the pasture in regular, measured rhythms, the bare spots began to fill in and rank weed growth succumbed to trampling, giving way to lower, more palatable forages. With impact spread more evenly over the pasture—and the calendar—the grazing season extended itself, beginning earlier in spring, extending into late fall and early winter, with increased drought-resistance to get us over the midsummer dry season. Poultry found more to eat, and better forage, in the mix of grass and forage species clothing the soil. Pigs preferred forage, fruit, and dairy surplus to a diet of dry factory pellets or crumbles.

And not only did we see an increase in the quantity of solar energy capture—grass, milk, meat, and eggs—their quality was improved exponentially. Grassfed beef and pork were leaner and better-flavored, grass-produced butterfat was yellow with beta-carotene, and on our improved pastures chicken egg yolks took on a deeper orange. Our purchase of inputs dropped dramatically, to such an extent that we no longer had to think our own farm hopelessly vulnerable to interrupted inputs, as our neighbors' farms were. We used little purchased feed and could go without if need be. The management-intensive grass harvest was improving our soil and pastures so that forage was of better quality, and greater quantity, for a longer season.

Prompted by the work of Eliot Coleman, we intensified our solar capture in the garden as well, adding new levels to our harvest. Succession planting and undercropping multiplied the size of our summer harvests, and adding fresh winter vegetables was as simple as some tunnels of 1-inch conduit and 6-mil plastic sheeting. Stumbling on Cornell University's online library of old farm books, the Albert R. Mann Core Historical Literature of Agriculture collection, we learned more about pre–World War II animal husbandry; as a consequence, when the human harvest was over, we filled empty

Disclaimer

Farming—modern industrial imitations notwithstanding—seems to us to be as much art as science, and to entail more commitment than either. Like marriage, it is affected by details that are peculiar to the individuals involved—people, animals, plants, time, and place—and, while similarities between farms are many, hard-and-fast rules are few. The best farmers we know or have read make the fewest claims for their knowledge: "I don't really know how to do it right, I just do the best I can," as one veteran grazier in our area often says. Allan Savory, founder of the Savory Institute and father of Holistic Resource Management, avoids the term "teach" as applied to his many workshops, preferring words like "demonstrate," "share," and "equip."

We, with our few poor years of experience, cannot teach anyone how to grass farm, not "teach" in the sense we learned in school, like explaining algebra. Nature's math, in any case, isn't like elementary mathematics: where 1 + 1 = 2 in arithmetic, in nature it is just as likely to equal 3, when the first two are a bull and a cow; or one, if one of the addends is a sheep and the other a coyote. Nor are we here to tell you How We Succeed and So Can You; just follow these twelve easy steps. If and when we ourselves achieve a resonant balance between weather, grass, ruminants, pigs, chickens, forage, table, and garden, maybe we'll write another book. But probably not—should we be so fortunate as to have a year in which all our resources are in balance, we'll expect nature to spring something new on us in the next.

The longer we farm ourselves, the fewer flat statements we are willing to make, having seen many a curveball come over the plate, and anticipating even more in the years to come. What we hope to do instead is to share the information we've picked up so far in our intense immersion in All Things Grass. Think of this book, then, not as a set of directions, but as the beginning of a conversation, like setting out for a point, distant and inviting, with someone who has traveled the first steps and is familiar with the early landmarks. But while we won't set ourselves up as experts on anyone else's land, garden, or animals, there may be much good to be had from sharing experiences *now*—not when we have all the answers, but while we are still deeply pondering the questions.

spaces in the garden with beans, turnips, sugar beets, and mangelwurzels; undersowed our heritage corn with field peas; and planted other areas to wheat and barley—and the pigs and chickens loved it.

We have come to believe that the secret to the success of the small farm is the capture and harvest of as much as possible of the solar energy falling on that land—each day, week, month, and year—by grass and grazing ruminants. Our farming efforts are focused by the application of this principle, whereby the farm is not a staging ground for assembling nutrients, but a font from which nutrients originate, when basic elements are assembled in the leaves of green growing things, consumed by herbivores, and converted into generous quantities of high-quality proteins and fats. This is the theme of a flourishing biodiversity, each species of plant or animal feeding and being fed, assisting and receiving assistance from the rest, and all, ultimately, fed by the sun.

No longer are we without personal recourse from a food production system that implicates even the unwilling in a widespread destruction of ecosystems. We can ransom food and agriculture on a small, individual scale, shifting our personal nutrient consumption to fresh, whole, local, responsibly grown foods—food produced with today's sunlight instead of fossil fuels. The careful, attentive management of our solar energy harvest brings benefits on every level: in the health of our soil, our pasture plants, our animals, our family, our community. To bring this about takes not only understanding and cooperation, but a good measure of dogged determination, an irresistible urge to play in the dirt, and iron control over your gag reflex. And it's worth every bit of the effort.

CHAPTER ONE

The Farm

There are two spiritual dangers in not owning a farm. One is the danger of supposing that breakfast comes from the grocery, and the other that heat comes from the furnace.

—*Aldo Leopold,* A Sand County Almanac

Before choosing a piece of land, putting up a shed, or bringing home our first dairy goat, it will be helpful if we take a brief course in plant science, and a penetrating look at the methods and principles of modern farming, both industrial and ecological. Knowing something about how healthy natural systems work, and identifying principles that promote or undermine those systems, will be essential as we begin building a homestead—like having a guidebook describing the pitfalls and giving a blueprint for building an integrated structure. That barnyard full of colorful laying hens, and green pasture groomed by sleek dairy animals, will begin to take shape more quickly if we know what we're trying to achieve—and avoid.

Figure 1.1. Awkward or neglected parcels of land often offer a clean slate for beginning a sustainable homestead.

Photosynthesis 101

We learned it in grade school. Sunlight falling on the leaves of green plants powers the solar energy collection and storage system we call photosynthesis. Carbon dioxide from the atmosphere, and water taken up from the soil by the plants' roots, are disassembled and reconfigured into carbohydrates—starches and sugars—and oxygen. This provides a source of energy for the growing plant, which in turn feeds, directly or indirectly, all animal life. When plants and animals die, the energy in their tissues is returned to the soil by the process of decay, where much of the carbon will be stored and the nitrogen and minerals released for the growth of the next generation of plants and animals.

But on the standard industrial farm of today, photosynthesis happens without the vigorous biological commerce that builds natural fertility. In a monoculture, nutrients are not efficiently cycled from soil to plant to animal and back to soil; instead, huge concentrations of a narrow range of nutrients are produced and then taken from the farm and exchanged for stored energy, in the form of money. As a result, money, not soil fertility, has to power the next crop with further applications of chemical fertilizers and biocides. Meanwhile the soil, deprived of its perennial ground cover, lies bare to the forces of wind, rain, and oxidation, bleeding away what little nitrogen, topsoil, and microbial life remains. Sunlight, rather than being banked in the topsoil to feed us today, tomorrow, next year, and for years to come, is reduced merely to the motive force in an assembly line, producing cheap commodities for conversion to cash. Gone is the elegant complementarity of natural growth and fertility that exists in nature.

Carbon Commerce

But as it turns out, there is more to photosynthesis than we were taught in elementary school. Scientific understanding of biological systems has come a long way, and we have today a much more detailed picture of the elegant and intricate commerce in carbon that is carried on between green plants and the living soil. Understanding of this commerce, along with an appropriate and corresponding husbandry of green things and livestock, gives us the power to rebuild our land and to harvest, store, and convert solar energy on a scale that makes petrochemical fertility assistance look like the poor crutch it is.

As plants grow and convert sunlight, air, and water into plant tissue, the parts below the surface of the ground grow in correspondence to the aerial parts; in other words, the root system is a reflection of the plant mass above the ground. This vast web of cellulose holds the plant in the soil and keeps it upright, while at the same time collecting water and air from the spaces between soil particles. When sunlight falls on the leaves, liquid carbon-based molecules are built and pumped not only to the aerial parts of the plant to fuel its growth and functions, but down to the deepest roots as well. There, microscopic biota in the soil receive it, in exchange giving the plant minerals from the soil that they have converted to forms the plant roots can take up. At the same time, the microorganisms build compounds that bind soil particles together, creating more spaces for the air and water needed for root growth—promoting plant vigor, which in turn pumps more carbon to the roots. This hidden symbiosis is the basis of soil fertility.

Add grazing animals and you have an even bigger powerhouse. Grazing signals a dieback in the root system correspondent to the pruning of the plant's aerial parts. This dieback leaves root cellulose deep beneath the soil surface where it can feed insects, bacteria, and mycelia, which die in their turn and contribute the nitrogen and minerals stored in their bodies to the further fertility of the soil. Spaces are left in the soil by the decomposed roots to make more room for air, water, and new root growth. The resultant increase in fertility promotes further plant growth, which provides more forage for grazing. At the same time, ruminants contribute to the soil the partially digested cellulose and teeming biological life in their manure; more topsoil is created, more fertility stored, more carbon-mineral commerce sparked between plant roots and mycelia, and so on, round and round. In other words, an ongoing net increase in topsoil, fertility, and available energy—all of which, if we farm in imitation of nature, means more food for human beings, too.

Extractive Farming

The problem is that we no longer have ready access to people and traditions that could teach us how to engage in a healthy, fruitful carbon commerce. Books and magazine articles draw their information from the industrial model, so that, deriving our goals from a milieu dedicated to industrial notions of productivity and efficiency, we employ—unknowingly, unwillingly—methods, tools, and plant and animal species inimical to the robust ecology we are trying to promote. Paradigms lifted from commercial agriculture lead us inevitably into the very imbalances we are trying to avoid. Without guides or guidebooks, we need principles by which to distinguish farming practices inconsistent with and destructive of the rich and vigorous ecology of the grass farm, earmarks not only to alert us so we can avoid dead-end roads, but so we can identify by contrast the methods that may be employed to build the sort of farm we are looking for. The industrial farm has many consistent characteristics.

First, *the modern farm is monocultural*, in excess even of what might be expected. Most crop farms today raise a single commodity, with corn and soybeans heading the list. Animal operations are similar: most raise a single species of livestock—often

Figure 1.2. A mixed herd of cows and sheep reap a generous solar harvest.

a single sex, generation, and genotype. Since these animals' value in the market is defined by a single product, all unrelated animal behaviors and propensities are suppressed as far as possible. Hence laying hens, desirable only for the greatest number of eggs that can be squeezed out of them in their short lives, are crammed into tiny cages where eating and egg-laying are virtually the only possible activities, and flapping, scratching in the dirt, and catching bugs are strictly out. Hybrid broiler chickens, designed for ultrarapid growth, may develop crooked leg bones and necrotic muscle mass if their lives are more than minimally active. Pigs are bacon and sausage, and get about as much exercise as if they were already converted into those forms.

Modern farms are also noteworthy for their *extremely large scale*—thousands of acres, barns the size of several football fields, and behemoth tractors. And since big machines have big price tags, in order to make the payments on half-million dollar tractors and implements, they must be used on big crops. Big crops, in their turn, place a big drain on local resources like water. They have to have big markets, and big trucks and fuel bills to get them there. But don't be misled: except for a few of the biggest farms, what *isn't* big are the net profits. The return on investment in farming at anything other than enormous scale is almost universally dismal; according to the USDA, most farms' operating costs exceed farm earnings, and the majority of farmers have to pay their bills from the proceeds of nonfarm jobs. Farmers as a class aren't greedy; if anything, the reverse. The majority of farm families work an off-farm job to pay for the privilege of growing food that doesn't pay them back in sales.

Farming is no longer done in alliance with natural systems. Well, after all, natural systems are so messy, and hardly what modern industry would call efficient. Aside from eating, growing, and reproducing, most natural animal behaviors are to be overcome or avoided. Seasonal variations, for example, must not be allowed to interrupt the steady flow of commodities into the market, so dairy cows must calve year-round, not just in the spring when nature intends them to. At least this doesn't upset the bull, since with artificial insemination there isn't one—or rather, there is *only* one, or a few, living far away from the herds of dairy cows needing to be bred and getting their manly satisfaction out of mating with a sperm trap instead of a cow. The ejaculated semen is then injected by the milliliter into plastic tubes, frozen in liquid nitrogen, and shipped all over the country to be placed in the mother cows by artificial insemination. The resultant thousands of offspring drawing their genetic material from just a few bulls is a situation now creating problems for farmers in search of genetic diversity.

Modern farming uses a minimum of human labor and human-animal interaction. At the scale of farming today, it would take hundreds of people to do all the farm work by hand or with animal traction, and naturally, if your profit margin is very narrow, you can't afford to hire a lot of help. Besides, what do you think all those machines are for? Mechanized agricultural systems are standardized for greatest efficiency; individual attention within the system, and temporary deviations from the established standard, are expensive and time-consuming. Hence, little allowance can be made for human creativity, regional variation, or individual problem solving.

Industrial farms are dependent on off-farm inputs. Efficiency according to the industrial model is incompatible with the complications of vegetable reproduction, animal life cycles, and natural maternity. Genetically engineered seed can't be saved from year to year but must be purchased annually—or you face a lawsuit. Chicks are shipped in from the hatchery, feeder pigs from the sow operation, semen in nitrogen tanks. In the field, dependence on mechanization means dependence on fossil fuels, and, with the resultant, unavoidable destruction of topsoil, natural fertility and pest resistance are replaced with fertilizers, pesticides, and herbicides.

Big farming is not local. How can an industry rooted, literally, in the soil, be anything *but* local? Although crop farming still happens in conjunction with local soils, and confinement farms are of necessity stationary, big farming is, nevertheless, *not* a local industry. Seeds, chemicals, and equipment come from big national or international corporations. Industrial farming does not raise food for local consumption; the bulk milk tanker and those tractor-trailer rigs loaded with grain, soy, or animals on the way to market aren't making deliveries at the local grocery store. In fact, most farming does not even put food on the farmer's own table, except indirectly; the vast majority of farmers buy their food from the store, same as you do. Farming does not generally use local money, either; loans and mortgages are held by big banks, while federal subsidies come from Washington—or, rather, from taxpayers all over the country. Nor does most of the money made in farming stay local. A great many American farm operations do only custom work; that is, the animals, feeds, seeds, fertilizer—and profits—actually belong to a corporation, while the farmer is contracted to do the work of growing the crop. Most profits end up in the pockets of big national or international ag corporations; in general, all that stays local in the end is a little money, a big mortgage, and toxic quantities of manure.

In fact, the net local effect of industrial farming is Loss. Loss of soil, as annual commodity crops cost us multiple bushels of topsoil for each bushel of crop harvested. Loss of fertility, organic matter, and stored carbon; loss of natural drought and flood resistance. Loss of biological diversity, as miles of land are cleared and plowed and planted to monocultures; loss of habitat for local wildlife. And who knows what invaluable natural services are lost as well, what beneficial interactions of great complexity eliminated, which might make the difference between ecological resilience and ecological fragility? The natural world suffers by this neglect of stewardship and, in conjunction with the natural order, local economies lose independence; local communities lose neighbors and neighborliness. Farming, according to this model, is as extractive as mountaintop-removal coal mining.

Regenerative Farming on the Independent Farmstead

Happily, large, monocultural farming-for-profit methods have little to do with the resilient homestead. There are principles of fecundity with which we may cooperate, to our own benefit and the benefit of the natural world of which we are just one part; local characteristics—animal and plant communities and behaviors, topography, watershed, day length, and climatic exposure—which may be allied to human observation and planning and careful, precise labor, to benefit the productivity and health of every aspect of local life, from the soil upward—and downward. In other words, we want to develop systems characterized by an approach opposite to extraction, with opposite emphases, opposite values: natural instead of mechanical, local rather than global, small rather than large, dependent on human labor and planning, exhilarant in diversity. We might call it *regenerative farming*.

Figure 1.3. In a mixed-species herd, animal requirements can be balanced and complementary.

The independent farmstead will be characterized by values integral to this regenerative principle. It will, first of all, be *diverse* and *integrated*; that is, it will naturally produce as many as possible of its own necessities. "Naturally" because that is how nature works: local, daily sunlight, as opposed to stored petrochemical sunlight, is, after all, all any natural system has with which to fuel its community of plants, plant-eating animals, predators and scavengers, human beings. Since the life of the system will thrive or decline with its plant life, the more diverse the plant community, the more diverse and balanced its withdrawals and deposits in the environment, the more sustainable its life as a whole, and the more numerous and diverse the animal life it will support. Dependence on imported energy and nutrients will be as unnecessary as it is undesirable: with species' impacts being myriad and dispersed, gaps that form in the food web close spontaneously. Likewise, life generates more life; resilience means that individuals have not only the ability to survive stresses, but also the life span and health necessary for generously adequate reproduction, thus guaranteeing the extension of the community over time.

Correspondent to its diversity of species, the diverse nature of regenerative farming will be seen in *its multiple levels of productivity*. Where one-crop farming derives its strengths from monetary exchange, regenerative farming, like nature herself, achieves flex with built-in redundancy, backup systems, and alternate sources. Sheep, grazing with cows, utilize plants not favored by bovines; poultry follow both and scavenge undigested seeds and pest larvae from their manure. Pigs forage at the bottom of the food chain, converting almost any kind of surplus nutrients into solid pig flesh and nitrogen-rich manure. In the garden, potatoes are backed up by the corn crop, beets with turnips. This stacking of species and enterprise ensures ample and varied nutritional sources for plant, animal, soil—and human—health.

Figure 1.4. Stacked enterprise.

Farming of this kind cannot be done from the cab of a tractor, painting in the broad strokes of combine and chemical sprayer; hence, the scope of a regenerative farm is determined by human limitations; it is by definition *human scale*. The level of observation necessary for judgment and decision making in regenerative farming requires time, proximity, intimacy, and the use of every one of our senses. Only so far as the farmer can see, smell, hear, touch, and taste, so far as legs can walk and so large as human-scale tools can reach, will he or she be able to make intelligent decisions, to take appropriate action. Such detailed attention to natural relationships means that regenerative farming will be noteworthy for its *limited use of technology*.

As a result, *human observation, judgment, and intuition are irreplaceable.* Even very small farms consist of multiple microecosystems, pockets of diverse plant species, areas more protected or exposed to sunlight and weather, each with its own set of requirements, each bringing different advantages to the overall community. No digital spreadsheet or computerized management program can take account of the subtleties in plant and animal condition, states of grass and forage, weather implications, and the flux of human time and labor available, as can an attentive, interested human mind; no machine can have the intuition necessary for evaluating varying natural conditions, nor share our intimate dependence on the soil for our every need. Human beings play a vital role in regenerative farming: in planning, timing, and organization; regulating food supplies for seasonal availability; protecting weaker species from domination by the stronger; harvesting, processing, and storing nutrients for increased food value or season extension; diverting rainfall from points of accumulation to points of deficiency; overseeing reproduction.

Where factory farming operates by importing animals, plants, and nutrients, converting them into forms with higher cash values and exporting them, *the primary inputs of the regenerative system are sunlight, air, and rainfall.* Properly managed diverse plant communities, in symbiosis with the animal communities they support, require little beyond their quotidian meteorological inputs to provide all the energy needed for health and growth—and even, often, some extra to spare. Farm needs are supplied through the interaction of farm species. Remember our photosynthesis lessons: herbivores consume sun-derived plant growth; omnivores feed on plants and plant-eating animals or on their products, such as milk and meat; and the soil is renewed and rebuilt through these processes from both above and below the surface.

In regenerative farming, *natural and farm ecosystems are seen as part of an interdependent whole*, not a collection of resources to be extracted and processed. The complexity of natural relationships and the compound mysteries of their successes—in which individual species fill many niches simultaneously and no part of the system, down to the last raindrop or tiniest grain of pollen, is without significance—incites respect, wonder, careful stewardship. In the face of such complexity, the regenerative farmer sees as her first responsibility the preservation of the sustainability of the community, defined simultaneously as the preservation and propagation of plant species, annual and perennial; the well-being of the animals; the renewed fertility of the soil; and the health and well-being of the farmers—with no part of the farm, whether directly benefiting the humans or not, allowed to compromise the existence of any of the others.

For the regenerative farm, the land and its indwelling life are the primary resources on which all living things depend, and, consequently, the first

Figure 1.5. Captured surface water irrigates without impacting groundwater or energy resources.

object of the farmer is the sustained health and fertility of the soil. It goes without saying that this kind of farming is rooted in one place, that it is essentially *local*. All other values—even the production of goods for home use or for sale off the farm—only come under consideration *after* the basic needs of the soil are met. The regenerative model must by definition be a viable, sustainable system. Only when we are generous with the farm ecosystem it is able in turn to be generous with us, filling to abundance our need for nutritious and delicious food, for health and fertility, for beauty, interest, elegance and simplicity *here*, in our chosen place.

Approaches to Resource Assessment

For methods to differ so widely, primary assumptions must themselves be in contrast. Thus, where extractive farming takes as the first question about any living thing to be, "What does it produce that can be sold for profit?" regenerative agriculture asks, "What are the characteristics of this organism, and where does it fit into the natural community as a whole?" Is it plant or animal; herbivore, carnivore, or omnivore? How big is it, and what are its physical requirements? How does it interact with the land—does it graze, root, scratch, trample? Does it eat other animals? Which of its needs is it capable of supplying for itself, and which may be supplied by other farm species, either wild or domestic? Which must be met by the farmer and which, if any, must come from off the farm? What benefits does this organism bring to the farm ecosystem, and how may it provide for the needs of other species? What liabilities arise from its presence—in the form of consumption or waste, destructive or aggressive behaviors, even legal restrictions—and can these liabilities be turned into assets?

Assets, in this case, are defined by value *to the farm*. Benefits and products of the farm community only leave if they are produced in excess of its needs. In other words, only surplus may be sold, "surplus" being calculated according to nutritional value, health enhancement, and increase of resilience to the farm, *not* as cash value in the marketplace. Thus, a gallon of milk that might bring ten or twelve dollars on the raw-milk black market may be considered even more valuable if it remains on the farm as an organic, high-protein superfood for a couple of feeder piglets, whose daily ration of commercial feed might cost only a couple of dollars but may include ingredients, known and unknown, of less than desirable impact on the pigs in particular and the farm in general. A couple of tons of hay in excess of what the livestock will require for winter may be of more benefit if fed on-farm to a few additional animals—in which case the minerals, nitrogen, and carbon they contain will be returned to the soil, with interest—than if sold for cash, with the resultant removal of those nutrients from the farm ecosystem.

Ecology of Scale

Small, as E. F. Schumacher reminds us, is beautiful; and intensive management of land for increased diversity and fertility means managing on the small, even the micro, scale. We read that there are as many as 1 billion microbes in a teaspoon of topsoil; an acre of land, then, is an empire for the intensive gardener. Most of us aren't in a position to buy a big piece of land anyway, but the same principles that make for a successful biological community on the scale of 20, 40, 60, or 100 acres can be applied even in a backyard. In many places, a ¼-acre garden can, under intensive management, provide all the fruits and vegetables for a family of six, year-round; two dozen chickens in movable tractors can apply spot cleanup and high-nitrogen fertilizer, increasing the fertility of the garden while providing the gardener with eggs and meat. Where rainfall is adequate, ¾ acre of grass may provide grazing for at least a couple of sheep or goats, building fertility and drought resistance while fueling a generous harvest of milk and even some meat. Five such acres will, in many climates, support a dairy cow, providing even more milk, for humans, poultry, even a pig, and, yearly, a calf to build the herd or fill the freezer. The owner or manager of just a small piece of land has the means of starting a new movement, beginning a revolution, founding a new country of carbon-wise, biologically sound, for-now-and-for-the-future fertility.

This isn't just an attractive theory. In various situations and climates, all over this country and the world, people are beginning to retake possession of the soil under their feet in just such ways, initiating biological processes and designing plant and animal communities founded on the principles nature herself uses; storing carbon and water in each swale and keyline, taking advantage of a tiny protected microclimate here, a south-facing incline there. New tools are being developed and marketed; methods of husbandry are being tested and refined. We are recovering lost information in volumes and sharing it in cyberspace, the best of old and new getting together for a small-farm barn dance. Hobbyists and farmers exchange their experiences with various animal and plant species, good ideas circling the globe at light speed. The consequence is that the person who is serious about becoming his own farmer has a wealth of tools at his fingertips.

The Steadying Hand

Wendell Berry, who has already said the best of just about all there is to say about responsible tenancy of the earth, points out that nature deals with abundance by a tip of the seesaw, spilling off excess of life on one side or the other, matching demand to supply, and so maintaining the overall balance. Human beings, dependent on that balance, may, with their powers of judgment, put a hand on the seesaw to moderate the spillage, to see that what might otherwise be wasted does not go unused. As in untamed nature, where life depends on solar energy and it takes many plant and animal species to reap and store the harvest, what is necessary to building healthy microfarms is diversity: animals and plants living synergistically, energy cascading from one life-form to another, abundance in one sector triggering new utilizations in another, death being just one of many manifestations of energy availability, and health—by definition—meaning the continuance of the community and the minimization of waste.

This is just as true on the small, sustainable farm of 5 acres as on 100 acres—and in some ways a lot easier to manage. Individual attention goes a long way toward preventing waste or nutrient buildups. Observation directs the grazier where to put his animals and dictates what number and species they should be. Is the green growth on this few acres grass, brush, or woody shrubs? Cows graze, sheep prefer coarser fare, goats browse thorny bushes—all three eat some of everything. Twenty rotationally grazed chickens can patrol a 1-acre homestead, eating pest insects before they migrate to the garden, salvaging undigested seeds from ruminant manures, scraping out parasite larvae before they can reinfect the host animal, and spreading valuable nitrogen-rich manure to fertilize the soil; on the other hand, they can also, without adequate oversight, rip out your garden, roost over the hay rack and defecate on the winter's stored forage, leave piles of poop on the front porch, and perch where they are easy pickings for raccoon, possum, or fox.

Figure 1.6. Intensively managed pigs store surplus forage as ham-on-the-hoof.

Likewise, two pigs under management can root out perennial weeds like Queen Anne's lace and the wild alliums that make your spring milk taste so rank it's only fit for pigs anyway, thus improving your pasture and making organic, grass-fed pork in the bargain. Unmatched storage units for surplus perishable foodstuffs, pigs may be raised on a forage diet supplemented with garden, kitchen, and dairy wastes, to minimize or eliminate questionable food sources like genetically modified corn and soy. Left too long in one paddock, however, they can also plow up 400 square feet of good pasture in about twenty minutes, and maybe ground their fence while they're at it so they can escape to dig up the garden, too.

Ours is the steadying hand on the seesaw. Management is paramount; the watchful eye and considering mind are indispensable.

Enrolling in the School of Nature

For those who are willing to roll up their sleeves and get dirty, to sweat and fail and try again, Nature herself is willing, even anxious to teach. Many's the prostrate corpse of a zucchini squash over which we have mourned, when we failed to pick off squash bugs assiduously, not to mention the hybrid broiler chickens that died in two hours of heatstroke because the calf that shared their pasture overturned their drinking water reservoir. Rescued baby dairy bulls transmit germs like lightning and die quickly until the farmer learns to keep them from sucking one another's ears and tails. Nature is not a gentle teacher, but she's consistent and impartial—reasonable, if unpredictable. For those who are determined to learn, Nature will give unlimited tuition, so long as failure is never an excuse for giving up. And she's a one-on-one instructor, ever at our elbow, every student receiving full and undivided tuition.

So start your own school. Beg, steal, or borrow a bit of land, never mind what condition it is in right now. Begin small, but diverse. Study local natural ecosystems so you can grow plants where they want to grow, where their wild cousins are thriving. Integrate animal species, ruminants grazing, poultry scratching and scrapping, hogs foraging. Let them produce offspring and breed to improve the strain. Encourage native forage species; learn to scavenge and conserve wild food sources. Use local energy: human power, animal power, wood heat, gravity. Minimize imported energy. Let symbiosis and natural animal behaviors replace artificial fertilizers and the myriad biocides. Reserve cash for the big things; relying on hand tools and sweat equity allows the homesteader to save money for those projects where the application of a little judicious mechanical- and petro-energy gives the farm a quick quantum leap away from purchased inputs. Natural solutions produce unexpected and happy natural benefits; money solutions, on the other hand, encourage extractive farming methods in order to pay for them. Whenever possible, solve problems with native ingenuity: get the "natives"—plant, animal, climate—to do the work for you. And there are no inputs more local than the land's own share of sunshine, rainfall, and topsoil.

The basis on which such an edifice rests is *grass*. Grass is our primary renewable natural resource, the first, largest collector of solar energy, the heat and life that fuels our planet. Grass power is a constantly renewed source of wealth, with ruminants to harvest and convert it into milk, meat and fiber, fertility, and more ruminants; birds to spread manure, fertilize the grass, and clean up pest and parasite species; hogs to graze, plow, and root while serving as four-legged storage units for myriad surplus perishable nutrients. Mankind provides the management: the *oversight*, directing cows onto fresh forage, protecting smaller animals from predators, directing nutrient surpluses to hogs; the *foresight*, planning ahead for seasonal availability of forage and water, for timely and advantageous breeding and birthing; and the *insight*, perceiving hidden connections and understanding how all the parts integrate into a harmonious whole, adapting systems to changing circumstances.

The results? Food. The best you ever ate. Lots of it. And security: invest money in stocks and watch the market fall; invest fertility in the soil and watch everything grow. Resilience: the ability to flex with changes in climate, economy, and policy. And what value can be placed on beauty, family, community, and culture?

All from a few acres of grass.

CHAPTER TWO

The Land

Anything worth doing is worth doing badly.

—*Gilbert Keith Chesterton*

Convinced of the necessity, even inflamed with the desire, to take up responsibility for our consumption and our personal well-being, what are we to do? If we are not already owners of a piece of land where we can begin our personal homesteading school, where do we go? What are we looking for? Which of the variables of soil, terrain, and climate are most important to our future venture? And won't a piece of land where we can realize our vision for a fertile, secure, sane future be far too expensive for most of us?

Granted that these are all points for consideration, the issue is not nearly as complicated as it may seem. All the Earth is, well, *earth*; practically everyone who is not presently airborne or on a ship at sea is standing over some kind of soil. And after all, our present quest is not to find fertile land somewhere and then enjoy the benefits of it, as long as they last, but to *build* fertility, somewhere, now. Where you are standing is not a bad place to start. Urban gardeners—great in number, if tiny in percentage of overall population—are converting city and suburban yards, porches, and balconies from chemically regimented, monocultural grass carpets or bare concrete pads to rich, biologically exploding microecosystems that overflow with almost unbelievable amounts of food. And while a studio apartment in a teeming metropolis, or a 1/10-acre backyard with restrictive zoning, may not be the place to start a grazing operation of any size, the landscape is just crammed with plots of land, many forgotten, neglected, or abused, which could become our perfect homestead. Finding one is a matter of asking the right questions.

Figure 2.1. Neglected land includes fertility in potential.

Where Do You Want to Be?

The ubiquity of dirt of some sort means that mere *earth* is not our limiting factor. So consider some of the others. First of all, you are going to live in this place; make sure it's a place you want to live. What do you want around you? Family? Look for land near home. Are you strongly attached to certain cultural surroundings? Know the flexibility of your aesthetic sense. Is it a sense of community that is paramount to your happiness? Maybe you should look in places where the language and culture are familiar to you. Climate matters: if you are a heat-loving person you may find you don't enjoy turning out in subzero weather to water the livestock or sweep the snow off your winter garden high tunnels. Habitat is important: if you are very fond of picking your oranges and avocadoes straight from the tree, northerly growing systems may not be for you; if you are determined to raise Scottish Highland cattle, residence in the southern states is probably a no-go. And of course, affordability is usually an issue; if you just *have* to own the land you are going to work, you'll have to work land you can afford to own. In any case, for most of us the question of where to live is not one over which we have complete control. The ability to jack up and move to the spot that appeals to us most is just not in our toolbox.

What Are Our Options?

Whatever the area where you plan to settle, the first step in choosing a piece of land is to study its individual characteristics. What does this earth want to do—that is, what are its natural tendencies? This may be hard to determine; widespread urbanization and commodity cropping have ousted or disguised the qualities and preferences of our ecosystems. Except to the eyes of love, Houston looks much like Atlanta, Pittsburgh like St. Louis, and the enormous cornfields of Illinois are remarkably like those of Kansas. The innate, natural tendencies of an area may be hard, at first glance, to see. Research in this area may help. Vermont, now stripped of much of its topsoil, once exported grain on a large scale. Eastern Ohio, its topsoil and native grassland destroyed by surface mining in the 1960s and 1970s, once was a thriving center of the lamb and wool trade. It is important to have a rough idea, at least, of the inclinations of the land where you hope to reside. Your agricultural plans need to be a good match, because when humas fight to make Nature do what she doesn't want to do, Nature always wins in the end.

Then, look at real estate ads. Land is expensive; there aren't a lot of pretty little farms on the market. Unless you're loaded, you probably won't find homestead here, but you'll begin to get an idea of what is selling in your area, and how the market is doing. Land prices in general are probably prohibitive, but don't, at this point, get discouraged: remember, land is everywhere under our feet, and you aren't looking for what real estate agents mostly are selling. Cruise back roads; make note of areas where people seem to be doing a little farming. Not a *lot* of farming; land already in row crops or commercial hay production, flat cleared land accessible to giant tractors and cheek by jowl with big agriculture is going to be gaspingly expensive, almost certainly already saturated with chemicals, and requiring for productivity more doses at frequent intervals. Look, rather, for weedy, overgrown fields; neglected woodlots; awkward, out-of-the way corners and choppy terrain. Go to the courthouse and find maps; familiarize yourself, at least rudimentarily, with the region's topography and watersheds; these things are going to matter.

And talk to people: vendors at the farmers' market, people at the local extension office, the man at the feed store, rural delivery people. Ask about services and organizations for farmers in your area. Attend local ag events like farm days, ecotours, and pasture walks. Visit farms—pick-your-owns with tourist appeal, organic farms that offer tours, orchards and vineyards that host wine-tasting nights. Ask questions, and listen carefully to the answers. Interact with people who are different from you; step outside your comfort zone. Volunteer; not only will you get a lot more from a person who can see that you are really interested than from someone who thinks you're just idly curious, but the conversation is likely to go on a lot longer if you're standing beside her helping turn the compost.

In the meantime, there is a great deal that you can be doing so that when your land suddenly drops in your lap—or you in its—you'll have some idea of where to start work first. Collect tools, especially low-technology, multiple-application tools; especially secondhand and hence lower-price tools. Don't go out and buy yourself a brand-new, imported, walk-behind mower with all the attachments—at least not yet. Collect books. Books about grass, about alternative energy, about gardening. Books about people doing what you think you want to do. Absorb the language, flounder through the unfamiliar terms, grow accustomed to the ideas they contain. Read, read, read.

Figure 2.2. Dutch Belted cow and calf. Courtesy of Anissa Wellington.

And plan what you'll do when you find your land. What do you think you want to grow? What does your picture of fertility, sustainability, and durable tenure look like? Are you in love with a particular species or breed of livestock? Study it, visit people who already have that animal, and imagine how you'll do things the same—or differently. Consider possible companion species, for guard, fertility, symbiosis. Make a twenty-year plan—but in pencil; keep everything flexible.

Buying Land Organically

When we are looking to buy land, we don't necessarily think first of real estate agents. There are some good reasons for this omission. First of all is this: most people don't want land for the same reasons we do, and since they don't want what we want, a real estate agent is pretty unlikely to be peddling it. Most people who use a real estate agent are looking primarily for a house, and the land on which it lies is of much less importance; we are looking for land, and the house, if there is one, is only of secondary significance. In addition, real estate agents in general aren't going to know much about the agricultural possibilities of land they are selling—its conditions, uses, or restrictions—so it will be hard for them to help us look at places that might suit us, even on the chance that they have a few.

Instead, we look at the local want ads. Check the newspapers; try the classified magazines, cheap or free, that are often to be found next to the checkout counter at the gas station or grocery store. Try craigslist or its local equivalent, by all means. Look at bulletin boards in the local grocery and hardware stores. If you have a regional agricultural paper, certainly take a look there; while most of the ads are likely to be aimed at conventional farmers or at sportsmen in the market for a hunting lease, you may also find small, inconsiderable parcels that are worth looking at. At any rate, it helps to get to know the area; even if you have lived all your life in the vicinity of where you hope to build your farm, you will find the neighborhood still has lots of surprises for you, little pockets the existence of which you never suspected.

Drive the back roads. Get off the beaten path. When you see a neglected side road, turn into it. Stop the car and get out; better yet, do your searching on a bike, or walk. Investigate curiosities; talk to people. Look for land that looks as though nobody loves it, junk-strewn, overgrown with weeds or weed trees, tucked back in corners difficult to access. If there's no For Sale sign, go to the local courthouse and look the parcel up; it may be that you can track down the owners and find out more about it. We know people who found a farm in just this way.

Assessing Properties—The Details

Specifically, what characteristics are we looking for in a piece of land for our future grassfed homestead? When most of us say the word *farm*, it evokes an image drawn from all our past experiences of farms—or, more likely, our past experience of farms in books, magazines, and advertising. Old MacDonald might just about sum it up for most of us: white board fence, a red barn with a gambrel roof, green pastures, plowed garden, a quack-quack here and a moo-moo there. As it happens, if that's what we're looking for in our search for land, we probably won't find much; most places that were ever like that picture have been swallowed up by agribusiness. In areas that are amenable to row cropping, whatever arable land isn't already owned by a megafarm is on their acquisition list, and the per-acre price is going to be high. Buying a farm that is already in existence is beyond the means of most of us, and maybe that doesn't matter, because buying an existing farm means, in most cases, buying land that has been under conventional systems for some time, complete with all the chemical fertilizers and biocides that implies. Not the kind of land we necessarily need for our smallholding.

Once we have found a piece of land that meets our requirements for region, price, and immediate neighborhood, it's a good idea to get a piece of paper and start making notes of its natural qualities. What sort of land is it? Flat, gently rolling, steep, sheer? Is there topsoil? Is it clay, sandy, or loam soil; are there many rocks, and if so, how big, and of what kind? What is growing there right now: grass, brush, weeds, trees, nothing? Is the land wet or dry? Is there surface water, be it pond, stream, spring, or marsh? Look at the microclimate of the parcel, including the limits of the climate zone—average first and last frost day, summer high temperatures and winter lows, average rainfall and days of sun; also slope and aspect, windbreaks and reflected sunlight. All of these things are going to affect what we can plan for this piece of land.

After the natural qualities, we look at the improvements (by which we mean the human alterations, be they improvement in our eyes or otherwise). Are there outbuildings, and if so, what kind, how large, and how are they placed on the property? Are there fences? Wire, mesh, wood, or stone? How much, and where placed? Is there access to

Figure 2.3. Weedy, overgrown fields now, but green pasture-in-potential. Courtesy of the Franciscan Sisters T.O.R. of Penance of the Sorrowful Mother.

the power grid, municipal water, natural gas? Is there a well? How deep is it, and has the well a working pump? Has the water been tested for toxic substances, like arsenic? The local extension office should know what the common water impurities are for the area. What are the other available resources on the property, things that might give direction or focus to a grassfed operation? What are the prevalent plant species? Are there orchards, vineyards, small fruit or nut trees? Are there specialized improvements, such as a gas well, RV hookups, or riding trails?

And then there are the human influences on a piece of property, in some ways the most important because they are the least susceptible to our efforts to change them. Who are the neighbors? What are their places like—modest homes or trophy mansions? Does their land use appear to be agricultural, recreational, or decorative? Do they have animals themselves, or look like they might have had any in the past? Do they appear to be potential partners in our grassfed interests, with empty fields we might be able to borrow? Or might they be in some way obstacles to our intended land use, as, for example, a fruit tree nursery next to where we had pictured our goat farm, or a market garden right up against the land where we were imagining free-ranging a large poultry flock? Don't by any means neglect to consider the qualities of the immediate neighbors.

Maybe the piece of property we are looking at doesn't have a lot of amenities; maybe, when we have narrowed down the possible parcels to something we can afford, we aren't sure it's something we want. Improved land, fertile land, and flat land all have a pretty considerable market value; after all, that kind of land is just the place to build a McMansion. For places like that we are outbid before the auction starts; land within our fiscal reach may not look so inviting. So what? Our mission is *improving* land, not buying land that is perfect already. Not to worry—once we get started, the place will soon be unrecognizable.

Good Enough Is Perfect

There are some real advantages to buying neglected land. Every patch of soil, regardless of market worth, has one intangible value to which no price tag is attached: *fertility in potential.* Where there is soil of any sort, where there is plant life, where there are sunlight and rainfall, there is the potential to build generous fertility, and, because the potential is hidden from most people, there's no charge for it. A couple of farmers we know in western Oklahoma found just such a place about thirty years ago, a bare, drought-ridden 20 acres with two oak trees, a sand-plum hedge, and a tiny shanty home. The ground was red and sandy, a sea of mud when it was raining and a dust bath within twenty-four hours when the rain stopped; lovegrass and horse nettles were the predominant groundcover. The couple moved in and began the process of regeneration: heavy mulch, responsible grazing practices and cover crops, rainfall capture, permaculture. Today the place is unrecognizable, a green oasis overflowing with food and beauty in a sunburned sea of eroded wheat fields and drought-starved pastures.

There is an added value, for people who like what they like, in land that has been allowed to decline: that is, a clean slate to work on. It really depends on the individual: would you rather own something you have created, or the product of someone else's creativity? Do you prefer to buy off the shelf or to customize? Not that we ourselves would turn up our noses in principle at land improvements, *if* we didn't have to pay too high a price for them; but it often happens that what appears to be an improvement needs, before long, to be undone and done over before it can be considered an asset. For creative people the neglected, abandoned acres may be the most promising.

Figure 2.4. Unimproved land can be altered to fit the homesteader's individual needs.

And those little chopped-up pieces of land that are left when all the "good" pieces are taken can offer some possibilities that are missing in the plots of level, smooth, cleared, and accessible land that look more like what we think of when we say "farm." We noticed this when we bought our farm in Ohio. Where we grew up, in Texas and Oklahoma, land was mostly (or absolutely) flat, dry, and featureless. It was easy to access, easy to plow, and easy to fence. Our parcel in Ohio, by contrast, consisted of a narrow east–west draw with steep sides of clay and slate, weeping moisture in summer and covered with icicles in winter, mixed second-growth trees on the upper slopes, and nowhere anything at all that looked even remotely like our idea of pasture. But although the state of Ohio has designated it "not suitable for agriculture," we find that our mixed and off-level pieces of property offer a wide variety of valuable variations and microclimates. East–west valleys, even when they are deep, capture sun and heat in their south-facing slopes and hold it against cool nights and prevailing breezes. Rocky outcroppings soak up sunlight during the day, offering an opportunity for growing less hardy plants than are normal for the gardening zone. The small streams and ponds that are often found on land too uneven for conventional use can be utilized for stock water and irrigation, and a damp place where sedges grow may be a natural seep—clean, soil-filtered water forced to the surface by an impervious substratum. Wet spots and woodland edges offer habitat to insect-eating birds and amphibians.

And neglected land is, after all, generally less expensive; on poor land you won't be paying for someone else's idea of improvements. If you're loaded—if your bank account is far and beyond "all right" and you have money with which to buy just what you want, develop it, stock it, and then live off the interest of what's left—maybe you don't care.

On the other hand, maybe you do: many a mistake is made, many an avenue followed that later turns the wrong way, because a person could afford to do things quickly. Haste, in some cases, does make waste. In our little corner of the culture—the sustainable ag world—there is a movement toward appreciation of slow things—slow food, slow money. We would like to propose that there is value in slow *farming*.

There is a universal value realized whenever a piece of wasteland is reclaimed. Each square yard of soil that is stewarded and encouraged from the condition of barren, sour, and weed-choked wasteland, sluicing rain and topsoil into local waterways, absorbing solar energy in the form of heat and releasing it slowly to warm up the microclimate, into perennial plant cover shading and cooling the earth, building and holding topsoil, capturing rainwater, and releasing oxygen into the atmosphere, is one more step toward reclaiming this place for fertility and resilience.

And consider: our information, in this new-to-us area of farming, is limited; our experience even more so. Our sources of advice, in local farmers, veterinarians, and in the vast majority of how-to books, are themselves limited—almost universally espousing methods that are just those from which we are trying to escape, suspicious or even downright scornful of the notion that agriculture may be practiced in any way besides the one they learned at ag school, or the one their seedsman or tractor salesman tells them is the latest problem fixer. Supposing we could afford to go out right away and buy our vision of what a grass-based, ecologically diverse farm should be, would it be a good thing? Would "our vision," however well we have done our research, be sufficiently well informed and all-inclusive to make its realization the farm of our dreams? In our opinion, the answer is "no."

Figure 2.5. Lady's thumb, a nongrass forage species, is edible for humans, too. Courtesy of the Franciscan Sisters T.O.R. of Penance of the Sorrowful Mother.

Experience is, after all, the best teacher—and experience cannot be bought. It has to be lived—and lived with. Especially experience of the sort a farmer needs: experience of living systems, dynamic, fluctuating; experience of vagaries in climate, water availability, pest population, rest, and renewal. Experience of changes in ourselves, our predilections, preferences and ideals, and in the circumstances of our own and our family's lives. Designing the perfect barn and having a contractor put it up for us certainly gets the job done quickly, but what if we discover once we put some animals in it that the loose stalls are just a couple of feet too small for the breed we want to house, or the gate swings in exactly the wrong direction for when we want to load livestock into the trailer? Closer to home, how will it be if we stretch miles of goat fence to keep in the herd of dairy goats we intend to raise, only to switch to pastured pigs after a year or two and discover that goat fence and hog fence are two very different things? And this caveat has application in small places, too; the hundred-dollar cobra-head hoe we pick up at the sustainable ag fair may be beautiful, but its application in the large potato patch that turns out to be the best use of our garden—because potatoes will grow there, and because we'll eat them—is going to be limited; what we may really need is an eye hoe and a broad fork.

How Much Land Is Enough, and When Is It Too Much?

How much land we need depends on a lot of things, but the answer is likely to be *less than you think*. First we have to ask ourselves some questions.

How much land can we afford? Heavy mortgage payments are going to mean more time away from the farm, working at our cash jobs. Who is going to live here and farm this land? What kind of time, realistically, will they devote to farm work—to grass management in the form, usually, of moving fences and water tanks for rotationally grazed ruminants, pigs, poultry, or some combination thereof? How interested are they, and how willing? What are their ages, potentials, capabilities, and skills? What are their time commitments, and are they willing to change them? Are we hoping to raise animals for dairy? For meat? How many people do we want the land to feed—ourselves, our family, maybe (sometime in the future) a customer base?

How brittle is the environment—does it receive rain only seasonally, or is the precipitation spread more or less over the year? How effective is the rainfall—that is, how much soaks in, and how quickly does it dry out after a rain? In what sort of condition is the primary ground cover, be it grass, brush, or trees; is the coverage general or spotty; are the roots deep or shallow? We might want to look up the conventional carrying capacity in our area of an acre of pasture for grazing cows, sheep, or goats—but we want to bear in mind that intensively managed rotational grazing gives an enormous boost to the fertility and resilience of land, so with time we are going to improve on the conventional numbers. Further, are there woods, and are they mature or young, dense or scattered? Is there surface water that can be utilized for livestock?

Too much land is when you have more than you can steward, more than you can improve, more than you can take responsibility for. A good rule for the smallholder is that you should be able to walk over every part of your land in the course of your regular routine, at least once a week, and preferably several times. For grazing land, as for gardens, the best fertilizer is the shadow of the gardener or grazier; the best feed supplement for livestock is the hand of the herder. A real farm, a regenerative farm, is not a machine that can be fired up, placed on a productive setting, and then simply maintained with periodic injections of inputs. Holistic management requires regular, attentive observation, frequent decision making, and thoughtful adjustment. Finances will dictate how much land most of us can buy in any case; chances are we won't be worrying about acquiring too much.

And fortunately, one hallmark of holistic land management is its multiple levels of productivity; fewer acres produce more nutritional calories, and more fertility, than a greater acreage of conventionally managed land. Holistic grazing means that, with adequate available water and sufficient commitment in time, a family-scale, grass-based smallholding can fit in a very small footprint; three years of holistic grazing on even a very small area can so improve forage production and rainfall retention, and so extend your grazing

Rental or Borrowed Land

Don't overlook the possibilities of land you don't own. In our neighborhood unused land is everywhere, and the owner who can't be bothered to mow it may be very grateful to have someone use animals to graze it off. Strangers have come down our drive just to inquire whether we would run some animals on their property; our home pasture, ours by title now, was for many years borrowed from a neighbor who was tired of mowing the steep hillside with a string trimmer. Because electric fencing equipment is so portable, it is possible to take advantage of borrowed land without putting money into someone else's property. If there's power and water available, even solar exposure and some form of surface water, we can take advantage of unused forage on a seasonal basis at least; the landowner gets improved land—the steaks, and the experience, are ours.

season, that the land will carry more livestock for more of the year, or even allow the possibility of year-round grazing.

Farmers don't have to have title to all the land they utilize. For over fifteen years we have harvested hay from land belonging to a neighbor whose horses aren't sufficient to keep his pastures grazed. Much conventionally farmed land is rented; 90 of the acres we manage actually belong to a Franciscan monastery 1.5 miles up the road, where the religious sisters are not presently using the property themselves and are glad to share the land and its produce. Not only does this arrangement provide us with pasture for our small herd of beef cattle; the sisters are saved the cost of having the land mowed twice a year, or having the cleared land go back to brush and scrub. A secondary, but no less important, result is the sense of community these arrangements foster in our neighborhood. Our rotational grazing practices improve the land and grow enough milk, meat, and other produce to share with everyone.

What Qualities Are Necessary?

While we don't have to start with perfect pastures and expensive improvements, there *are* qualities the land has to have before our agricultural projects are possible. Number one, unquestionably, is water. It must be considered the homesteader's primary limiting factor, without which we could not even begin using rotational grazing to regenerate the land. But while this is a limitation, it's one that can be met in most places, one way or another; the trick is to meet it in the most sustainable way.

Water

Pressurized water—water from a municipal system or domestic well—is the form of water supply with which we are most familiar: turn a tap handle and there it is. We don't think too much about where it is coming from, or even how much we are using; for most of us, tap water is just a fact of life. But city water is expensive, and often chemically treated; well water is effectively nonrenewable—fossil water being pumped out of natural reservoirs deep underground by means of electrical power. And even shallow wells, while they are more readily recharged than deep wells, still require some sort of power to raise the water to ground level.

The regenerative farmer will look for other water sources for his farming endeavors, emphasizing especially various means of detaining rainfall on the land, of using it and giving it a longer period in which to percolate into the soil. Does the land have ponds or streams; does it border a river or lake? Are there low spots where rainwater collects after a storm? In hill country, like our part of Appalachia, springs are frequently to be found high on rocky hillsides, providing an opportunity for directing water to points where it is needed lower down. The rate of flow need not be considerable to collect sufficient water for a small number of livestock. Seeps—springs the flow of which is so slow that they don't create a rill, just a soggy place—can be captured and directed into a holding tank; even a mere drip provides tens of gallons a day. Elevated water sources are especially valuable; whenever possible, look for opportunities to use gravity, rather than electricity, to move water on the farm. Even a slight variation in the surface of the land can offer an opportunity to move water without a pump.

Figure 2.6. This primitive improvement of a seasonal spring provides hundreds of gallons of stock water every year.

Use local information. Ask the neighbors about the seasonality of any spring you find; they may know whether it tends to flow all year or only during wet seasons. Don't make the mistake of one of our local grass-based farmers, who put thousands of dollars into an extensive spring improvement program only to find that the water he intended to feed his series of descending tanks ran only from March through June. Ask at the Natural Resources Conservation Service (NRCS) office about local toxicities, and have the water tested; the biggest spring is useless for stock water or food crop irrigation if local mining activity has contaminated it with heavy metals.

Soil

After water comes soil. We place it second because, while we cannot do without at least *some* soil, even a little will do, and its condition at the time of purchase may be—in fact, most probably *will* be—poor. No matter; we are going to improve and build the soil, and by doing so we will improve and increase the water-retentive abilities of the land. An example we have seen are the Benedictine monks of Our Lady of Clear Creek Abbey in far eastern Oklahoma, building soil on 1,000 acres of undulating land where second-growth oak and maple trees cling to limestone outcroppings and the soil is in many places so thin the white bones of the hills show through. Well-planned rotational grazing, combined with cover cropping and fermented whey biofertilizer from their small creamery, is resulting in deeper topsoil on their rocky hilltops, and more forage for their flocks of sheep and their dairy cows. Not to make your own first efforts too difficult, however, you'll want to look for land with at least some depth of soil. It need not be *good* soil when you buy it: in his book *The Small-Scale Poultry Flock*, author Harvey Ussery includes before-and-after pictures of his rusty, sandy soil, improved to black loam with the aid of rotationally grazed chickens.

Thin, poor soil is one thing; *contaminated* soil is another. The means by which humans have polluted the surface of the earth are many, and not all of them leave signs that are readily visible. If the land has been mined or used commercially, it may be good to have the soil, like the water, tested for toxicities. Inquire locally if you are unsure of previous land use; some forms of abuse leave few signs. In our township, a piece of cleared creek bottomland with an east–west aspect—land apparently adventitious for homesteading—was owned, up to a couple of years ago, by a scrap dealer. Broken appliances from which he wanted to separate the scrap metal from the plastic parts he would haul to this plot, douse with gasoline, and light afire. The black smoke that billowed up on these occasions was enough to convince anyone that the local representatives of the EPA must be sleeping on the job; what toxins this practice deposited in the soil we can only imagine. We would be unwilling to eat anything that was grown or grazed on such land, supposing it could be induced to produce much at all.

One of the qualities of a piece of land that could be easily overlooked is *slope*. If the place we are looking at is not mesa-flat, we need to pay attention to the degree and angle of its inclines. These may be advantageous or otherwise: very steep slopes can make access to some parts of the property difficult for humans or animals, but a pasture on a south-facing slope warms and dries sooner in spring, while one with a northern aspect resists dormancy longer in the heat of the summer; an orchard facing north may blossom later than one in more direct spring sunlight, an advantage in the event of a late frost. Furthermore, it may be important in some areas to get a topographical map and investigate the slopes of our proposed purchase to find out who are our uphill neighbors; their land use or waste disposal practices can have devastating effects on the quality of our water.

Figure 2.7. Land with slope allows the farmer to put gravity at the service of the homestead.

Road Access

Access is a point on which land values pivot, and here you may have a considerable advantage over the generality of land-seekers, who mostly want something with a nice drive or with commercial access. Homesteaders can be more creative, but we must have access of *some* kind. How are we going to get onto this piece of land? This must be investigated carefully. Does the land have road frontage with permit to build a drive, and if so does the part of the property contiguous to the road allow vehicular access? Our own farm is cut in two by a well-maintained county road; nevertheless, the steepness of the hillside means that visitors to our farm hesitate, hearts in mouths, before they take the plunge down our precipitous lane. How will we bring in building materials, livestock, or any large movable? What about access for maintenance or emergency vehicles? Our neighbor to the south is without any road frontage at all, only a right-of-way over our property, so he depends on us to maintain the lane, which we do, but only as far as our first barn. From there the lane continues downhill, crosses North Creek and Jeddo's Run and climbs steeply up the south side of the valley, where the narrowness of the way prohibits access to any really large vehicle, including fire trucks—a fact which resulted, forty years ago, in an overfired woodstove burning his first house to the ground. Road access matters.

Future Uses

A great deal of trouble, frustration, and disappointment may be avoided by doing a little more prying, this time into the future of the neighborhood into which we are thinking of moving. Apply at the courthouse for information on the zoning in the area of your piece of property; find out especially what permits have been granted or applied for. The best piece of land may lose much of its value when the neighborhood goes bad. When a construction and demolition dump was permitted in a county next to ours, neighbors in the area were soon complaining of foul odors. Up to several miles away the air was almost constantly rank with the odor of decay—we spent an anniversary weekend at a bed-and-breakfast in the area, which lost some of its pleasure due to pervasive dumpster smells. Obviously not all the garbage coming in on boxcars from New Jersey was from construction and demolition, but none of the signees of that contract live within nose distance of the new dump, and so far no one has been able to do anything about the breach of permit. In a similar case, friends upriver, purchasing 10 beautiful hilltop acres with an old farmhouse and a bank barn, found, within months of moving in, that the local power plant was constructing a 5-mile long conveyor belt along the border of their property, the purpose of which was to carry fly ash from the coal-burning plant to a new ash dump only a mile from our friends' farm. The constant hum and clatter of the conveyor serves to remind them daily of the known and unknown health risks associated with dioxin-laden gypsum. We all live in compromised environments; it is nice to know about hazards ahead of time, so we can limit their extent.

As for your own future use of the land, check out the zoning laws before you buy. How restrictive are the requirements for building? Are there restrictions on the number or type of livestock that can be kept there? What do the laws say about septic systems, and do they require installation of access to municipal water, sewage, or gas lines, whether or not the landowner desires or intends to use these services? Better to discover these things beforehand than to find a sheriff on the front step with a cease-and-desist order in his hand.

Power, of One Sort or Another

We will assume for the purposes of this discussion that you are not planning to do all your farming and building—all of your cooking, cleaning, and food handling—with nothing but hand tools, wood fires, and springhouses. Most of us are going to want some sort of access to electricity. If there is already grid power on your land, it will save you several thousand dollars in installation fees, if you chose to go that route. Being belt-and-suspenders people—sometimes—we consider electricity-on-tap an asset, even if it never gets turned on. Options matter. If there is no power on the land as yet, find out what it will cost to have it run all the way out to your future farm site: the longer the line, the higher the price.

Don't want to tap into the grid? Then pay special attention to the qualities of the plot of land you are planning to buy. If you hope to use hydroelectric power, check out the elevation, rate of flow, and seasonality of your surface water. For wind power, look up average wind conditions in your area for all seasons. Look around; are any of your neighbors using wind power? Ask how well it is serving them. And if you live in an area like ours, pay attention to the average number of sunny days: while we do use solar power for some things—lighting one of our big barns, powering electric fence in our more remote pastures—the number of sunny days in this part of Ohio is more or less equal to those of Seattle, Washington, which makes for a good many weeks when our solar batteries run low. Be especially attentive to the aspect of your land—its north-south-east-west exposures—if you are thinking of using solar power. A north-facing slope in our latitude may get only minutes of sunlight a day during the winter months, shivering the rest of the time under an unbroken blanket of snow and ice.

Figure 2.8. In January our neighbors (*far side of the valley*) receive less than an hour of sunlight per day.

What Not to Pay For (Unless You Have the Money)

Naturally, there are lots of amenities that are desirable on a farm, some of which are downright necessities; but while we may not be able to do without these things, we don't necessarily want to pay for them. Things we think of when we say the word *farm*—things like fences, barns, and cleared pastures—may be well within our ability to provide and may suit us better if we do the work ourselves.

Fertility is number one on the list of things *not* to pay for. If you do find a stretch of virgin land, unpolluted, rich of soil, for a price you can afford to pay, buy it by all means; you will be in the minority. Most land of that sort has a hefty price tag, if you can find it at all. Fortunately, fertility is not a *fixed* quality, but a *fluid* one—that is, it is mutable; it ebbs and flows. Committed attention to intensive grazing is going to set a spring tide of fertility on your land, poor though its condition may be at the outset. We don't have to pay for fertility up front.

Cleared, flat land and existing pasture with tight fences are great if you can afford them but, like fertility, aren't necessities. There is no rule that says pasture has to be flat, and since flat land is more desirable for building and for mechanical cultivation, it is generally more expensive as well. Existing pastures in any case, if they have been managed conventionally, have in all likelihood been chemically fertilized and planted with just one or a few species of pasture grass; soil biota will be less flourishing, and the forage spread less versatile than native pasture, as it is less diverse. For the same reason, it will be less resilient to seasonal fluctuations in temperature and rainfall. Look for unlovely parcels instead. Brushy land can be cleared with goats, which seem to prefer their pastures hilly anyway. Weedy, overgrown lots may only need to be grazed properly to become pastures of highly diverse native species.

Fences, especially perimeter fences, are good things, but not absolute necessities, and you can put up your own, where—and when—you want them. The marginalized land that will best repay your stewardship is not usually going to have nice fences. Existing interior fences may—in fact, probably will—be in places you *don't* want them. Permanent fence of any kind, as opposed to the inexpensive and lightweight temporary electric fences of the rotational grazier, may have to be mowed under regularly to keep it from becoming overgrown—a thankless and endless chore. Don't make the mistake of thinking your land has to come with fences.

Barns, corrals, and inground stock watering systems all raise the price of a piece of land without necessarily increasing its value to the smallholder. Grazed animals spend nearly all their time in the field, not the barn; consequently the grazier's barn, if he needs one, will probably be small and not elaborate. Small numbers of livestock that are accustomed to close proximity with human beings, as are those of the grazier, generally require less handling than conventionally managed livestock, and are easier to handle when they do need it, so corrals and cattle chutes may be extraneous. And because animals are moved to new grass regularly and the land is rested often, parasite cycles are interrupted; hence the need to corral and chemically worm animals may occur seldom or never. Inground and frost-free water systems are great, but not absolutely necessary to the small-scale grazier, whose stock water, ideally, will follow the animals over the pasture rather than being stationary; in any case, we will be looking for opportunities to use captured or surface water sources, rather than municipal or well water.

What other seeming necessities might be dispensed with? Your flexibility will determine some of these. What sort of shelter do you require for yourself? Our home, the Sow's Ear, was derelict when we bought it. The previous owner told us if he couldn't sell the land for what he was asking, he would bulldoze the house and try again—meaning that in his opinion the value of the land would go *up* if the house were demolished. Outbargained, we paid his price. The house, if terrifically unattractive, was structurally sound; we stripped it down to the studs and remade it slowly, while living in it. Saved us a ton of money. Can you live in a yurt? Manage with an outhouse and solar-heated outdoor shower? Live in a secondhand house trailer for a few years? The simpler your requirements, the more various the types—and prices—of land you can consider.

Still, if You Can Get Them

Trees—for shade, fuel, timber, nuts, maple syrup, temperature moderation, oxygen—are good things and good to have if you can afford them. Cattle may be wintered in our climate with no other shelter than the edge of the woods, and in summer a loop of fence thrown into the trees gives them access to cooling shade. A belt of trees as a windbreak is good shelter for most livestock, and can mean big savings in energy for the house as well. And browsing tree leaves, summer and winter, provides herbivores with minerals brought up by the plants' deep roots.

While surface water is not a sine qua non, access to a lake, pond, or (better yet) a creek or stream may mean you can water livestock without an electric pump or city water. Moving water is especially valuable because it so seldom freezes. Springs and seeps, coming from the constant temperature zone beneath the frost line, can also provide frost-free stock water. Watering livestock in winter can be a battle of wits—and endurance—in cold climates, and God be praised when Nature will do it for us. A well is a good thing, too, for the people on the farm, especially if it can be rigged with a secondary, hand-operated pump to make it independent of power availability.

Verticality—ups and downs in the terrain—is a very good thing. It allows us to use gravity as a motive force, for one thing. Water captured in a high place can be made available to lower points on the farm energy-free. Hills can provide shelter from prevailing winds, a fact we remember every time our hilltop neighbors lose shingles in a storm. And even a slight, south-facing slope will greatly increase absorption of solar energy; gardening on only a gentle southern slope effectively raises the microclimate half a weather zone. Other variations in terrain—a rocky outcropping; a swale; a low, marshy spot—create variations in microecosystem, microclimate, and habitat that add up to more opportunities and greater biodiversity.

Figure 2.9. Temporary fence makes it possible to shelter livestock in the wood's edge. Courtesy of the Franciscan Sisters T.O.R. of Penance of the Sorrowful Mother.

Access to grid power is generally valuable; even if you are planning to rely on alternative energy sources, grid power may get you over the hump before you have your other systems in place. After all, you may want to use a power drill for putting up your wind generator.

And don't fail to look over the fence, or past the property line. What is next door? Vacant fields; unused barns; a long, unmown right-of-way—all are potentially useful to the enterprising smallholder. There may be lots of grazing in a roadside verge or ditch; the low profile of rotational grazing, where all or nearly all of the fences are temporary, makes renting or borrowing land a real possibility. And not land only, but other resources might make this neighborhood appealing. Do the neighbors keep a pet horse? It may be a source of manure for your future garden. One of our neighbors up the road operates a small sawmill; he lets us take as much sawdust as we can carry off, especially when he finds gifts of home-cured bacon and fresh sausages hanging from his doorknob. Friends in Minnesota harvest bushels of apples every fall from trees in a pasture near their farm, where they are unvalued by the owner. Our country is full of resources that go unused for want of someone to value them.

Nontenancy

And what if you are not going to live on the land? Not everyone who wants to increase his independence and improve the environment can, or will, live in the country. Can a nonresident hope to practice rotational grazing? As farmers often temporize, "It depends." Lots of people who own livestock don't live where their animals are grazed; indeed, given the size of the average ranch, even for resident farmers there is often a good distance between the home of the grazier and the pasture on which the livestock are presently grazing. Nevertheless, nontenancy has its own set of challenges. "My last good night's sleep," Shawn's dad, for some time a city dweller with a hobby ranch, used to say ruefully, "was the night before I bought the farm."

Fences are not perfect; animals with wanderlust can be clever and cunning. Escaped livestock are a worry even for the farmer who lives on-site; for the nonresident, they are a bigger problem. First of all, the lapse of time before the security breach is discovered will be longer, perhaps very long; second, recovery, even then, is going to pose extra challenges. And fences are permeable in both directions: if the neighbors' livestock breach the fence and gorge on your forage, you may lose a week of grazing before you discover them; the ingress of predators into the pasture may only be revealed by the vultures circling when you next visit the farm. Human predators are a bigger risk for the nonresident farmer as well: our own first little farm, ½ mile from our home at the time, lost two chickens a week to a "fox" of great resource and ingenuity who carefully refastened the door behind him, until the last chicken was gone.

Still, it can be done. Naturally, if the land is not to be your home place, your requirements for it will be different, but there are amenities you will want anyway. A place to be comfortable—somewhere for shelter from sun, rain, or cold—will make your visits to the farm more enjoyable, more frequent, and longer. Good, tight fences are of especial importance to the nonresident grazier. A place in which to store tools—extra fence posts, reels of polywire, hoses and the like—under lock and key is going to be a necessity. And if you plan to make your venture a family enterprise, or even if you don't, consider buying land that provides its own reasons for you to go there: beauty, shade, fishing, bird-watching, and so on.

Durable Tenure, Regenerative Agriculture, and the Next Right Thing

Land regeneration is something that all concerned people should be able to collaborate on, be they wealthy or of modest means, liberal or conservative, religious or atheist. Whether we are motivated by abstract virtues like moderation or good stewardship, or concrete issues like the toxins in our food, air, water, and soil, the fate of the earth, and the people who live on it is everybody's problem.

Ultimately, we need to look not at what the land presently is, but what it can become under good stewardship. Proper grazing is a part of the solution. Instituting this, we can build soil, capture and securely store atmospheric carbon, retain rainwater in the landscape, lower local temperatures, and purify air and water. We all eat; if we want to eat in the future, it's time to start considering how we may regenerate fertility and diversity in our food production systems.

CHAPTER THREE

Water

Now the earth was formless and empty, darkness was over the surface of the deep, and the Spirit of God was hovering over the waters.

—The Book of Genesis

More than temperature, weather, topography, or native plant and animal populations, water determines where human beings can live, and under what conditions of truce we can live there. Cultures are shaped by water—its abundance or scarcity, and the forms in which it is most prevalent in a given place. The movement of nomadic peoples and migratory animals is dictated, directly and indirectly, by water availability; it has a bigger effect on plant and animal populations than either forage or prey scarcities, these reserving their direct effects for a limited range of species, whereas flood and drought—too much water, or too little—affect all the living things in an ecosystem.

Figure 3.1. A small stream is a big advantage on the low-input farm.

Short- and long-term deficits in precipitation, falling water tables, and increased pollution in sources of water for human use are moving concerns about water up in the priorities of individuals, communities, businesses, states, and nations. The diversion of regional water sources from rural to urban uses, and from local to distant systems, is marginalizing communities, ranchers, and farmers all over the country and even the world. Where you decide to begin your homesteading efforts should be largely affected by the availability of water.

Finding Water

When we think of water availability, most of us think primarily of three sources: *surface water* like oceans, lakes, and rivers; *subsurface water sources* for well water; and *atmospheric water*, that which reaches the earth in the form of precipitation. From these we get our domestic water for drinking and cleaning, industrial water for manufacturing uses, and water for irrigation of crops and watering livestock. When we think of the dangers to our lifestyle or well-being resultant from desertification/global warming/climate change—whatever you wish to call it—the scarcity of water for purposes of this kind is our primary concern, and fear sparked by an effective inadequacy in the supply of surface, subsurface, and atmospheric water occupies most of our attention. But we are overlooking a reservoir right under our feet, and it is one that holds an estimated five times the volume of all the water contained in the atmosphere. Ecologists call it *green water*, and it may have more to do with the conditions of our local climate, and the ability of our land to feed us, than most people realize.

Green water is the water contained in the soil under our feet. The soil itself is a huge sponge, and it can absorb and retain, under favorable circumstances, the majority of the water that falls on it. It is estimated that up to two-thirds of the precipitation that lands on the soil is absorbed before it reaches a waterway. The absorbed water is stored in the structure of the soil and released in various ways. Some percolates down, being filtered through layers of soil and rock, and either exits the soil gradually via springs or is eventually deposited in subsurface water reservoirs, the sources of our deep and shallow wells. But where the soil is bare of plant life or plant litter, much of the precipitation absorbed by the soil is lost almost immediately through surface evaporation, as in our own native Southwest, where heavy rainfall turns sandy soil to quicksand, then dries to dust in twenty-four hours.

The ability of land to absorb and retain precipitation is largely a function of the amount of stored carbon—read *organic matter*—in the soil. In the first place, organic matter content affects the soil structure, helping to determine the number and size of interstices in the soil—spaces that are available for water retention—and hence its permeability. Decomposed plant matter acts as a sponge, soaking up water to be released slowly, over time. Land with plentiful plant life also receives the water more gradually, as drops are slowed by impact with plant leaves, and ground cover decreases the rate at which surface water travels, giving it more time to soak in. On bare land, or land with sparse plant cover, runoff happens quickly, leaving little time for absorption by the soil. What moisture the land does soak up is rapidly lost through evaporation, the rate increasing when sunlight heats the bare soil; the temperature of soil bare of plant cover can vary by 20 degrees or more from nearby ground occupied by growing plants. Land with sufficient plant material in the soil and on the surface, as either living plants or litter, shades the soil and keeps it cool, allowing absorbed water to remain longer.

Figure 3.2. Five years of intensive grazing has transformed compacted clay soil and sparse ground cover to lush pasture.

The quality and quantity of water on a piece of otherwise habitable land will have more effect on its uses than any other single factor. Intensively managed grazing will increase, over time, the surface plant cover, organic matter in the soil, and consequently the degree to which precipitation is absorbed and held in the soil. Increasing the ability of a plot of ground to absorb and retain rainwater gives land the potential to become more fertile, more habitable, and therefore more valuable.

Pressurized Water—City and Well

The most obvious water source, to most of us, is the one we've grown up with: tap water. Be it from municipal systems or private sources, pressurized water—water in pipes—is the only water most of us consider trustworthy for drinking or cooking, and probably even cleaning as well. But city water and well water, while they are conveniences, are not the only way to get water to animals, nor are they necessarily the most reliable. Municipal water is an imported resource—an off-farm input—and reliance on its use decreases the overall self-sufficiency of the farm. And if one's water comes from a deep well, drawing water off the aquifer, it is tapping into an effectively nonrenewable resource. Availability of pressurized water can be an advantage, but for livestock water other sources may prove perfectly adequate.

City water costs money, too—another off-farm input. In a neighboring town we have seen water costs rise by a factor of five over the past twenty years. According to the EPA, the average American uses 100 gallons of water *per day* for personal use. Livestock are less consumptive, but large animals, dairy cows for instance, may drink 10, 20, or even more gallons of water per day. A small herd could require hundreds of gallons of water a week, significantly increasing the cost of keeping livestock. Pressurized water from municipal sources is also liable to expensive leakages. The first indication our neighbor up the road had that the pipe supplying water to his barn had sprung a leak was an unexpected eight-hundred-dollar water bill, a big expense to tack on to the cost of the pigs he raised for bacon that year.

Moreover, municipal water comes from a wide variety of sources, and contamination is not unheard of. Industrial waste, agricultural runoff, mining disturbance, and airborne contaminants are more and more universal; by some estimates there is nowhere on the continent now free from some form of waterborne contaminant. And the processes by which water is purified for domestic use are not always acceptable to individuals concerned about environmental or human health. Local and regional water sources are subject to pressures beyond the influence of small landowners; a new business or industry in the neighborhood may affect how much, what quality, or when water is going to be available for small-scale agricultural or domestic use, and other interruptions of water service do happen, leaving the owner of livestock with an immediate and urgent need for an alternative water source.

On-site well water—water derived from subsurface sources, whether shallow subterranean streams or deep wells piercing to the aquifer—is a real asset to the landowner. Given a power source—grid, solar, wind, or generator power—a well on your property is a source of water that may be unregulated, for the most part, by industrial or government interests. If your land has an existing well, have the water tested for potability; the local extension office will have information about water testing services in your area.

Other questions about well water may require more work to answer. Determining your well's rate of flow will be important to decisions like the size, number, and species of livestock your land can carry. High-volume wells provide an effectively constant water source, meaning that the well replenishes at a rate at or beyond demand, keeping water before the livestock at all times. A low-volume well—one with a slow replenishment rate—is not necessarily a bar to keeping animals, but may mean that you will have to install a cistern or holding tank, so that water collected gradually will be available in volume when there is a demand.

Wells, like surface water sources, may be seasonal; that is, they may provide adequate water for only part of the year, being slow or nonexistent in certain seasons or under adverse conditions. This is a factor that can only be determined on the spot and over time, although it is possible the neighbors will be able to tell you something helpful. If they don't know the history of your well, any information they can give you about their own well may indicate something about others in the area. If their well depth and location make it possible that your wells share a single source, there is a good likelihood that they will have similar qualities. And even if your well does not supply water for all twelve months of the year, don't despair: seasonal wells aren't necessarily a bar to keeping livestock; they just mean that other sources of water will need to be explored as supplement to the well system.

A private well is only as reliable as the pump that moves the water. The depth of your well is highly significant to its reliability; a really shallow well can be tapped with a hand-operated pump, but many if not most subsurface water sources are too deep to be pumped manually. Very deep wells require a high-powered pump to raise the water to surface level; pump size determines the amount of power necessary. Friends in Washington State, who draw their house water from more than 2,000 feet below the ground, tap their well sparingly for farm use, relying as much as possible on surface water and precipitation to irrigate crops and water livestock.

Grid power to run a pump costs money and is unquestionably an off-farm input, and therefore a gap in the farm's sustainability, but nevertheless a good option to have in your back pocket. But it may be subject to periodic interruption, temporarily cutting off water supply to homes—and livestock—that rely on an electric pump. Investigate alternative sources of electricity. Your options for other sources of power will be dependent in part on the land itself, and whether the climate and aspect make solar or wind power an option. Prevailing local wind conditions, the elevation of your land, and the inclination of slope—exposure to north, south, east, or west—will have a lot to do with whether it will be cost-effective to explore these options. Our own little farm, tucked deep in a hollow of the Appalachian foothills, is too protected for wind generation to be reliable, and most of it receives too few hours of direct sunlight for generating much solar electricity. Fortunately, we have plenty of other ways to get water to the livestock.

Surface Water—Ponds, Streams, Rivers, and Lakes

It is a considerable benefit if you have, and can make use of, sources of water already on the surface. They have, after all, the great advantage of being already in place. Surface water—flowing or standing, in permanent or temporary catchments—is where most wild animals get their drinking water, and using it to provide for your livestock, if you can, is going to be the cheapest, as well as the easiest

Figure 3.3. Verticality adds greatly to the value of existing surface water.

and most sustainable solution. Making this water available to livestock in a rotational grazing system can require some thought, but with a little ingenuity and simple technology it can be done, increasing the sustainability of the farm as a whole.

Riparian Banks—Flash Grazing

Naturally, the easiest way to give your animals water from your pond or stream is by allowing access to the bank. This has its drawbacks, however. Prolonged animal activity on a stream or pond bank breaks down the edges and tears up plants living next to the water, causing soil to wash away. Over time the erosion damage can become so extreme as to decrease pasture size—a situation not uncommon on our Ohio creek bottoms—and muddy the water moving downstream. On a bermed pond bank or earthen dam, animal trailing can seriously compromise the integrity of the reservoir; in our native Southwest we have often seen an artificial pond, or "tank," lose depth, and therefore holding capacity, when animal trails incise the soil bank, creating an unstable spillway for overflow during heavy rainfall. A sudden influx of water, breaking through the bank at this point, carries soil away with it, deepening the cut and reducing the pond's holding capacity or even destroying the dam altogether.

Erosion isn't the only damage caused by stock animal access to creek and pond banks; cattle wading or wallowing in ponds and streams increases contamination and can transmit pathogens and parasites. Water fouled or muddied by stock access becomes inhospitable to aquatic and amphibious animals, biological indicators of water quality. Fortunately, there are options that can enable the homesteader to avoid these problems and still use the surface water that, under some circumstances, may be the most ready and plentiful source of livestock water on her land.

One option is to limit the area and duration of animal access to waterway banks. Temporary fences that allow access for short periods to only limited areas of bank at a time can actually have a beneficial effect on creek and pond banks. The wearing down of abrupt slopes and edges around the waterway, which can happen with short duration livestock access, actually creates conditions that under proper management can increase the living vegetation on the bank, thus increasing its stability. The reduced slopes and exposed soil surface are conditions favorable to seed germination and plant growth. Creek banks clothed with living vegetation are protected against erosion during times of spate, while a grassy bank filters runoff, improving overall

Figure 3.4. Sheep flash graze a section of creek bank; brief access followed by a long rest helps clothe slopes with vegetation.

Figure 3.5. Portable stock water tank with Jobe float.

water quality. In this model, "flash grazing" is accomplished by moving livestock gradually around the pond, or along the creek bank, with short grazing access and lengthy recovery periods for each section of bank, during which plant life is able germinate and grow. The temporary fences of polywire or polytape that the rotational grazier uses are ideal; electric netting can also work but poses more of a challenge to erect.

Running Water

The ideal, in a rotational grazing system, is to have the stock water travel with the animals from paddock to paddock. In large, professional operations this is usually done with buried or surface pressurized supply lines to hydrants scattered over the grazing area; but for most small-scale homesteaders on limited means, supplying stock water to individual paddocks is obviously going to be least costly if the animals are followed with a small tank supplied by a portable hose. Under many circumstances, this can be done from existing surface water sources.

The location and elevation, with respect to your pastures, of the pond or stream to be accessed for stock water will have a lot to do with how you make it available to the animals. Gravity, in this situation, is often your best friend. An elevated pond, or a sufficient drop in the line of a creek bed, can be tapped using a simple intake and a garden hose or other tubing, flexible or rigid, and the water moved to points downhill.

A pond is especially easy to divert for stock water if it already has an overflow pipe or *drawdown*. The latter is a drain pipe, either vertical or horizontal, the opening of which is placed at the highest desirable water level of the pond to provide an outlet for overflow water, thereby preventing the erosion damage that would be caused by overflow across the pond bank. In the case of a pond where a steady replenishment rate means that there is always water moving through the overflow, this excess can easily be captured for constantly available stock water.

It is *essential* that any capture of overflow water be done in such a way that there is no obstruction of the drawdown pipe. This is because the volume of water that needs to be carried away to prevent pond overflow during a heavy rain may be—almost certainly will be—many times the carrying capacity of the hose or tubing used to tap the source for stock water. If your overflow pipe is, say, 4 inches in diameter, and your diversion hose is ¾ inch, there is obviously going to be some backup when the 4-inch pipe fills up. Under these circumstances, water unable to exit the pond through the overflow provided will have to find other ways out, with potentially destructive results.

To get around this problem, when we tap a pond overflow we place open tanks or tubs *under* the exit pipe to catch runoff. Half a 55-gallon drum works well, but even a 5-gallon bucket, if it is firmly anchored to prevent it being knocked over by a rapid outflow, will do the trick. We plumb these tanks just below the rim with a hose coupling, to which our stock water supply pipe is attached. Being kept constantly full by the overflow, this setup provides unlimited water to our stock tanks. Excess water simply flows over the side of the catch tank and back into the spillway or waterway.

The same system can be used to divert creek water to a stock water hose. A half barrel, plumbed for a hose fitting and set under a culvert or waterfall, or partially buried in a creek bed, will be constantly refilled. Rocks in the bottom of the tank keep it stable when the current is strong, while the overflow, being confined to the creek bed, poses no threat to the integrity of the bank. Another option for creek water capture, damming the creek, is not only more labor intensive, and nearly always more costly, and improperly constructed or poorly located dams are subject to washout during spate. On a large impoundment, a compromised dam is a danger to downstream neighbors; and even when the dam is of adequate strength, it may present an erosion hazard if the spillway has been improperly constructed. A bucket or small tank captures more than adequate water for the smallholder's purposes, while posing no obstacle to natural water flow. To prevent the ingress of waterborne leaves, twigs, rocks, or silt, cover the inlet side of the pipe with large-mesh screen. This will also keep out the aquatic animals that may creep into the hose and be difficult to dislodge; our local crawdads and water snakes seem to find that little black hole fascinating.

Figure 3.6. Simple, low-technology water capture.

Stationary Water Sources

A pond that is filled seasonally (its refill therefore only periodic) or a pond without an overflow pipe

Figure 3.7. Drawdown for pond overflow.

can still be tapped to fill a movable water tank—it will just mean a few additional steps. A hose or pipe, the end of which is protected by a screen to prevent clogging by aquatic plants, animals, or muck, can be anchored just above the bottom of the pond and the water siphoned over the bank; alternatively, the inlet to the pipe can be suspended just below the surface of the pond by means of a float, while an anchor to the bottom of the pond keeps the hose end well out from the bank.

The system will have to have a full hose to jump-start the siphon, a ticklish business if you don't have a pump for the purpose. Submerging the entire hose and evacuating the air will fill it with water; keeping the ends closed while the hose is dragged over the pond bank is trickier. A sprayer attachment closing the male end of the hose—the one away from the spigot if the hose were hooked up—will serve to close that end, which can then be carried over the bank and down to the desired point of delivery. The upper end must remain below the water surface; when considerably more water-filled hose is below the level of the pond than is inside the pond, the greater weight of the water trapped in the hose outside (and below the elevation of the pond) will—provided air cannot get into the upper end of the hose—drag water up over the level of the bank. It is necessary that the upper end of the hose remain underwater, so that the siphon pulls water, and not air, into the hose to fill the vacuum.

Ponds may be of almost any size, from a few dozen square feet to acres of surface. Whether you hand-dig a pond or hire excavation equipment for the job, the water you capture is going to increase the resilience of your land, holding rainfall that can percolate gradually into the soil. Collect all the local and technical information you can before you build your own pond to avoid disappointments; contamination from higher elevations, leaky ponds, and broken dams make expensive mistakes. Your local NRCS can be a valuable source of material, but remember your sustainable farm goals, and analyze their recommendations from the individual perspective of your homestead. Agricultural extension advice is almost always going to reflect the values and trends prevalent in the Department of Agriculture this year, and even with grant money (a thing that always comes with strings attached), their solutions may not be cheap.

Building ponds is like planting trees, a gift you give to the future. Holding rainfall on the land in ponds slows runoff and prevents soil erosion, as well as giving precipitation an extended period over which to percolate down and replenish the moisture in the soil. Surrounding pastures receive slow-release rainfall for deep root systems and increased drought resistance. Pond banks provide ecological edge spaces, habitat for plant and animal species that bring benefits both tangible and intangible to the small farm ecosystem. Stock water is only one of many reasons for the homesteader to consider building a pond.

Rainwater and Runoff—Roofs, Foundation, and Pavement

Captured rainwater, in many instances, represents the most abundant free water available, so when we can impound the rain from a roof or pavement and hold it for use on the farm, we not only slow the passage of precipitation into local waterways and increase the drought resistance of the land, we save money while we are doing it. A single inch of rain falling on 1,000 square feet of horizontal footprint can shed more than 600 gallons of water; in our area, where average precipitation is relatively high—over 35 inches—this translates to over 21,000 gallons of water shed yearly by a small roof or paved lot—much of which can be captured at very little cost and put to work on the farm. Rainwater capture systems don't have to be expensive or elaborate; practically any means you can devise for channeling clean runoff into a catchment will do the trick. Our own rainwater capture systems are a case in point.

Roof water is usually the first rainwater we think of for farm or garden use. There are 50-gallon rain barrels available from the local home improvement store or the Soil and Water Conservation District (SWCD) office, for a price, but why stop at 50 gallons? After all, this would hold less than a twelfth of the water that might be harvested from even a small roof in a 1-inch rainfall event. We ourselves favor the 300-gallon plastic tanks known to industry as IBCs, or "intermediate bulk containers." These are the large square plastic tanks with exterior aluminum frames used to hold and ship liquids of all kinds, from beverage concentrates to detergent,

Figure 3.8. This intermediate bulk container holds six times the water of the rain barrel next to it.

food grade iodine to glue. In our area right now they can be had relatively inexpensively, the cost depending in part on what substance the tank originally held, and in part on how hard the seller has worked to clean it out. Obviously, since we are proposing the use of these tanks for relatively long-term stock water storage, the cleaner the tank and the more innocuous the former contents, the better.

IBCs are designed for ease of fill and drainage, just what you want in a rain barrel. They have two large openings, a fill opening at the top, and another, governed by a spigot closure, at the base. Diverting rainwater from a roof gutter system to the tank is easily done with a length of flexible plastic pipe—or more elaborate arrangements can be assembled that include valves either to divert the first flow of a rainfall event, which will contain the most impurities, away from the tank; or to shunt away the overflow once a tank is filled. The spigot in the lower opening has a strong plastic ball valve to open and close it—very handy, but not the ideal diameter for the average garden hose. Since we can't always find an adaptor for just these dimensions, we solve this

Figure 3.9. Rubber coupling and hose clamps make it easy to adapt the wide IBC opening to any diameter pipe.

problem by mating the spigot, with an arrangement of flexible 4-inch rubber couplings and hose clamps, to graduated PVC adaptors, reducing the 4-inch diameter to ¾ of an inch. No doubt someone else may have come up with a more elegant solution, but for us this does the trick.

Roof water is easily delivered to a tank or cistern by means of gutters and downspouts. Collecting water at ground level, as from a paved area or a French drain, is trickier; but where there is some variation in topography we find it can be done. We water the animals in the monastery's back pasture from a French drain system that exits on the hillside. It isn't safe just to extend the exit pipe and run it into a tank; as with a pond overflow, it is imperative that the outflow of either a roof system or a French drain never be impeded. Any compromise to the speed with which a roof can drain is a compromise to the integrity of the roof itself, a fact brought vividly to our attention one winter when ice plugged the downspouts at the monastery and snowmelt came pouring through the refectory ceiling.

Catching water from a ground-level exit without covering the end of the pipe took some thought; eventually, we set a bucket rim-deep in the ground below the exit pipe, to collect the water as it came out but still permit unimpeded overflow. The bucket is plumbed on one side to take a 2-inch hose (we'd have made it bigger, but the most convenient bucket to hand at the time was the tank from a defunct shop vacuum—might have been made to order for the job). We lengthened the vacuum hose with some rigid drainpipe, and the captured water is delivered to a series of IBCs lower down on the hill. This array captures nearly all the volume of a moderate rainfall but allows the overflow from a sudden spate to run away without any danger of its backing up into the basement or onto the roof.

Nonelectric Pumps—Nose, Sling, and Ram

There are ways to drag water up small slopes without use of electrical power, and involving technology only a little more complicated than simple gravity. Several devices are available commercially for this purpose.

Nose Pumps

Nose pumps are really ingenious: piston pumps that operate when livestock push a lever with their noses.

Don't confuse them with the nose-operated switch valves you'll see for use with pressurized water; nose pumps are actual *pumps*, which can raise water up to 20 feet over a 200-foot run. Although the ones we have seen in use are for larger animals—cattle, horses, bison, and so on—we have read that they are now being made in smaller sizes for use by sheep and goats, as well as calves; these, however, since the animal itself is the source of motive power, probably have a lower lift height and shorter run than nose pumps for large animals.

A nose pump is generally used to bring water from a fenced-off pond or creek bank to a pasture only a few feet away. An intake pipe anchored below the surface of the water is laid aboveground to the nose pump itself. The pump is firmly attached to a solid base inside the pasture, which is raised somewhat to decrease the chance of animals stepping or defecating in the pump reservoir. The permanent base is a disadvantage, since it makes the pump less convenient to move—a base has to be set up or built for each paddock location. And because in most applications these pumps will freeze in cold weather, they are not usually a solution for winter watering of rotationally grazed livestock. Despite our own heavy reliance on surface water for livestock, we have never used a nose pump and know of no neighboring farms that do; partly a function, no doubt, of our hill-country location, where gravity is ever ready to provide motive force, and partly due to the cost of a pump—the ones we have seen run about four hundred dollars.

Sling Pumps

Sling pumps—also called propeller pumps, or rotary pumps—are another use of simple technology for a nonpressurized stock watering application. Sling pumps require an active stream flow—at least 2 feet per second—for operation. Basically, a sling pump is an encased Archimedean screw operated by a propeller. Anchored in the flow of a stream with the propeller toward the flow, a sling pump will lift small amounts of water over short distances; since it runs constantly, it can move a couple thousand gallons of water daily. Constant flow means that a return pipe is necessary to carry overflow water back to the stream. For filling a permanent stock tank located near a fenced stream sling pumps are a good option, but the complications of setup—particularly the necessity of anchoring the pump where stream current holds the pump in active flow, rather than pushing it toward the bank—and the necessity of watering animals within a few feet of the stream bank, or managing a lengthy return pipe, make sling pumps less popular among graziers in our area than gravity or pressurized water systems. The cost of a sling pump and fittings starts at about seven hundred dollars.

Ram Pumps

Where nature has not made our job easier by locating springs in elevated places, a spring or stream of sufficient flow can, with some assistance, be made to lift the water for us. In simple language, a ram pump uses the greater portion of a flow of water to pump the remainder of the flow to points desired. The speed of a descending column of water pushes about 10 percent of that water up to several hundred feet above the ram (though somewhat less than that above the water source, since the pump needs some vertical drop to create speed of flow). We are fascinated with ram pumps, and plan to install one on one of our larger springs. Water raised from this spring could be delivered to a tank on the highest point of the pasture, and daisy-chain from there along multiple ridges to cascading tanks located

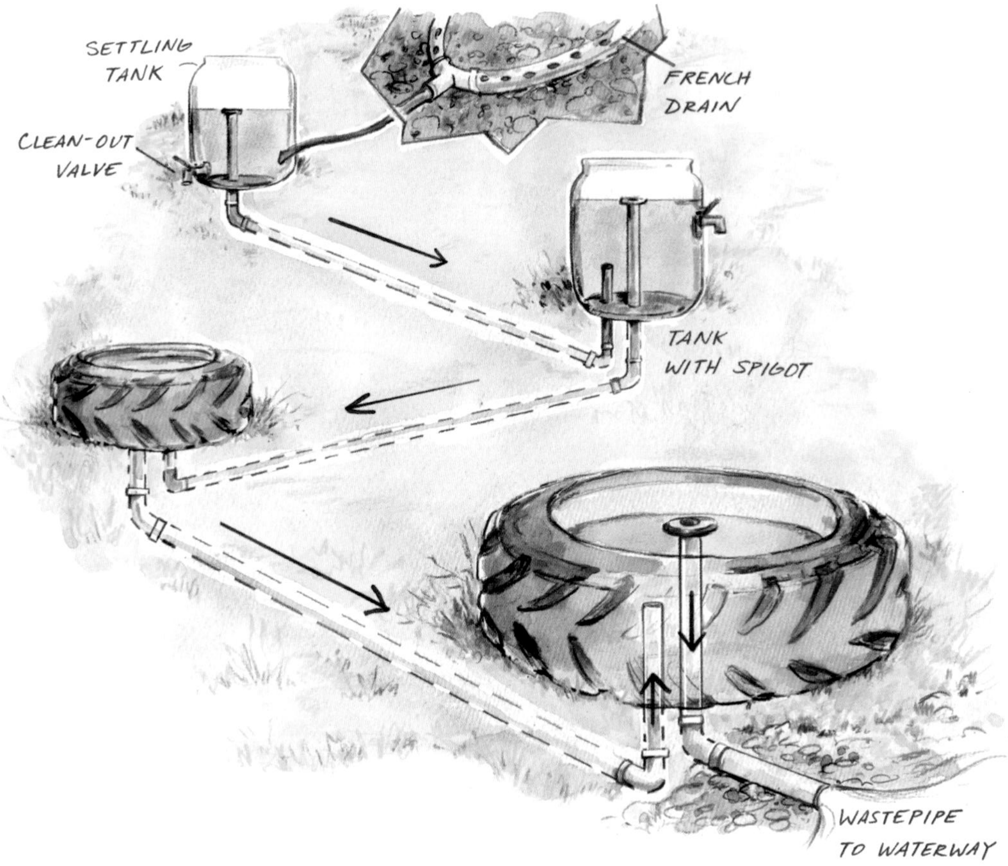

Figure 3.10. Captured water cascading through a series of tanks for constantly available and frost-free livestock and irrigation water. Illustration by Elara Tanguy.

over most of the area. Because the water is constantly moving, danger of freezing is minimal in even our coldest winters, when livestock water is arguably our biggest problem.

Rams do present a couple of disadvantages, though: first of all, a pump big enough to raise the water to the point we have in mind will cost in the neighborhood of one thousand dollars, an amount not in every budget; second, ram pumps are rather noisy affairs. In operation, a ram pump cycles between twenty-five and fifty times per minute, each cycle including a *click* and a splash: small *click*, big splash. People for whom the silence of nature is paramount don't want a ram pump too close to their home or favorite haunts. We haven't given up the plan, though, as the nonelectric power in a ram pump has considerable benefits for the self-sustaining homestead.

Developed Water Sources—Springs and Seeps

Some groundwater is less obvious than ponds and streams. Springs, and their shy sisters, seeps, are places where subterranean water makes its way to the soil surface in either concentrated flow (springs) or diffuse wet spots (seeps). For natural reasons they are more common in hill country, where layers of rock or clay are exposed on a slope. Water percolating from above encounters an impervious layer and runs horizontally until it reaches the surface, either emerging as running water, or just saturating surrounding soil.

Springs and seeps are not always immediately evident when you are getting to know an area, but they can often be located by paying attention to small details. A place on a hillside where the soil is constantly wet, especially lower down on the side of a slope or swale, may be a good indication that there is a source of constantly or at least seasonally available water there. In our shale hills these spots are often right out on the open hillside. We have found springs by observing the plant composition in an area; when sedges, reeds, or cattails—wetland plants—grow on an open hillside, it's a sign that water is constantly available at that spot. Low spots, footprints, or cattle tracks that are always full of water are signs, too, as are accumulations of icicles on canyon walls in winter. Sometimes our detective work may have to extend over several seasons, but less than a year on our hillside farm was enough to give us a good idea of several places that might reward a little digging for water; some of these, now developed, provide us with hundreds of gallons of stock water a day.

Don't overlook small opportunities to harvest water. Even a very slow seep should be investigated, since a flow as inconsiderable as a drip a second can deliver nearly 10 gallons of water in a day—enough to supply a couple of goats or sheep. And slow seeps directed into large cisterns can make it possible to water far more livestock than their per-minute flow would seem to indicate. We're not suggesting that you invest time or money to improve a seep with a maximum flow of one drop per second, but do recognize that what may appear to be an inconsiderable flow may carry more than enough water to be useful for your smallholding.

Figure 3.11. Toad eggs. Aquatic and amphibious animals can be indicators of water quality.

Seasonality of flow is important to observe and consider. Not long ago a farmer west of us bought 100 acres of grassy hillside. It was autumn, and he immediately put several thousand dollars into improving the generous spring that rose in the upper pasture, installing a series of four or five descending tanks, each filling from the overflow of the one above it. This was a rotational grazier's dream, since spigots in the side of each tank could be hooked up to movable hose/tank arrays and led to paddocks anywhere in the pasture. The hitch in his system didn't appear until the next June, when his spring dried up completely just when those pastures were wanted for grazing. The captured spring didn't begin to flow again until rainfall recharged the soil in the fall, limiting the advantages to be gained from the system. However generous a spring flow may be, it is of little use in the wrong season.

"Spring water" is not synonymous with "clean water," let alone potable water. Generally speaking, spring water quality is going to be high enough for stock water use, but this should not be taken as a hard-and-fast rule. Local land use can contaminate groundwater to the point where it is not safe to use. Strange smells in your spring water may be naturally present sulphur, or they may be biohazards from the septic tank overflow of the neighbors up the valley. Mining in the area can cause waterborne toxins such as heavy metals to reach the surface, too. If you are in any doubt about the quality of your water, by all means have it tested.

The degree to which a spring can be made of use in a rotational grazing system depends on several factors, and as with surface water, elevation is high among these—no pun intended. In an operation as small as ours, especially, a spring's rate of flow is of less importance than its elevation. A small spring filling a 50- or 100-gallon cistern, if it is located on a hillside, can be carried through a hose to points below where it may be used for quite a few animals; whereas a spring of greater flow will be far less useful if that water comes to the surface at the lowest point on the farm, where moving it to follow the animals is going to be energy- and labor-intensive, or require expensive and specialized equipment.

Figure 3.12. Water from this captured seep (*blue tank just above barn roof*) has to be pumped to the high IBC before it can gravity feed to the sheep and calves in the middle pasture.

Basics of Spring or Seep Development

The technology for improving a spring or seep located on a slope is simple and straightforward; your local extension office will have diagrams explaining the process in detail. Briefly, in a place where you have located a spring or seep, a ditch is dug on the uphill side deep enough to intercept the flow of water above (behind) the present point of egress, and a French drain is installed. This gathers the collected water and delivers it to a settling tank situated lower than the ditch, where particulate matter is removed by settling. From the settling tank, the clear flow is directed into a tank or cistern of large capacity, from which it can be drawn to serve any point below the elevation of the cistern.

Such spring collection systems can, but need not, be an expensive and high-impact project. Backhoes and bulldozers will cost a lot to rent or hire and make a big mess while you're using them, but why use them? For a small project it's surprising how much can be done in an afternoon by a couple of people with picks and shovels and a purpose. The spring supplying water to our big barn and 5 acres of pasture

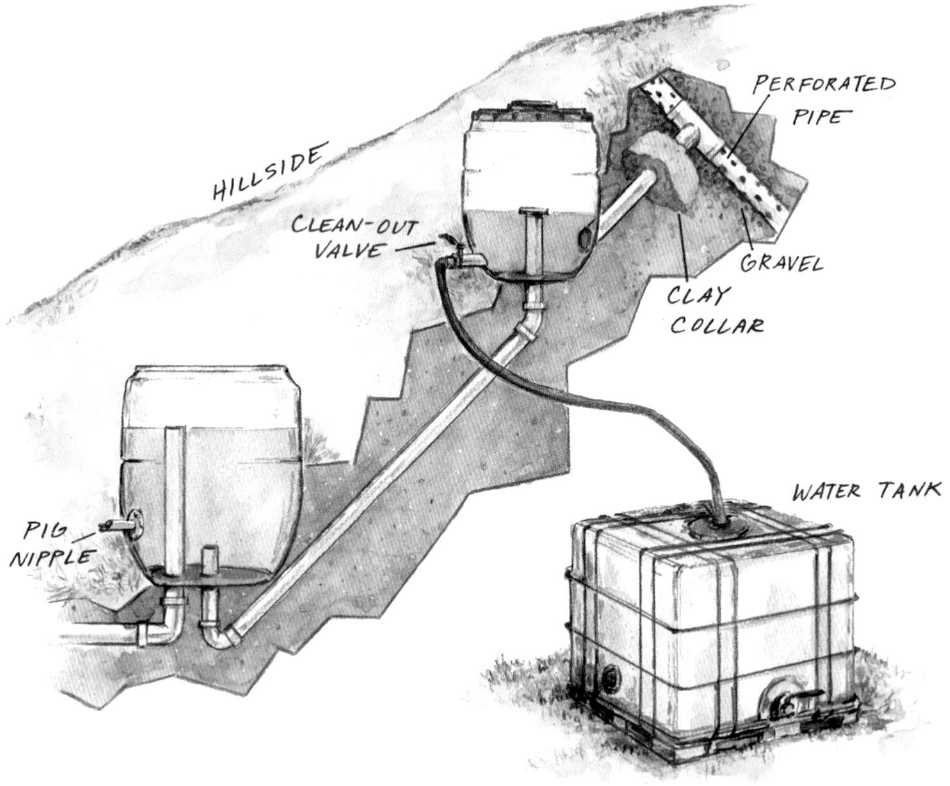

Figure 3.13. Water from a spring or seep can be captured with simple technology: pick, shovel, gravel, perforated pipe, and a settling tank. Illustration by Elara Tanguy.

is a slow-flowing seep with a French drain no more than 60 feet long; most of the time it runs in a stream smaller in circumference than a pencil. The weather, the October we built it, was rather wet and dismal, and the soil on that hillside is mostly big pieces of shale and sticky, shovel-clogging clay, but in a couple of weeks of on-and-off labor we dug the 60 feet of 3-foot-deep ditch for the French drain and another 40 or so feet for the delivery pipe that carries the water, first to the settling tank, and then to the holding tank. We also dug another 20 yards or so to bury the overflow pipe that carries runoff to the creek, a choice we consider wasteful now—we should have directed the overflow into a series of tanks descending the pasture. Ah, well—a project for another day.

The settling tank on this particular spring turns out to have been an unnecessary precaution, since very little silt finds its way into the water flow, but settling tanks are generally a good thing to have on a spring development: spring water is directed into a preliminary tank before being piped to the point of delivery—in this case, the stock tank at the back of our barn. Slowing down the flow of water in the first tank allows silt and other heavy matter to settle to the bottom, while clarified water is drawn down through a short vertical pipe extending up from the bottom of the tank. A valve low on the side of the tank allows us to drain the tank occasionally and clear it of silt, although, as we say, in practice we haven't seen enough silt to make this worthwhile.

Stock Watering Systems

Differentiation should be made between permanently installed tanks with belowground, frost-free fill pipes, and smaller, movable stock tanks with aboveground pipes or hoses.

Stationary Stock Tanks and Storage Cisterns

Ideally, rotational grazing systems are built around a movable stock water system with portable tanks, but undergirding this array it is frequently useful to have one or more permanent tanks, particularly if the grazier wants to avoid off-farm inputs like city water or grid-pumped well water. Water from ponds, streams, seeps, or springs may be easier to access, and provide four-season water availability for a low-pressure, gravity-fed system, if it is first captured in a permanent tank or cistern.

Permanent tanks are most useful as part of a water capture system where some of the water will be used on-site, and some piped to other points for use in rotating paddocks or movable pens. The captured water is directed first to a permanent tank via a buried pipe coming up through the bottom; overflow can be drawn through a second pipe back below the frost line and travel underground from there to another tank, or series of tanks, daisy-chain fashion —or it can be accessed aboveground by means of a spigot or coupling plumbed into the side of the first tank at a point below the normal water level.

Plastic barrels can make good permanent bottom-fill tanks so long as the construction is sufficiently rigid to withstand earth pressure on the outside. The deeper the tank is bermed, the greater the pressure the earth will exert, and this is doubly true where

Figure 3.14. Via the drawdown, water can flow from this spring-fed tank to other tanks lower down in the pasture.

livestock water directly from the tank. Don't underestimate the horizontal force exerted by 1,200 pounds of cow directed into the soil through four smallish hoofs; a tank of insufficient strength will cave in quickly under such pressure. The same rule holds for any pipes you are going to bury in or around livestock use areas: make sure you get pipe of sufficient strength that it won't be crushed by the largest thing you plan to run over the surface above it, be it livestock or equipment. The cost of heavy-duty pipe will spread itself out over years, even decades, of service, whereas the cost in time, inconvenience, and delay when you have to dig up a system and re-lay it properly is hard to overestimate.

One disadvantage of plastic in-ground tanks will be seen when you are plumbing pipe through the bottom of the tank. Adhesives and caulks especially formulated for use with plastic make it possible to cut two holes through the bottom of the tank, run up a fill pipe and a drawdown pipe, and then seal the joints so they hold water; but later, with livestock pushing and shoving around the tank, the seals around those stationary pipes are going to

Figure 3.15. Overflow from this spring-fed tire tank feeds a second tank in the woods behind it before emptying into the creek bed.

experience some stress. This is not necessarily a big problem; a small leak under an earth-bermed stock tank is not likely to create a mud hole or otherwise compromise the tank installation. But where the volume of flow into the tank is very small, the compromised tank may end up with a refill rate rather slow for the livestock use it receives, meaning that livestock spend more time around the tank and less time grazing. Worst-case scenario, if the leak is big and the inflow is very small, it will drain as fast as it fills. In that case, you'll have to stop the water coming into the tank and try to reseal the joints.

Used tractor tires can be used to build very good stock tanks. The USDA considers tire tanks safe for livestock use, but of course some of us look beyond government regulations when it comes to food and water safety concerns. Tires can be cleaned out with detergent and a strong scrub brush, but there is the tire material itself to consider; tractor tires, so far as we know, are not food grade. We minimize this concern by using our tire tanks for water from a spring with a good flow; where the water is running constantly, as with a spring capture, it is nice to think that no chemicals from the tire composition are building up over time.

To make a tire tank, we select a site that is below the spring we are developing and directly over the trench, dug to below the frost line, where we intend to lay the pipe carrying the captured water. This spot will be dug out to about 6 to 12 inches below the soil surface, leveled, and bedded with crushed rock or gravel. The inner portion of one sidewall of the tire is cut out (this can take a while and use up a couple of reciprocating saw blades) and the tire is then set, cut side up, over inlet and drawdown pipes (temporarily plugged), centering the pipes as much as possible. Dry ready-mix concrete is packed into the tire to a depth of several inches, so that it fills the center hole and covers the lower wall of the tire in a single pad. The inlet pipe is then opened and the tank permitted to fill with water. As the tank fills, the concrete mix is hydrated and, curing slowly below the water surface, will be even stronger than if air-dried. Earth is back-filled around the tank for additional bracing, and ½ ton or more of limestone may be spread around the tank for a mud-free stock pad.

The same method can be used to turn a short length of strong culvert into a stock tank, making sure to berm the installation deeply since there is no lower flange, as there is with the tire, to lock the tank sides into the concrete bottom. For making a stock tank, plastic or concrete culvert sections are preferable to metal, since they will keep the water cooler in summer and warmer in winter. Additionally, rough edges on a metal culvert could easily prove a hazard to watering livestock. Culvert 24 inches in diameter is adequate for all the livestock a smallholder has room for, and such a tank's smaller surface area can help keep water temperatures moderate. As with the tire tank, ideally the area surrounding the tank will be rocked to minimize puddling and pugging.

There may be a temptation, while you're at it, to use a longer length of culvert than the depth of a conventional livestock tank, up to 6 feet or so. Although to do so makes the task of digging it in a lengthy one, the deeper tank takes advantage of geothermal heat to moderate stock water temperatures year-round; in fact, using this dodge allows Canadian cattlemen to set up nose pump waterers that operate in the coldest northern winters. We have so far resisted the temptation to install an extra-deep culvert tank, for two reasons. First, because in this case, the inflow and drawdown pipes would have to be plumbed through the side walls of the culvert, a ticklish business for us seat-of-the-pants farmer types; and second, because deep tanks create a real and unnecessary risk to people and livestock who might fall into the tank. We like to think that our tanks are shallow enough that if anything alive falls into one of them, it will be able to get itself out again.

Figure 3.16. Plumbed together at the spigots, these IBCs fill and empty simultaneously.

Volume

Bigger is better when it comes to capturing water, and an IBC holds five times the water contained by a standard rain barrel. Still, even a 300-gallon tank holds only half of that inch of rain falling on our theoretical 1,000-square-foot footprint. Are we to let the rest run off into the local streams unchecked? Where one IBC is good, two, three, four, or more are even better. On our back pasture, roof water is directed into four IBCs plumbed together to make a 12,000-gallon catchment; in a summer of normal rainfall, this usually provides stock water for eight to ten cows, from April through late November (when we drain the system to prevent freezing).

How to connect these tanks together took some thought; eventually we ended up with an arrangement we think works optimally. For these multitank arrays, we set the tanks side by side on a flat, stable earth base and plumb them all together at the spigots, using Fernco couplings to mate each spigot to a length of 4-inch PVC pipe, with T-couplings at each except one terminal spigot. At that end of the array we attach the PVC pipe with an L-coupling, rather than a T; at the other end, we use PVC adaptors to reduce the pipe diameter to a standard ¾-inch copper hose fitting. With all the spigots open, all the tanks fill—and drain—simultaneously, getting us around the necessity of moving the hoses from tank to tank as they fill and empty.

Stability

It is important for the stability of a rain barrel of any size that care be taken to divert overflow water away from the tank, preventing erosion of the base on which it stands. The bigger the tank, the more important an issue this is—300 gallons of water weighs over a ton, and you want that mass to be stable. If there is an overflow diverter in the downspout above the tank, be sure this water will be delivered to a point at some distance, preferably to a gutter or waterway (for example, the overflow from one of our tanks runs into a creek bed). If you do not have an overflow diverter, the worst thing you can do is to let the water simply run over the sides of the tank, where it will wash away soil from under the IBC. Instead, make an opening as high as possible on the side of the tank and plumb it for a hose or pipe to carry away superfluous water once the tank is filled, remembering that the overflow pipe diameter needs to be sufficient to handle a real downpour.

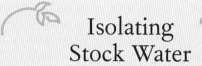

Isolating Stock Water

When we need to isolate stock water for any reason (usually because we want to treat the animals with a natural remedy in their water) we close all the tanks but one, wait until use has drained it of all but the amount of water we want to treat, then add the remedy and let the animals finish the tank. Once the treated water is gone, we simply reopen the other spigots and let the water return to level. The disadvantage to this method is that when we get an unexpected rainfall while some of the tanks are closed, that space is not available for capturing precipitation.

Purity

Anywhere you create an impoundment of water you will have to guard against contamination. Fortunately, with livestock water we need not be quite as particular as with water for human consumption. On our farm, we filter rainwater to prevent leaves

Contaminants

Rainwater, whether collected off a roof or pavement or from a foundation drain, has a number of potential vectors for contamination. It is important, therefore, to know as much as possible about places from which you intend to collect precipitation.

On an older house with painted wood siding or treated cedar shakes, French drain water could carry leached lead or arsenic; it may also be contaminated by chemical fertilizers, pesticides and herbicides applied to foundation plantings, and by commercial mulches treated with fungicide. Lead roofs, or roofs with a zinc strip to prevent algal growth, will leach into runoff as well; you may want to find out the extent to which you can expect your runoff water to be affected. Air pollution dissolves in falling raindrops; if you know your local sources, you can disconnect your rainwater collection system when weather patterns put those sources upwind. Likewise, you can check the Environmental Protection Agency/Clean Air Markets Division (EPA/CAMD) site for local contamination alerts and their sources. Pavement runoffs can include pollutants like gas, motor oil, heavy metals, or polyaromatic hydrocarbons (PAHs); the homesteader might consider diverting initial flow before capturing pavement runoff.

from gathering in our roof-filled tanks, either by placing leaf guards over the roof gutters or putting a screen over the intake of the tank. We set the lids loosely over the top-fill openings of the lower tanks in a multitank array (the water enters only through the first tank), since the water entering these lower tanks through the spigots must be able to push the air in the top of the tank out through the fill hole; otherwise, the tank will pressurize and be unable to fill. Either the lids must be set loosely over the fill holes, or some sort of screen used in lieu of the solid lid. You could, of course, leave the top opening uncovered, if you don't mind the occasional drowned bird, rodent, or amphibian in your collection tank; we prefer covers.

Algae buildup and insect larvae may become issues wherever water is static for any period of time. Placing tanks in a shady spot means less solar energy to heat up the water and fuel the growth of aquatic plants; the north side of a building or slope, under a tree, or covered by an arbor of trailing plants are good places for locating a water storage unit. A bundle of barley straw floated in a pond or tank is said to cut down on algae growth; a few goldfish or koi are a natural and attractive way to manage unwanted resources like mosquito larvae and duckweed.

Cisterns

Cisterns may be buried or located aboveground. Inground cisterns used to be traditional in many areas, where earth-sheltering protected water from freezing in the winter, and many American homes once drew their water from a cistern filled by the house and barn roofs. Even today, there are some good examples of inground cisterns in our area; on one two-hundred-year-old farm the roof of the road-front barn, which enjoys the highest elevation of all the farm buildings, drains into an old brick-lined cistern. Pipes from this run below the frost line to connect it to the house and lower barns. Due to the slope of the hill, in this case the advantages of elevation and of earth-sheltering are happily joined. The farmer, tenth generation of his family to hold the place, knows the value of this free water and keeps the system in good repair; sadly, far more often an old cistern is allowed to deteriorate and is eventually filled in.

A new cistern is a valuable improvement on a farm, if you can afford it. They can be constructed of a variety of materials—stone, brick, cement, metal, plastic. It need not be elaborate or overly expensive; one farm near us uses a 500-gallon polyethylene septic tank to hold water for its four horses and small herd of sheep. Because the tank is located at a low spot on the farm, gravity feed isn't possible, so a small pitcher pump lets the farmer fill a stock tank for the livestock.

Water in the Pasture—Pumps, Hoses, and Tanks

So much for water capture and storage; now to carry the water to where it is needed, out in the pasture. Where the permanent tank or cistern can be located at a point elevated above the desired point of delivery, the situation is ideal. Gravity-fed water not only saves us the energy inputs of pumping, pressurizing, or carrying stock and irrigation water, its availability is independent of off-farm energy inputs like electricity or petroleum to power a generator or pump. But where circumstances make gravity feed impossible, we look at other options: some require electricity and some do not. Pipes and hoses to carry water to pasture and paddock can be rigid or flexible, inground or aboveground; stock tanks may be small or large, stationary or portable. Details of land, finances, and personal skills will determine which options the rotational grazier selects for use on her farm.

Inground Delivery Systems

The system favored by many farmers in our area is an inground, pressurized water system with multiple hydrants in the pasture. Either stock tanks are set next to the hydrant or a hose is run from the hydrant to a small, portable tank fitted with a stock water valve, which can be moved from paddock to paddock over the area around the hydrant. This is an easy solution to the problem of providing water in each paddock, and certainly it has its advantages. Pressurized water flows fast enough to water large numbers of livestock at once, let alone the small number of ruminants on a smallholding, and it is not subject to short-term dry spells, unless you count power outages. And if the hydrants installed are frost-free—that is, they close and drain below the frost line, reducing or eliminating the incidence of frozen water lines—you can use them to fill tanks in winter.

Figure 3.17. This rain barrel holds about 600 gallons, several weeks' supply for a small flock of sheep or goats.

But they have disadvantages as well. Installation is labor-intensive and expensive, and the pipes, once buried, must be *un*buried if they need work. A neighbor of ours, installing an inground system for his herd of thirty brood cows, spent several days digging trenches with a rented Bobcat and laying snap-fitting pipe to three in-line frost-free hydrants. He backfilled the last trench about sundown on the third day and christened his labors with a well-earned lager. Next day he came home from work to find a swamp in his pasture; one of the snap fittings had blown under back-pressure once the stock tanks filled. Out came shovels (the Bobcat had gone back to its lair), and 15 feet of trench had to be re-excavated to find and repair the leak. Next evening there was a new leak at a point farther out in the pasture. Our friend had done enough research into the plumbing materials he was using to know that systems like his are common and not usually subject to materials failures, so instead of ripping the whole thing out and changing pipes he dug up the second leak and fixed it as well. Problem solved (apparently he had just been unfortunate enough to get two compromised pipe joints), and there were no more leaks—*until* one day in January, when one of the frost-free hydrants froze (despite its name), cracked, and made an ice slick over the county road in front of the house. Luckily no passing cars wiped out before he got the water shut off to the system and the ice melted. His herd spent the rest of the winter in the home pasture, where a spring-fed tank kept water before them regardless of the weather.

If you have the money, need, or inclination to install an inground pressurized water system, there are plenty of sources for information to help you do the job right. The local NRCS or SWCD office would be a good place to start. Their staff is usually experienced in the design and installation of stock water systems and can help you with technical advice for your application. They will know what strength and pipe diameter will be appropriate for your system, and help you avoid the experience of one of our local farmers, who ran a water system out to the hill pasture above his house and got a basement full of water when the back-pressure from the uphill column of water in his supply pipes blew the fitting in the basement. Thinking to improve his water supply, he had substituted a larger diameter field pipe than the system design called for, and the greater weight of water exceeded the strength of his pipe junctions. A constant supply of water in the field is a real plus; just make sure you do it right.

Figure 3.18. Both fences and water supply may be portable and temporary, a real advantage on borrowed land.

Gravity-Feed and Surface Pipe Systems

But pressurized water is not the only way to get running water to the pasture, and inground systems, while they have many advantages, are less flexible in application than pipe laid over the surface of the ground. Many graziers, on large operations as well as small ones, have come to prefer overground pipe systems because of the lower cost of installation and ease of setup and adjustment. On a large farm, semirigid pipe branching from a main line, with quick couplings at intervals along the lines, allows the grazier to move a single small tank with the animals from paddock to paddock, hooking up to the water line with a short hose and water valve. This makes it possible to use a small, even very small, stock water tank; because the tank is in the paddock, animals drink whenever they are thirsty, instead of bunching and making the trip back to water as a herd. This means that the tank volume and refill rate need not match a sudden surge in use but will be draining and refilling more or less steadily over the day. Otherwise, when livestock visit the tank as a group, as they will if the tank is at a distance (bunching for water hole visits is an instinctive protection against predators), the sudden demand for water may easily exceed the rate at which the tank can be refilled. Larger, more dominant animals drink first and then return to the pasture, with younger and less aggressive animals going without a drink rather than being left behind. Small paddocks

Figure 3.19. Hudson valve for low-pressure regulation.

with water always available in the paddock avoid this situation.

Permanent overground water pipe is often installed under permanent fence lines, this having the dual advantage of putting the pipe out of the way of regular traffic and in the shade of the ungrazed forage under the fence. Shaded pipes keep the water in them cooler, an advantage in summer. Warm stock water is not as cooling; hence animals drink less, resulting in less efficient grazing and forage use (read lower milk and meat production). In addition, warm water is more apt to grow algae and harbor pathogens.

But on the smallholding, it is possible to run water out to the pasture without even the formality of a permanent rigid overground pipe arrangement. Our systems use simple garden hose, hundreds of feet of it, to carry water where we need it in the pasture. This is delivered to half barrels plumbed with low-pressure stock valves set wherever the animals are grazing. This simple, low-tech, quick installation system serves us from April through some time in November or December.

Low-Pressure Stock Water Valves

A number of different valves are available for use in low-pressure stock water systems. Among these are the Jobe float and Hudson valve, both of which we use on our farm in gravity-feed systems with low pressure and a slow refill rate. Jobe floats are simple to plumb into a half barrel, where they are not subject to interference by curious animals; the Hudson valve simply screws to the male end of a garden hose and can be dropped into any tank you want to fill, provided the livestock can't pull it out again. We generally fix a 2-by-6 brace across the top of a tank we want to regulate with a Hudson valve; with a hose-diameter hole in the brace, we can suspend the Hudson valve where we need it and prevent animal interference at the same time.

There are few problems with it; occasionally a hose will spring a leak and need to be mended, or one of us will get careless and drive over a fitting and bend it. These repairs are easily made and take only a few minutes and a couple of dollars in parts—perfect for our small operation. No pipes to be dug up; no expensive PVC fittings. It does require a little more *oof* to pull 70 feet of garden hose where you want it to go, and we do have to carry additional hoses out to the field when we need to lengthen a water line, but then part of our love of farming is a love of the simple labors involved. We don't always feel like hauling hoses on a hot day in August, but the effort, even the annoyance, slight and brief as they are, are some of the spice in our farming life, intensifying the pleasure of the moment when,

Figure 3.20. Tools of the trade. Courtesy of the Franciscan Sisters T.O.R. of Penance of the Sorrowful Mother.

soaked in sweat and hot as Hades, we stop for a few minutes under the pasture apple trees to cool off.

A simple garden hose arrangement is within the budgetary limitations of even the most impecunious grazier; we have often collected old or damaged hoses from friends or curbside trash piles and, for the cost of mending them, had an extra 50 feet of water line. Other possibilities suggest themselves: PEX, or some other flexible hose, would serve the purpose just as well. For us, plain garden pipe, easily lengthened or shortened by adding or removing additional hoses, is the simplest solution we have found so far. The biggest chore involved comes when we drain the cisterns at the onset of cold weather and coil the hoses for storage.

Stock Water Tanks

Stock tanks come in many sizes and compositions. On the smallholding, anything bigger than about 100 gallons is probably going to be unnecessary, the number of livestock being generally small and the frequency of our visits to the paddock, and chances to check on water supply, being great. Black rubber and galvanized steel tanks of 100 gallons or less are common at the farm store, and other, repurposed containers can make perfectly serviceable stock water tanks as well.

The size tank needed for your grazing animals is going to depend on the number, age, size, and species of your livestock, as well as the refill rate of your water source. Basically, your stock tank needs to hold *more* water than your animals will need over the period of time expected between refills. If you are filling your tank by hand (with a hose or bucket), this means that it must hold sufficient water so that when something delays your return to the pasture, the animals don't run out of water right away. This is important; livestock can be distressed very quickly, even die, when they cannot access water on a hot day. The tank needs to be stable, as well, sitting solidly on a wide base so that when animals choose the tank for a scratching post, or begin jockeying for position around it, it can't be tipped over, even when the water level is low. Its sides need to be low enough for the smallest animals to drink even when the tank is less than half full; calves, goats, sheep, or pigs that can reach the water when the tank is full may well go without a drink once bigger animals have lowered the water level. One can find—at a price—stock tanks with side reservoirs designed especially to allow drinking at two levels, very handy for the multispecies herd including both large and small animals.

Stock tank materials can vary, but they need to have certain qualities. Tanks that are going to hold static water for an extended period (as opposed to, for example, a tank filled by a constantly running source—from a spring or ram pump, for example) need to be made from materials that will not leach into the water. There is some latitude here, obviously; tanks made from tractor tires are not food grade,

Figure 3.21. Lightweight stock tanks go wherever the livestock are.

but the USDA considers them acceptable for stock water use. (Our own tire tanks are not used for static water storage but run constantly, so that any undesirable chemicals in the rubber have little time to contaminate the water.) On the other hand, our small tanks made from plastic 55-gallon drums cut in half may hold water for a number of hours. For that reason we avoid using barrels that have contained anything incompatible with our farm principles and with the uses to which we will put the barrel. We have turned down drums that previously held windshield wiper fluid, for example, due to our uncertainty about what method of cleaning would be necessary to make the barrel fit for use—or if any cleaning at all could do so.

It is important to use durable stock tanks, both so they will stand up to the abuses of livestock—goats or hogs propping their feet on the edge, yearling calves shoving one another for position at the tank—and because of the frequency with which they will be dumped and moved in a rotational grazing system. Buy, beg, or manufacture tanks that will stand up to regular tipping, and will not develop leaks from being pulled or dragged over rough ground. With that in mind, they should be as lightweight as is compatible with the necessary volume and durability required, since you will soon get sick of moving an overly heavy tank all around the pasture. Not too light; a tank that is easily tipped will give you more headaches than will the periodic necessity of moving a heavier tank. Rubber and plastic tanks have worked well for us, being relatively lightweight, conducting little heat, and being less likely than galvanized steel to develop cracks or open seams when water freezes.

Stock Water Valves—High and Low Pressure

In recent years there have been some simple but handy innovations in stock water valves for low-pressure use; these, with the old standard float valves for pressurized water that have been around for generations, give the grazier a number of options for keeping water in front of the animals all the time. Other types of valve for watering individual animals on demand are available, too. Which one or ones we find most useful depends on our situation.

Valves should be selected first of all for appropriateness to the water source. Relative water pressure, and the presence or absence of particulate matter in the water, have a lot to do with determining which valve will work best in a given application. Jobe floats, Hudson valves, and similar low-pressure regulators may pose insufficient opposition to the flow of water under high pressure; and while we do use some high-pressure float valves with our captured water, gravity flow (low-pressure) systems, not all high pressure valves may serve in such a situation. Check out what is available at your local farm store, or better yet, look online at a reputable farm catalog to determine the best valve for your situation. While well water and municipal water should have few particulate impurities, pond, creek, or captured rainwater sometimes carry solids that hinder the function of a stock water valve, either blocking the valve so water cannot flow, or jamming the valve open so that flow does not shut off. We have had both problems, but we find, by and large, that the openings to our low-pressure valves are large enough for small detritus to flow through without impeding the function of the valve; however, periodic cleanout is sometimes necessary, as bits of algae or leaf build up over time. Your own situation may be different; some filtering of collected water may be necessary at the intake end of the hose.

Small Animals: Cows, Swine, and Sheep

Hogs are especially susceptible to overheating, and it is important to make sure that pastured hogs have water available at all times. Stem valves, also called pig nipples, and nose-operated valves are a handy way to present water-on-demand to pastured pigs. These are valves designed for low- and high-pressure systems, and can be a real asset when dealing with large hogs that have both the strength and mischief to turn over an ordinary water tank. Nose-pressure valves can be mounted on the side of a large tank; galvanized or rubber stock tanks with a valve already in place are available from many farm catalogs. Pig nipples can likewise be plumbed into

Figure 3.22. This pig nipple on a step-in post lets us put water anywhere on our tilted acres.

the side of a tank, or welded to a two-prong step-in post with attached hose. We use both with equal success; we are real fans of the pig nipple. For one thing, it is resistant to leakage: a tipped stock valve or slanted tank can drain off 1,200 gallons of captured rainwater in about thirty minutes, as happens when our miniature horse Bridget gets bored and yanks the float valve off her water tank; stem valves, on the other hand, are only open when the stem is depressed, as by the mouth of the animal taking a drink, and close automatically when the pressure is removed, preventing extended leaking. Stem valves can even be installed at an angle without leaking, so that in winter, water does not remain in the delivery trough to freeze and prevent the valve from being depressed.

Leaks of any kind are the bane of captured water systems, dependent as these are upon uncertain and intermittent rainfall for refilling; and even for a pressurized water system, a leak in a water line can be a real problem and result in a real loss of efficiency. We check our system often for leaking hoses, and stay especially aware of areas where lines run underground. Contamination of stock water can occur when a line cracks beneath the surface, especially if proximate to the barnyard, and the muddy spots above an inground leak are breeding places for flies and vectors for parasite transmission. And even a small leak in a hose joint can result in the loss of many gallons of water over a few days, water we may need for our thirsty livestock.

Winter Water

There is no doubt that the single biggest headache we face in our cold Appalachian winters is keeping running stock water in front of our pastured animals. Not infrequently, with nighttime temperatures below zero, we have had frozen "frost-free" valves to thaw, and ice-clogged hoses—despite the care we take to drain them every time we use them. An inch of ice on even our spring tanks is evidence that the cold can sometimes penetrate below our usual 2-foot frost line. Hog nipples freeze at night and have to be checked multiple times to make sure they thaw long enough to give the hogs access to water during the day; and hens watering from Jeddo's Run have to have hot water carried to them when the creek freezes over. Keeping winter water available can be a big chore.

Pressurized house water can be accessed from outside the building when we install frost-free spigots that shut off the water flow inside the house. In our experience, these only freeze in the coldest weather—but when they do they can make a real mess (experience speaking). With inground high-pressure systems installed in field or barn, some of the work of winter watering can be lessened if stand-alone frost-free hydrants that drain below the frost line are installed, but even these are susceptible to freezing during extended or extreme low temperatures. Electric stock tank deicers, either floating or submersible types, can be installed on a tank to warm water above freezing point, but since these depend upon access to electric current they have limited use while our animals are out on pasture. And just as inground pressurized water comes with a price tag, so does operating a stock tank deicer—at present electrical rates around here, about a dollar a day. No kidding. So other, lower-tech options are definitely worth checking out.

Nature's deicer is the first and best in our experience. Moving water, so long as it keeps moving, resists freezing even at temperatures below zero, if the cold is not extended over very many days. Jeddo's Run and North Creek run all winter long, seldom icing over completely and thawing quickly when daytime temperatures rise above 32 degrees

Figure 3.23. Moving water seldom freezes.

for even a brief time. The tank behind the winter calf shelter fills from the overflow of a spring-fed tire tank farther up the hill. This water runs through an aboveground pipe delivering to a point about 6 inches over the lower tank; the surface disturbance of the falling water keeps that tank ice-free even when the spring tank above it is half iced over. Another device for keeping water from freezing is the pond mill, a small, floating, wind-operated mill that creates turbulence to prevent pond water from freezing; back in Oklahoma, we used one of these on our farm pond with good success. And there are various other ways to keep a full stock tank from freezing, including many home-construction projects like insulated half covers, black tanks for collecting solar energy, or setting a tank within a tank and filling the space between with heat-generating compostables, like chicken litter, a method one of our local farmers has found satisfactory. A warm drink on a cold day goes a long way to improve any animal's comfort and well-being.

Behind our big barn, the spring-fed water barrel has continued to run all winter, each of the four years since we improved that spring. The sheep, when they are finished grazing their stockpiled paddocks and go into the barn on hay before lambing, drink from the top of this barrel, using the earth-filled tire berm surrounding it to give themselves a leg up. The barned hogs drink from a pig nipple in the side of the same tank; this, being metal and extending a few inches out from the side of the tank, will freeze in extreme weather even though the tank does not. But there are remedies for a frozen stem valve like a pig nipple. We have been told that a copper nail with the head clipped off can be inserted in the back of a pig nipple so that it extends a little way into the body of the reservoir, to act as a heat conductor from the (relatively) warm water in the tank to the inside of the valve, and this seems as though it should work in most cases. But since in this instance the valve is on a 4-inch steel extension pipe, other methods have so far been necessary. An ordinary, floating stock water heater is inadequate to keep the valve unfrozen since it heats the water at the top of the barrel only, but we have found that a submersible heater, suspended so that it hangs just behind the valve, will prevent freezing on most cold nights and thaw the valve in a reasonable time if it does freeze. What we have in the pipeline for warm water this winter is a Jean Pain mound, a large pile of composting wood chips used to heat slow-moving water in a coil of pipeline. A well-built mound is reported to have the potential to heat water to 115 degrees or higher at the rate of around a gallon per minute, which is more than enough to warm the incoming water in the big barn tank.

Hand-Filling Tanks

In the worst weather, when we move the dairy animals up near the barn for a few weeks, we have recourse to the labor-intensive but low-tech option

Figure 3.24. Filling stock tanks by hand requires frequent visits to the paddock or pen. Courtesy of the Farm School, Athol, Massachusetts.

of filling stock tanks by hand with a garden hose run from a frost-free spigot on the house or an outbuilding. Good quality hoses are necessary; cheap hose breaks easily in sub zero weather, and plastic hose ends will sometimes split under water pressure alone. Alternatively, a neighbor of ours buys several "pocket hoses" (quality uneven, she says, so buy an extra one or two) and uses them carefully. Several will fit in a 5-gallon bucket, which means she can take hoses indoors between uses, obviating concerns about freezing. An electric stock tank deicer keeps tank water from freezing, and we have to refill usually once a day, draining the hoses as conscientiously as a firefighter so they don't freeze and become unusable. We don't move water tanks very often in winter but instead create lanes from the animals' paddocks back to water. Normally, of course, we avoid stationary water tanks, because of the pugging, trailing, and puddling that happens around them, but once the ground freezes we seem to be safe from impact of this kind; even when we have used a stationary tank for a number of weeks, we have seen no significant loss of forage production in those places when spring arrives.

Finally, it is possible, and used to be common, to water animals twice a day, filling tanks only as much as the animals drink so there is no concern about freezing. Many people manage poultry this way, carrying out hot water morning and evening, and dumping any ice that remains from the previous watering. Mounds of bowl-shaped ice blocks outside a chicken house door attest to the simple practicality of this method on the small scale, and there is no reason it can't work for larger animals. Our reservation is this: making access to water only periodic increases the urgency of our being extremely conscientious about chore time. When every drink of water comes directly from our hands, there can be no skipping a trip to the barn, and no delay; and no possibility, as there is with, say, hay-fed livestock, that you can put out an extra feed allowance and skip a watering. Watering by hand means chores must be done like clockwork.

Record Keeping

Water is the single most limiting factor for when, where, and how we may graze our livestock. Record keeping—keeping track of the movement, timing, and performance of livestock over the whole farm, for the whole year—is an essential tool for helping us work out efficient, year-long use of our grasslands. Seasonal availability of water from the various sources on the farm dictate to some extent what kind of forage grows there, how much, and in what season, as well as how many livestock may be grazed there, and for how long. This matters, even on a 5-acre homestead.

In our own grazing schedules water comes first. On the back pasture, where stock water comes from roof-capture storage tanks, we get our most complication-free grazing during the warmer months. From April to November, the sky sends the vast majority of our stock water and keeps it from freezing; in the winter, we have to run hoses and extension cords to keep tanks filled and open. In the west pasture, on the other hand, we stockpile forage (leave mature forage ungrazed) for winter use, knowing we can use the (mostly frost-free) spring tank in the winter months. At the bottom of the home pasture, a tiny captured seep—really just a shallow depression lined with stones—fills an IBC at a slow drip, which is adequate to water six ewe-lamb pairs from April to July most years; but later in the summer, we may have to run a hose from the big barn spring to provide water in those paddocks. Keeping track on paper of these and other seasonal variations helps us every year as we try to plan for improved efficiency of forage management.

CHAPTER FOUR

Grass

You'll gain more produce by sowing and planting what the land readily grows and nurtures than by sowing and planting what you want.

—*Xenophon*

According to the Food and Agriculture Organization (FAO), 40 percent of the world's land mass is grassland; arguably, it is the largest solar collection force on the planet. Grasslands provide food and habitat for the greatest number of the world's herbivores, not to mention the smaller

Figure 4.1. Fall pasture, weedy but palatable. Courtesy of the Franciscan Sisters T.O.R. of Penance of the Sorrowful Mother.

lives dependent on grasslands for nutrients and shelter. With its pattern of growth/grazing/root dieback/regrowth, grass builds topsoil, providing nutrients for soil microbes and mycorrhizae and long-term storage of groundwater, while keeping soil temperatures low and modifying microclimates with the cooling action of transpiration. Grass, clearly, is of vital importance to the health and stability of the planet.

And yet, what is grass?

Most of us, if we think of grass, think first of a lawn. Front yard, golf course, playing field; Kentucky bluegrass in the south, St. Augustine grass along the Gulf coast, Bermuda grass in climates drier still, these are mowed areas of generally monocultural ground cover. In a rural setting we see other, larger expanses of grass, often with grazing cattle; this growth is taller, rougher, dotted with the large broadleaf plants we think of as weeds or shrubs. Sometimes we may see a hay meadow before cutting, tall grasses and legumes—timothy and alfalfa, maybe—making a gray-green carpet to the tree line, or a wheat field in early spring, the closely spaced young plants a smooth, verdant carpet; all of these are grass. And yet when we begin to think as graziers—*grass-iers*—we find that we have to give grass a new definition.

Grass, Pasture, Forage

Pasture, in a conventional grazing system, is often a limited polyculture, fields seeded to forage species that the farmer, or more likely the extension agent or seed salesman, says are best for that area. Agriscience has over the years developed many different grasses and legumes for pasture, specialized plants bred to provide forage in wet season or dry; cool or hot; for cattle, sheep, or horses. Our own fathers, raising cattle in the Southwest, planted sorghum Sudan, johnsongrass, and lovegrass; elsewhere, the rage was for fescue—a silver bullet, farmers were told, for nearly every grazing situation. But the fact that seedsmen are still developing grasses at a rate, if anything, greater than that of forty years ago, and that farmers are still plowing up their fields and replanting, is demonstration enough that the perfect pasture grass is yet to be developed, and evidence that planted pastures are not a low-maintenance, fix-it-once-and-you're-done enterprise.

Even what we mean when we say the word *pasture* varies according to our goals and methods. In a conventional grazing operation, animals are given the freedom to wander and forage over large areas for extended periods. Water is provided at a fixed point, and whatever the land produces may be grazed and regrazed at any time, and in any state of maturity, for as long as the stockman deems there to be adequate forage and water available. When prolonged, unlimited grazing inhibits or kills the existing ground cover to the point where it will no longer support livestock, pastures must be reseeded, usually with one or a very few plant species, which will in their turn be grazed down and planted over. Grass, in this picture, is a conventional crop sown for harvest with ruminants, the profit on whose sale being hopefully sufficient to pay the cost of the grass seed and drill and finance the resowing of the pasture next season.

In pastures under intensive rotational grazing, on the other hand, native species are preferred and are generally not sown but rather encouraged. This is good news for the smallholder whose goal is to manage natural systems for an increase of fertility and diversity. Introducing intensive rotational grazing on your own plot of land doesn't need to be—*shouldn't* be—an operation entrusted to the local extension officer and the recommended protocols of the state department of agriculture. Whatever may be growing on the land initially, it is, in all likelihood, potential food for herbivores. We want to learn all we can about the plant species represented there, but as an observer of the ecosystem, not assuming the responsibility of bringing in new species or voting others off the island.

We have observed in our own area that rough, weedy, and rocky pastures, such as produce lots of perennial broadleaf plants like ironweed, ragweed, and goldenrod, will, after a year or two of intensive rotation, exhibit a rich diversity of grass species and legumes that we can only account for by saying that, as pastures grow in fertility, they *want* to increase in biodiversity. Seeds of diverse native species that have lain dormant in the soil for decades, or sprouted and born seed as an inconsequential percentage of the pasture makeup while conditions were unfavorable, germinate and spread as managed, timed grazing opens up space for them in the plant community. Many respected graziers, including Allan Savory, Joel Salatin, and Greg Judy, encourage a wait-and-see approach to pasture forage composition, noting that biodiversity is favored by this hands-off attitude. Fertile soil is by definition richly diverse in soil biota, and the diversity belowground will be reflected in diversity above. If we trust our pastures, trust our ecosystems, we can put our animals on grass and watch the array of forage species unfold.

Elements of a Good Paddock

A good paddock has five basic elements: grass, water, shelter (shade), minerals, and a charged fence. The first of these, grass, is the constant study of the grazier: it includes the species, consistency, and condition of the forage; its state of maturity; and the duration of the grazing period for a given paddock. Learning to judge these well is an ongoing education, but even in the early stages of acquirement, management-intensive grazing is a practice

When Is Good, Good Enough?

Total digestible nutrients, number of tillers, Brix, stock density—these are words, phrases, and ideas to help the grazier know how well he is using—and building—his forage. But so many new, unfamiliar terms can be overwhelming. Fortunately, our goal is to rejuvenate a parcel of land while we produce food for our home and our farm, so the level of exactitude necessary is much lower than for operations whose success is measured by monetary profit. We are trying to manage our animals for the conversion of sunlight into forage, forage into milk, meat, and manure—to build a homestead where energy cascades from organism to organism with a net gain for the ecosystem as a whole—and we want the system to feed us while we do it. Whether our harvest achieves the absolutely greatest possible ratio of human-available output to fertility input—the most meat and milk relative to litter and topsoil—concerns us but little, at least at this stage of the game. For measuring success according to homesteading standards, our eyes, hands, feet, and memories can give us all the information we have to have.

Figure 4.2. Early summer pasture is high in protein.

that can only improve your land. Which is to say: *There is no such thing as bad rotational grazing; only good and better.* It helps if we can keep this principle firmly in mind, especially in the beginning, when we don't really know what we are doing.

The goal of the intensive grazier is to imitate the effect of large herds of feral herbivores on permanent grasslands. These animals migrated across rich and abundant savannah or prairie, grazing, trampling, and defecating, leaving behind them a heavy layer of litter and manure. Hence, the forage in a paddock to be grazed needs, first of all, to have good growth on it. You don't want to begin your rotational grazing on a freshly cut lawn or field; it is necessary that the forage have sufficient height, such that, when the animal or animals in the paddock have full bellies, there is still plenty of plant material on the ground. This serves to protect the soil from erosion by wind or rain; to shade the ground to prevent moisture loss through evaporation—with subsequent overheating of the soil, killing soil microbes and sending plants into dormancy; and to provide organic matter for building topsoil. Remember, also, that leaves are your solar collectors, the dynamos that power the farm; the grazier wants to leave plenty of solar collectors for prompt and efficient regrowth.

Keep It Simple

This is one reason we try to build paddocks that are fairly consistent. This doesn't mean that we don't mix types of forage—native pastures are polycultures, after all—but we don't include such widely varying topography in a single paddock that the area will be grazed inconsistently. An extreme example from our own farm would be a paddock including a weedy patch thick with blackberry canes (which we may be trying to suppress) with an area of rich clover; the grazing pattern you would expect to see in this case, of course, would be untouched canes and overgrazed legumes (if your animal is a cow), or browsed canes and untouched legumes (if you are grazing goats). The same advice holds true in the case of extremes of terrain: if we were to build a paddock in which were included an area of our deep grass bottomland with a section of the steep shoulder of the south hill, the animals could be expected to graze, lounge, and defecate in the level area that is easiest to access, while the hillside, in greater need of the remedial effects of grazing than the bottomland, would go untouched. The exception—there are always exceptions—is when we are grazing mixed species; in this case, we may include briars for the goats, browse for the sheep, grass for the cows. Our goal is to achieve consistent impact; keeping this in mind helps give direction to our forage decisions.

Take It or Leave It

How much to take and how much to leave is a matter as much of art as of science, a question answered differently by different graziers, and even by the same grazier at different times and places. Species of animals to be grazed; duration of grazing period; pasture composition; season; weather conditions past, present, and future—all of these go into the decision making process when animals are grazed holistically. Nevertheless, improving land and building resilience through rotational grazing is possible even for graziers with no previous experience, even on a very small scale. Some basic rules of thumb are handed around the rotational grazing world:

Take half, leave half. This should be self-explanatory; nevertheless, it takes some experience and attentiveness for the infant grazier to get a sense of what "half" of an area of forage looks like. "Half" is half by volume or mass, not half by height; keep in mind that pasture plants are denser closer to the ground. Also, ruminants are not lawn mowers; unless a paddock is so small the animals scalp it, preferred forages will always be grazed more closely than less favored species. Remember the goals: well-nourished animals leaving plenty of litter (and manure). Rough, trampled areas with manure piles every few feet is about right.

Sixty/thirty/ten. Percent, that is: 60 percent of the forage eaten, 30 percent trampled, 10 percent still standing. It takes a while to get where you can estimate the percent eaten—after all, that part is in the animals' stomachs—but you get the hang of it. We use this one a lot.

Graze only the tops. This means more rotations during the growing season, less impact per rotation; or, in other words, bigger paddocks, shorter rests. The idea is to postpone the stage where grasses form seed heads and stop growing, keeping forage in a vegetative state for a longer grazing season.

Duration of rest. This rule helps the grazier determine paddock size according to the period of time he intends to let pass before grazing this paddock again. The desired period for rest and regrowth varies for different times of the year and under different conditions. There are no really hard-and-fast rules, but some generalization is possible. In our area most graziers agree, for instance, that in the early spring animals should move rapidly over the pasture, leaving no hard grazing lines and hitting the whole area lightly over a period of two to three weeks. In early summer the rate of movement slows, and grazed paddocks are built smaller, increasing the time it takes to get around the farm to thirty, sixty, or even ninety or more days. In arid regions, a paddock may be grazed as seldom as twice in a year; sometimes not even so much.

Plant mass. Your local soil and water conservation office may have available tools for helping the beginner to judge the mass—literally the weight and volume—of the forage available in a pasture. The grazing stick is marked on one side with a grid on which there are small dots in various numbers and arrangements. The tool is slipped into plant growth at ground level, and the grazier, standing upright, counts the number of dots visible through the foliage. With this number, and information printed on the remaining three sides of the forage stick relative to different types and seasons of pasture composition, it is possible to estimate the amount of forage standing. Further calculations relative to the number, ages, size, species, sex, and state of gestation or lactation of the animals to be grazed can give the grazier a ballpark idea of how to determine his paddock sizes and durations. This method has the advantage of lots of objective calculation, which may give the infant grazier confidence in his decision making process. It is also (comparatively) lengthy, complicated, and (for people without essential math genius) fraught with possibilities for error. We own a grazing stick; we have never used it. If you don't just love math, you can leave this tool alone.

Number of tillers and growth points. For the science lover. Your soil and water office will probably have literature available describing "optimum" grazing impact for various forage species in your area, or you can find this information online.

Every plant eaten, stepped on, manured, or urinated on. Self-explanatory.

Figure 4.3a. Showing 60 percent grazed, 30 percent trampled, 10 percent standing.

Figure 4.3b. In early summer, impact may be very light.

Figure 4.3c. Still a desirable level of impact.

It will be seen from the above list (just a selection from almost unlimited possible systems) that while the goals remain the same—that is, increased solar energy capture, deepening topsoil, greater fertility, more and better food from the land—the means adopted can vary considerably. The multiplicity of approach might be intimidating, if we thought that for successful grazing we needed to understand all the possible intensive grazing methods so that we could be sure to apply the best one to our own homesteads at any given moment. A better view is that, with so many approaches to one goal, it is easier for the amateur to succeed, and harder for our efforts to produce negative results.

Modern technological and educational systems condition us to look for "right" answers and to expect consistent practices to produce consistent results, but you can forget the idea that there are hard-and-fast rules for best grazing—or for practically any other aspect of farming, either. It has been important in our own grazing efforts to realize that nature is not static. In a thousand years graziers will

Changes over Time

There are books and magazines, print and online, just for practitioners of intensive grazing, and they make great reading and are consistently inspiring. Following an author over several years, though, we find his style changing, his preferences for before-and-after grazing height, duration of grazing period, or extended rest varying from year to year; and, since all the advice might sound good, we find ourselves scrambling to adjust to each new quirk. It's all good, but we wish we could keep up with the learning curve! And, we wonder, do the new methods mean our old ones were wrong?

More experience on our own farm has taught us to see things differently. Intensive management of ruminants has been around for a long time, yes, but most of the people at the front of today's grass-farming movement have only been doing it for a few decades at most. This works out fine for us, because as they help people new to the art, they remember what the questions are; they are still aware of all the things we don't know intuitively. It usually means something else, too: that the variables they are working with are changing all the time.

Most people—nearly all—start out grazing on land that has been badly managed, at least for the purposes of all-grass farming, for as long as the present owners can remember. The land in these cases is as badly in need of remediation as we beginning grass farmers are in need of tuition; and for at least the first few years that we graze a piece of land, the soil (its depth, water retention, and biodiversity) and the forage (both in composition, health, and resilience) are changing even faster than our growing sensibility to the nuances of intensive grass management. Naturally, the best-use methods for a changing landscape are also going to be changing. Heavy pressure from perennial weeds, a disadvantaged understory, and the need to keep the ground shaded might mean mob grazing the first year or two on a piece of property that, three or four years down the line, with tall perennials less prevalent in the pasture community, might better reward the farmer if she runs the dairy ruminants over it in large paddocks and then follows them immediately with the meat animals. Or, starting out with a lot of bare soil, we might initially build small paddocks on a fast rotation and offer hay to make up the balance of the diet; then, when increased soil depth, water retention, and forage production make it possible, we dispense with supplemental feeding and shift our management practices to increase the percentage of this or that forage species, or for a longer active growing season.

Which is all to say that it is natural that your methods, anyone's methods, are changing over time, not just because we are learning new and better ways to care for the land under our management but because the land itself is changing all the time. And after all, isn't that what we're farming for?

still be discussing what constitutes "best"; goals differ, and success is defined in large part by desired output. For the smallholder, whose intention is to increase the fertility of his land and to build security and resilience into his food production system, the desired outputs are food and fertility, not cash flow; by this estimation, if we and our animals are eating, we're still in the game.

Estimating Forage

How much grass we should put in a paddock is a moving target. A lot depends on the plants themselves; even more on our livestock's varying needs at different times of year, of life cycle, of gestation or lactation. Often, adequacy is a thing determined after the fact, by looking at the animals and forage in question. Are the animals restless long before it is time to move them onto the next paddock? Are they waiting at the fence before we are ready to move them; do they challenge their fence often? If so, we want to observe the paddock more closely, and if the forage is nibbled off very short, we make the next one larger. A paddock in which a large proportion of plants are untouched while certain species are grazed hard would tell us to outline a smaller area for the next grazing period. Paddock planning, like everything else about rotational grazing, is learned by doing; there are, however, some tools and rules intended to make the process a little easier.

As a tool for estimating paddock size, after observation of the pasture and animals, there is understanding the forage species themselves. How high are they when we put the animals in? When we take them out? How does this compare to their expected height when mature? In what state of growth are they—active growth, in bud or bloom, vegetative or mature, green stems or brown? Does the forage look more like a lush jungle or more like a collection of broom straws? Very young, tender green growth with no signs of flower or fruit (seed head) is distinctly different in composition and food availability—what the commercial cattlemen call TDN, or total digestible nutrients—from coarse, brown, mature grasses. Paddock size will be largely affected by the growth stage of the forage species.

The grazier who keeps a dairy animal has a daily, even twice-daily, means of knowing how well he is gauging his paddock size and composition: milk in the bucket. Milk quantity and fat content will rise and fall with the forage available to the lactating animal, and when the dairyman fills jars, or skims off the cream from the previous milking, he will know something about the grass in the last paddock. Making daily trips to the pasture, handling animals once or twice a day, and recording their milk and cream production gives the dairyman a distinct advantage in learning the ins and outs of rotational grazing.

Simple record keeping is a must in an intensive grazing system, and an important tool for determining when and how long to graze a paddock. One simple aid for judging forage is leaving stationary measuring posts, against which forage height can be estimated. The fiberglass posts we use most often have black plastic clips visible even from a distance; these allow us to judge the level of growth or regrowth in a paddock. Note is made of the forage height when animals are turned into a paddock, again when they are taken out, and at intervals during the rest period. In this way it is possible to keep a grazing history for individual areas in our pasture, allowing us to predict future growth patterns.

Best of all, over time one learns to see the success of rotational grazing in the livestock themselves. Shiny coats and clean tails indicate overall health and low parasite load. The body form appropriate to your species and breed of animal—meat animals blocky, ribs not showing; dairy animals with more prominent but still padded rib and hip bones, and full but not overfull udders—let us know that our livestock are getting enough nutrition from their forage. Walking behind the younger dairy cows as they come up to milk, we notice the fat around the

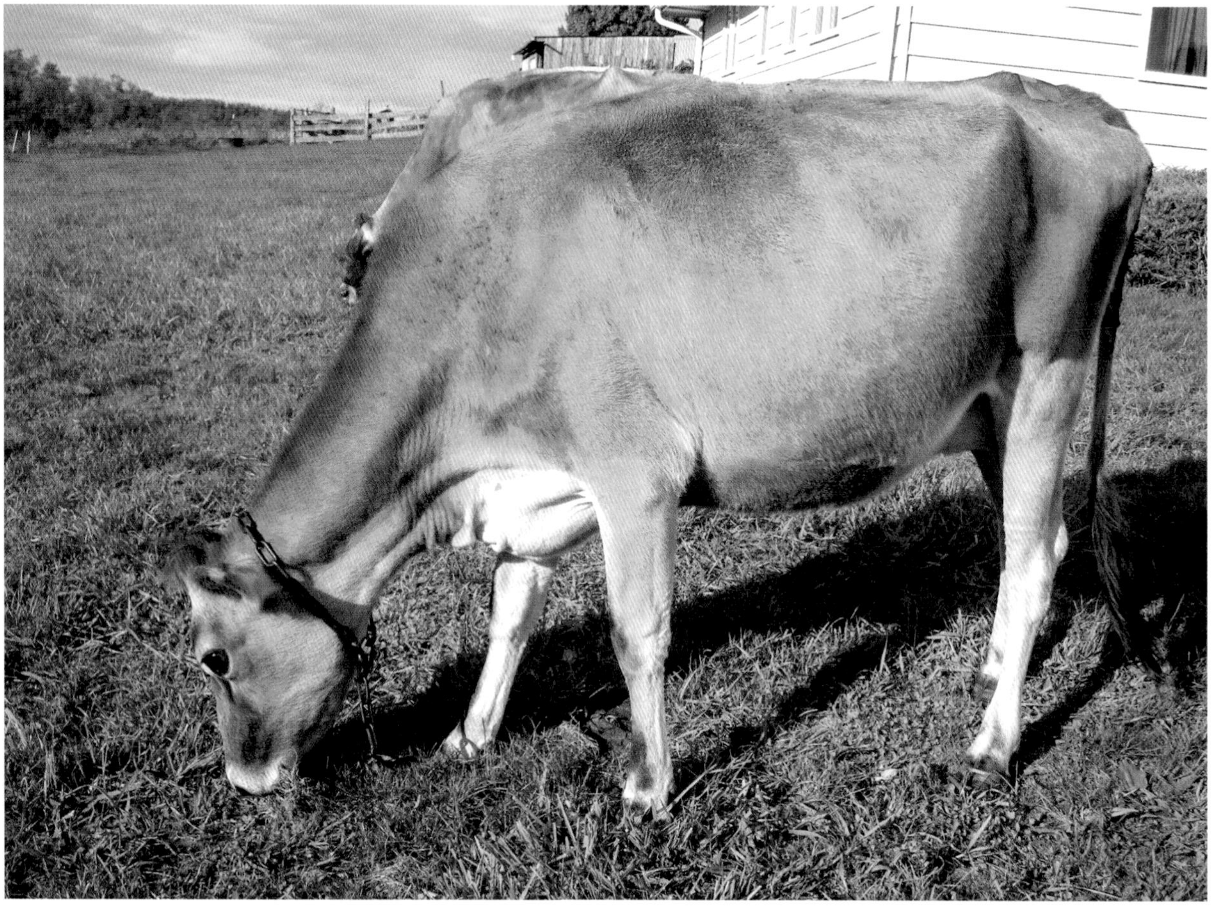

Figure 4.4. The area under the left hip bone is well filled out, indicating a full rumen.

Figure 4.5. New growth showing through litter.

base of their tails, giving them a comfortably filled-out look; if you approach a cow from the left, you see that the space under her hip bone is well filled and taut, indicating a full rumen.

Ultimately, our method of determining paddock sizes, patterns, and durations needs to work on the practical level; it should be neither complicated nor require lengthy calculation. So we look at the forage, walk the land, count our days, step out our paddocks, and make a decision. Then we observe the results. If, after the grazing period, there are areas taken down so far the soil is exposed, we needed a larger paddock or a shorter

Poisonous Plants

We have found the presence of toxic plant species to be of far less importance than we once imagined. Our animals in general have a much better idea about what they should eat than we do, and of course they have far more opportunity to study the composition of our pasture. When we put them out on a limited area, so long as the quantity of desirable forage is adequate, they will find and eat what they need for health and growth, leaving the bad stuff alone. There are some poisonous weeds common to any part of the United States, and while it is good to be aware of the ones in your area, chances are your animals will avoid them naturally. Don't be afraid that, because intensive grazing paddocks are small, livestock will be so hungry as to eat toxic weeds out of desperation. And in any case, it is much more likely that, in the beginning, you will err on the side of making your paddocks overlarge, not the reverse.

Jimsonweed and nightshade, both toxic, are very common in our area, yet we have never known or heard of livestock being poisoned by eating them. The same goes for poison ivy; in a new pasture there is often a good deal of poison ivy mixed into the forage, but we have never seen ill effects from animals grazing it, and goats seem to prefer it. We even think that when goat milk was our primary dairy food, the apparent immunity of most of our family to this plant's allergic properties might have derived from drinking the milk of animals that grazed on it. Some plants are only poisonous part of the time; wilted cherry leaves, for example, contain a toxic substance quite dangerous to herbivores, so if we are cutting wild cherry (*Prunus serotina*) around the pasture, we always clean up after ourselves. The leaves are not supposed to be toxic when green or when autumn-dry, but only in the wilted state. Better safe than sorry; we haul cherry slash into the woods, out of reach of our grazing animals.

Not only are some of the plants that you will see listed as toxic widespread in our local pastures; some, we find to our surprise, are even favorite forages. We've asked our local extension office experts about hemp dogbane, for example; herbicide companies list it as a toxic weed (and sell expensive and potent herbicides to wipe it out), but it's the candy of choice for local cattle, and they eat it first when they go on any new paddock. It is an invasive plant, so if you have it, you have a lot of it; if it's toxic, why are our animals eating it and thriving? One possibility is that, attractive as the forage obviously is, maybe in a conventional system cattle eat more of it at once than is good for them, whereas the total amount of dogbane in any single paddock is too small to do any harm. Who knows, maybe moderate doses of hemp dogbane provide an unknown benefit, some necessary mineral or natural antiparasitical; hence, its attraction for cattle. Many things that, eaten in moderate amounts, can be good for us, may be harmful when eaten by the bushel.

grazing period. Lots of standing forage, with some patches of favorite species hit hard, mean our paddock was too large; we make the next one smaller. Getting it "right" isn't really important; after all, in a few weeks we'll be back at this point in the rotation again.

The grazier's primary goal is *no second bite*; that is, while the actual amount of impact at any given

Figure 4.6. On this neighboring pasture, conventional grazing has reduced the plant cover to defensively recumbent growth and woody forbs. The hayfield to the right (*distance*) is ready for a second cutting.

time may vary, the regrazing of new growth is always detrimental to the strength of the plant; that is, to its ability to renew its root mass, hold soil, shade the earth, capture sunlight, convert carbon dioxide, and grow new leaves for the next time our animals are on this paddock. Even if we think a paddock has been undergrazed, we don't hold animals over unless we are sure that the total duration on that paddock will not exceed the period of time during which the pasture can be expected to show regrowth, usually a few days at most.

Aspect

Predicting how much impact a grazing period will produce is another part of paddock construction. Sometimes this will have less to do with forage height or species than with other aspects of composition. In our hill country, the degree and direction of a slope—steep or shallow; north, south, east, or west—will have a lot to do with the speed with which forage regrows at various times of the year, as well as the rate at which the soil firms up in spring, or dries to concrete in high summer. On a north-facing slope topped by a tree line, for example, we know to expect cool-season grasses to produce the longest, and warm-season forages to come on slowly. A southerly slope heats up early, sending plants into dormancy; deeper ground cover or litter on the surface will be necessary to keep the soil shaded and cool and the forage plants productive.

Grazing Patterns

Rotational grazing, as with every other instance where humans partner with the environment, is a blend of art and science. No two graziers have identical methods; every theory gives way before different conditions of forage, grazing species, weather, season, rainfall, insect population, atmospheric purity (or otherwise), the neighbors' farming practices, the prevailing wind direction, and so on. We use many terms for rotational grazing because there are many nuances to the art.

Under hot, dry weather conditions, on our farm we may do some *strip* or *line grazing*, moving a fence line out in front of the cattle to offer fresh forage, but putting up no back fence, so the animals have unimpeded access to shade at the wood's edge. In early summer, when grass is growing faster than the animals can graze it, we may skip some paddocks, letting areas of forage grow tall and mature; later, if forage regrowth has been limited, we can *mob graze* these paddocks, bunching the cattle tight and moving them frequently. The resultant heavy impact will deposit a thick layer of litter and manure, shading the ground to keep it cool and delay or prevent pasture plants from prolonged dormancy. Sometimes *two passes* over a paddock may be indicated: catering to their greater nutritional needs, we may graze lactating animals over a paddock first, following them immediately with the rest of the herd, whose needs can be satisfied with the higher-carbon grasses the milk producers didn't graze. Often, we run a mixed species herd: ewes with weanling calves or rams with dry cows, taking advantage of their different needs to make grazing more efficient, while separating males and females outside of breeding time. Each of these approaches may be beneficial for our grasses, our soil, our livestock, and our farm; all are *rotational grazing*.

And it is important always to remind ourselves: any rotation is better than no rotation. Hence, the first rule of intensive grazing: *Never open all your gates. No conventional grazing.*

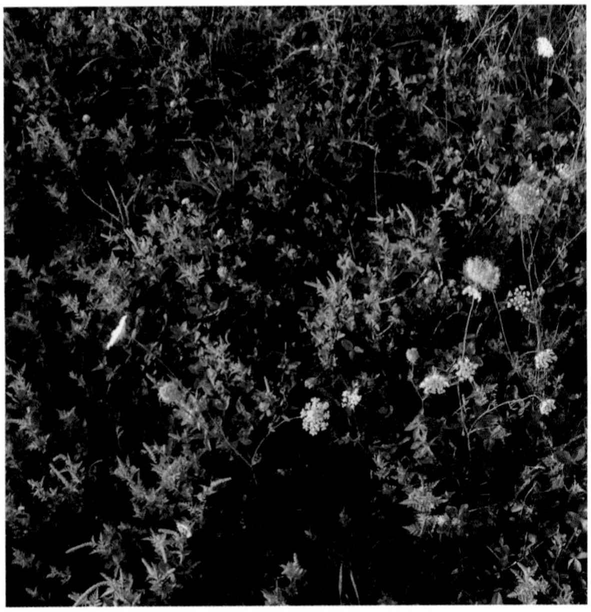

Figure 4.7. Bare soil and weeds disguise fertile pasture in potential.

Grazing with the Seasons

Rules and indications governing rotational grazing decisions vary with the time of year and the state of the animals. Any attempt to make a comprehensive list would be contrary to the complexity of ecosystems and the innumerable variations in climate, plant communities, and grazing herds. However, since we have to start somewhere, it's good to know some basic grazing patterns that are common in nonbrittle environments.

Spring Grazing

Mid-April is always a period of high suspense in our part of Ohio, as grass farmers keep one eye on the height of the first brilliant green in the pasture and one on the condition of the soil. Farmers meeting in town or conferring on the phone ask one another, "You put your cows on grass yet?"

There are multiple considerations at stake when making the decision to start your animals on their first grazing pass of the season. We want, of course, to turn the animals out onto that beautiful, rich, high-protein forage as soon as possible once they have calved or lambed. Mamas who have spent the last of the winter on stockpiled (standing dormant) forage or hay are hungry for the nutritionally packed green growth of spring, the perfect food to help them make rich milk for their babies. But best to serve the farm—and, ultimately, every animal and plant on it—our first priority must be to protect the soil. If livestock are turned onto wet, cold soil, the resultant pugging—pressing soil particles into tight clumps, excluding spaces for water, air, insect, and root penetration—and trailing may have a negative impact on the turf that will take months, even years, of careful stewardship to undo.

In addition, we don't want to start the animals on their first pass over the pasture until we are pretty sure that its rate of growth—at this time of year just beginning to accelerate from the almost imperceptible crawl of late winter to the not-to-be-kept-up-with sprint of early summer—will be such that in two weeks, or at most three, the paddocks first to be grazed will have regrown so as to be ready for the animals' second pass. In fourteen to twenty-one days we plan to be back at point A, having made the entire round of the area available for grazing, so by that time we will need the forage to be tall and lush, preparatory to summer grazing.

There are other dangers in putting animals out on new grass too abruptly, and these must be avoided as well. Diarrhea, bloat, and tetany are some of the conditions that may arise when livestock go out on spring pasture. In the case of diarrhea, a common effect of grazing "washy" (fast-growing, moisture-laden) spring grass, the result may only be that the animals fail to "get the good" out of their forage as it passes through them all too quickly; but extreme cases of diarrhea are as detrimental to animal as to human health. Additionally, sudden change of diet can cause a shift in rumen flora that may result in bloat, when the rumen is filled with gas that cannot be evacuated by natural means. Internal pressure in these cases can become so severe as to prevent the lungs and heart from functioning, resulting in death. It can become necessary to trocharize—puncture with a sharp tube—the rumen through the abdominal wall, or to run a tube down the animal's throat for the evacuation of gas, or to drench the animal with vegetable oil to break up the gas bubbles that are causing the problem, all procedures the infant grazier would rather postpone learning. Tetany, or staggers, as it is sometimes called, is a potentially fatal condition in grazing animals on green, high-moisture spring forage. Actually a magnesium deficiency, it is less common in intensively grazed herds than in conventional operations, where high-nitrogen fertilizers bind up magnesium in the soil, making it unavailable to plants—and, consequently, to the livestock that graze them.

All three of these conditions are less likely when animals under frequent observation are introduced gradually to the new spring grass. Offering a few flakes of hay in the paddock, to slow digestion, is usually enough to ward off problems; we ourselves have never seen a bad case of bloat caused by spring grass, and we have never had a cow with staggers.

Summer Grazing

As the weather warms up, the rate of forage growth, accelerated by spring rains and intensifying sunlight, increases dramatically. The mature grasses of summer are higher in carbon and lower in protein than the tender grass and legumes of spring; after the early spring, two- to three-week, large-paddock passage over the grass, paddock sizes are reduced to reflect the density and height of the forage. This increases animal impact, breaking down coarse grass clumps to expose growth tips, pressing grass

Figure 4.8. Unpalatable "hot spots" of high nitrogen growth around manure piles, like this one, are avoided when poultry following ruminants spread pies.

seed into the soil for germination, and breaking the soil surface for the better absorption of rain.

This slower rate of passage, coupled with the increased growth rate of forage, means that the number of animals adequate for grazing the pasture when we were moving fast will be inadequate to keep up with summer growth. Early paddocks will be ready for regrazing before the livestock have made the complete round of the pasture the second time. We often wish that we could "grow" our herds at this time of year, doubling our stocking rate (or, rather, their grazing impact) in order to utilize the abundance of forage. Later, when winter comes and we are anxiously watching weather patterns and calculating the number of "cow days" left in the stockpiled forage, we are grateful that our herds are no larger. Early summer, though, finds us with more grass than we can utilize in real time.

At this point, the decision has to be made whether to mow, make hay, or stockpile. As grass moves to maturity and then beyond, it declines somewhat in nutritional value. Halfway through the second rotation, the regrowth in paddocks grazed early in the second pass will frequently be more nutritious than the paddocks that are left, and the grazier may decide to skip the overmature paddocks and start over again with the earlier ones. The skipped paddocks will not receive the benefit of impact, manure, and urine from the second pass; however, they will not go unutilized.

Stockpiled forage is a godsend when the long, dry periods frequently experienced in midsummer to late summer send even warm-season grasses into dormancy. This will be the time to graze those paddocks skipped earlier in the season. Alternatively, skipped paddocks can be cut for hay. This doesn't have to mean the purchase of expensive miniature farm equipment, although it is out there. Early in our farming years we cut hay with a BCS walk-behind and a Gravely, helped out as necessary—the Gravely tended to rattle so hard it shed bolts like a cat shedding fur on a plush sofa—with a line trimmer. We raked and turned the grass by hand and brought it to the barn loose, loaded in the bed of our F-150. Later we used a rheumatic Massey Ferguson and a baler with hiccups to do the same job. Technology is not necessarily an improvement. Give me the walk-behind every time; the work is harder, but the headaches are smaller. Or hire the baling out to a neighbor on shares. Hay in the barn is our insurance when the weather turns really nasty, and anyway, there is always need of a truss of hay here and there on the farm: dry fodder for a calving cow, a bit of grass for pigs confined to the pigpen, bedding for everyone. Baled or stockpiled, no grass should go to waste on the independent farmstead.

Figure 4.9. Clover and timothy—midsummer pasture.

Animal Condition

Since summer grass decreases in protein as it increases in fiber, our animals' needs by volume will vary over time. We find it important to keep an eye on the condition of our animals, to assist us in judging paddock size and composition. The bucket is still a great indicator for the keeper of dairy animals, but it's important to remember that with breeds selected for high milk production—read all modern "improved" dairy breeds—most individuals will continue putting pounds in the bucket even as they begin to lose condition. Heritage breeds and grass-efficient bloodlines are less inclined to these excesses, placing survivability over such merely human concerns as specialty production.

You can watch the consistency of your animals' manure piles to help you form a rough notion of their forage nutrition. Cow pies that splash and spread way out are indicators that grass is passing through the cow quickly, as with washy early spring grass. Cow pies that stand up as high as the crown of a farmer's ag-tech-logo-embellished cap indicate high-carbon, low-protein forage. The perfect cow pie, in the median position, has proportions and the consistency of an unbaked pumpkin pie, indicating a good carbohydrate/protein/fiber balance. Keeping hip bones well covered may require that you skip some overmature forage in a few paddocks, or build paddocks large enough for a little selective grazing, sacrificing for the time being the heavier impact that is most desirable. In any case, an inadequately grazed paddock can be "clipped" after grazing—that is, gone over with a scythe, trimmer, or mower set for 8 inches or higher—to cut off tall weeds and seed heads before they mature, leaving more litter on the surface and a cleaner paddock for the regrowth you want to see in your next pass over that area.

An important consideration when building paddocks in summer is the near-constant sunshine and increased heat that can be expected at this time. In hot weather, it is vital that we build some shade into our paddocks; even when the animals' health is not compromised by high temperatures, their comfort, well-being, and performance may be. Building shade into a paddock does not mean giving the animals a lane back to the barn—not, at least, when there is any other alternative to consider. Grazing animals will only travel just so far between grazing area, water, and shade. If the distance is too great, animals tend to settle down in the lounge area rather than returning to pasture. When everything is located in proximity, on the other hand, more time is spent on the pasture grazing and less just lounging. Ideally, some natural shade in the pasture—mature trees widely spread, as on a savannah—would be about perfect. Including a shade tree is all that is necessary for our climate. Failing this, we have to utilize the wood's edge for shade, throwing a loop of polywire fence into the trees here and there (chapter 5 will go into fencing layouts in more depth), but making sure not to destroy all the species-rich edge space where the two ecosystems of forest and open field meet. Usually our daytime paddocks are built with shade

Figure 4.10. Prolonged access to shade areas results in bare soil and pugging.

access in mind; for the late afternoon and overnight, when shade is unnecessary, livestock can graze in the high part of the pasture where there are no trees.

Water, to which animals should have constant access if at all possible, is of great importance during the hottest part of the year. Keeping a self-filling tank in the paddock is ideal; we make sure that the ratio of its size to its refill rate means that every animal gets a chance at the drinking fountain. We run hoses through high grass or under fencerows whenever possible, to keep them shaded and avoid solar-heating the water. If the animals must return to a stationary tank or pond for their water, we keep the access lane as short as possible and plan our grazing so that hot weather finds us building paddocks in proximity to the tank. It is important to keep salt available in the paddock; our plain salt block (kelp is offered at the dairy) follows the cows in a cut-off water heater tank. Another local farmer mounted his mineral tub on an old STOP sign, which he pulls forward from paddock to paddock with his all-terrain vehicle.

Summer grazing gets your animals ready for fall butchering or breeding; careful attention to paddock composition now makes sure our animals are in the best condition when autumn arrives.

Fall Grazing

Fall weather usually means a return to regular rainfall and a second flush of cool-season grasses. Generally, by this time we are already stockpiling forage, skipping paddocks of mature forage to hold them for winter grazing. In our part of eastern Ohio, July is not too early to begin thinking of stockpiling for winter, and by the end of August everyone who has plans to winter his livestock in the field has grass maturing toward this end. Stockpiled paddocks won't be grazed until the grass has gone dormant; meanwhile, the animals are pastured in other areas. For very small operations like ours, stockpiling is often the natural result of our very slow rate of rotation over dense summer growth; with no commercial pressures on our production rate, we can afford to graze more thoroughly those high-carbon, mature grasses other farmers might brush hog or bale. We can take the slight decrease in milk production; it will pick up again as the weather cools.

Deciding which paddocks to stockpile requires some planning ahead, especially for the small-scale grazier. Watering animals in the field with a hose from the rain barrel is simple in spring, summer, and fall, but with freezing weather all that is going to change. No longer will we be able to leave a hose out to a Jobe float in the pasture. Maintaining water availability in winter means planning to do our winter grazing on those paddocks closest to our best frost-free water source, be that a frost-free valve on the house or in the barn or pasture, a bucket filled in the utility room, or (if you are very fortunate) a creek that stays open in cold weather, a pond with a mill to prevent freezing, or a spring tank that runs all winter. Pastures stockpiled for winter should be in proximity to frost-free water if at all possible—and, for the convenience and comfort of the dairyman or -woman, to the dairy barn as well.

Winter Grazing

Stockpiled forage is the perfect food for herbivores in the coldest weather. We remember the first year we kept cattle on pasture all winter. We had four or five steers on grass at the monastery while several more animals were at home eating hay in the barn. All winter those field-kept steers—exposed to whatever the sky wanted to throw at them in the form of cloud cover, sleet, and snow—ate the standing frost-cured grass and stayed in perfect

Figure 4.11. A dry autumn means less stockpiled forage going into winter.

Figure 4.12. This manger, attached to skis, can be moved across the pasture for better distribution of waste hay.

condition, right up until a few weeks before green-up, when grass condition finally deteriorated and the animals began to show signs of thinning out a little. At the same time, the barn-kept animals—sheltered, on deep bedding, with water just a few steps away and fair- to good-quality hay always before them—took on what, up until then, we had considered just "the winter look"—leaner, bonier, rougher-coated, less blooming. It was an eye-opener for us; until that time we had not realized the degree to which grass loses quality when it is cut, raked, baled, and stored. Shattering alone—that loss of nutritious, leafy material under the heavy handling of rake and tedder, baler, and repeated moving—accounts for a good deal of nutrition lost between the field and the barn.

Paddocks usually have to be larger in winter, when animals must eat not only for nourishment but to keep themselves warm. Planning with this in mind helps the grazier predict how long his stockpile will hold out, and how much hay he needs to have in the barn. Snow, even a foot or two of it, is not an obstacle to grazing; a thick coat of ice, on the other hand, will lock down cattle and prevent their digging down to the forage, or even from moving to a new part of the paddock. Thick ice on top of snow is slippery for animals to walk on and, when broken through, cuts or bruises the leg that has been forced through. We find that when heavy ice makes it difficult or impossible for animals to graze, feeding hay is best done in the field, where waste hay, manure, and urine remain where nutrients are most needed. Winter hay feeding in the field, in fact, is an excellent tool for pasture remediation.

Round bales are often unrolled for feeding, avoiding the heavy ring of waste hay (and manure) that builds up, concentrating nutrients in a small area around the feeder while at the same time laying down a heavy mulch that will suppress the next year's forage growth. When only a few animals are fed a round bale, it can be unrolled gradually, and a line of polywire used to keep the animals off the balance of the hay. Where unrolling is difficult, we may reduce waste by pushing the bale over on its flat side and binding the lower half with a cut-down stock panel. These welded fence sections, usually 16 feet long by 3, 4, or 5 feet in width, can be cut down to about 30 inches in width and tied tightly around the bale with twine or cord, leaving only the upper half of the bale available for feeding. The layers of hay are held in place by the wire panel, even when partly eaten through, instead of falling out and being trampled, as they would be were the bale in a standard round bale feeder. This arrangement allows the animals to get the most food value out of the bale, while preventing the accumulation of hay that will otherwise build up into a smothering mulch.

Shelter in very rough weather is a consideration, but there is no need to overdo it. Cattle and sheep are range animals and except in sickness or extreme weather do not require any shelter more elaborate than a windbreak or the shade of a tree. Rain and snow trouble them little; but when wet conditions are followed by intense cold or wind, or when unusually cold temperatures prevail over an extended period of time, it is good to provide them with a place to get out of the worst of it. In this part of the country many cattle are wintered at the edge of a wood, having access to the trees when snow or wind is severe. This was our own policy until the winter of 2013–14, when in two weeks of temperatures below zero, we lost a six-months calf. We were monitoring the animals two or three times daily and carrying hay and corn late at night to where they bedded in the woods behind the spring tank; it is our belief that this calf was simply too cold for too long. Consideration now tells us that the young stock should have been put in the barn for the duration of the cold snap. Hindsight is 20/20. The last few winters there has been a run-in shed in the field adjacent to the spring tank, and when the weather

Figure 4.13. On a hot day, soil temperatures may be as much as 15 or 20 degrees cooler under the shade of good pasture plants.

turns bitter we run the young animals in there for the duration.

Special Cases

In very wet, very cold, or very dry weather, special effort may be necessary for planning paddocks and erecting fences.

Wet Conditions

Soft ground and deep mud usually mean you should plan to keep the animals off the pasture until the ground has firmed up. On the other hand, the smaller heritage animals many homesteaders favor are less liable to damage turf in periods of heavy rain, and the constant movement of intensive grazing means that impact in any given place is kept to a minimum. Our animals—mostly Jersey or Jersey cross cows, and Katahdin sheep—remain on pasture year-round, regardless of weather, and although our climate frequently includes long periods of heavy rain in spring and fall, the only pugging we see happens when we make overfrequent use of a lane for walking the cows back to the barn. One spring, during an unusually long rainy period, we pugged the back lane pretty badly; the compacted soil came up in ragweed and Queen Anne's lace and grew little forage for most of the summer. Still, with long rests between grazings, by late in the season the action of root, insect, and rainfall had gone a long way to restoring that part of the pasture.

Dry Weather

When the soil heats up and sends plants into dormancy, leaving the grazier with short, ragged pastures or fields of dense woody overgrowth, it may be tempting to turn the animals loose on the whole farm to scavenge what they can. This is *never* a good idea; when forage is scarce is exactly the time when intensive grazing becomes most necessary. Limited forage should be met with tighter paddocks and faster paddock shifts; the rule of thumb is "hold them tight and move them fast." When the dry weather relents and grass begins to grow again, your paddocks will be found to have increased in fertility and abundance as a result of good grazing practices. In the event that it really becomes necessary to supplement forage, we feed hay in the field, spreading it out and letting the animals trample what isn't eaten; in this way we can shade the soil to conserve water and add organic matter where it is most needed. Indeed, the graziers in our area insist that when you feed purchased hay on the pasture, you not only provide your animals' nutritional needs of the moment, you get equal value in the minerals—phosphorus, magnesium, calcium, and so on—which are added to the soil via the waste hay and the animals' manure.

Hot Weather

In hot weather we are careful to make sure animals have access to shade; on a sunny day a black cow may overheat quickly if she cannot get out of the sun. If you ever see a cow standing with her neck arched, mouth open, saliva dripping from her jaws, respiration about eighty breaths per minute, get her out of the sun right away—but at a walk, not a run. And this is not just a summer issue; early spring days can hit the animals harder than higher temperatures will do in July, when they will be acclimated to it. We avoid situations of stress or exertion—unusual handling like shots, castration, or dehorning, for example—when temperatures are high. And while water is always an important consideration in any paddock, during hot weather we check stock tanks extra often; even a short time without adequate water can compromise livestock well-being.

Pasture Management

Weeds, especially early in the process of reclaiming a pasture or regenerating degraded land, are bound to be an issue. Unfortunately, the conventional solutions to weedy pastures are generally not conducive to land health, biodiversity, or resilience. According to popular wisdom, weeds are wholly negative in impact and should be eliminated, either by mechanical or chemical means or by the use of fire. Invasions of euphorbia in the western part of the country, for example, are being addressed with Roundup and bulldozers. On a neighboring farm here in eastern Ohio similar methods were used to remediate a pasture seeded with fescue—silver bullet of agronomists twenty years ago but later discovered to host an endophyte that causes elevated temperatures in cattle, reducing their heat tolerance in summer and seriously compromising their ability to conceive and carry offspring. In our native states of Oklahoma and Texas, fire is a frequently used tool to fight cedar seedlings in pastures—with the unfortunate side effect, however, of leaving bare soil exposed to desiccation and erosion, and ultimately compromising the very pasture grasses it is intended to assist.

In holistic rotational grazing systems, weeds are part of the community. What role they play in the natural grazing picture, and how we help them to play that part, develops over time, as we use the tools at our disposal for the overall increase of fertility and biodiversity in the ecosystem. It is true that land low in fertility and overgrown with nonforage species may need a little help before we can begin to use it for the animal species we have in mind. Our own most frequently used tools are grazing and clipping; which tool we use in which situation depends largely on our resources at the time.

Figure 4.14. A member of the carrot family, Queen Anne's lace will be grazed when young. Courtesy of the Franciscan Sisters T.O.R. of Penance of the Sorrowful Mother.

In our part of the country, when fields are abandoned, common pioneer species in the early stages of reforestation include briars, seedling trees, and wild cane. These can invade cleared land so as to make it unavailable for grazing by cattle, with their large bellies and udders exposed to briars and thorny growth, and for sheep, whose thick fleeces snag in the wiry tangle. There are other options, though. This is prime land for remediation by goats: forage that is too tough and thorny for cows and sheep is delectable to goatkind. A single goat, or a few, or an entire herd, depending on the amount of land in question and the means you have for containing the animals, is the best remedy we know of for land overgrown with such coarse plants. Secondary-growth forest made impenetrable by thorny undergrowth can be opened up by allowing goats to browse it in small sections, moving them forward as areas are cleared. Whether or not your long-term vision for your land includes goat raising, these animals may be the best tool for improving your land right now; later, cattle or sheep may move in to join or replace them as the prevailing plant life becomes more favorable to these species.

Land primarily given over to those "Band-Aid species"—perennial weeds that tend to germinate well on bare, infertile soil—presents a different challenge. Here cows and sheep can be used right away to rebuild fertility. Broadleaf weeds like ragweed, ironweed, dock, thistle, goldenrod, and asters, to name only a few, often become primary species on unused and wasteland. An hour's drive through our local countryside reveals uncropped fields and vacant areas abandoned after being used for logging, equipment storage, or staging ground, acres thick as the fur on a dog's hide with coarse growth of very limited value as forage. Your local extension office will probably recommend eradication with herbicides, but cattle and sheep can turn those fields into pasture in cooperation with the unwanted plants themselves, the soil biota, and the grass species living hidden lives under the broader leaves of the "weed" species.

Our experience teaches us that in tight rotations—small paddocks moved every twelve to twenty-four hours—cattle will graze many of these plants while they are young and tender. If we are beginning land remediation in the spring, we expect our cows to eat much of what would later be too tough and unpalatable, trampling the rest. Subsequently, grass plants, given a greater allowance of sunlight, have a chance to put on some height and get ahead of the weeds. A second or third pass later in the season, when paddocks are smaller, will continue to disadvantage the tall weeds in favor of more nutritious grass species. After the animals have gone over, we often come behind them with a machete, eye hoe, or string trimmer and cut off the large weeds at the root. If we are starting on a weed-ridden pasture later in the summer, when the broadleaf plants have matured, we might pasture sheep, whose browsing strips tall weed stems of their leaves. Then when winter comes, the woody stems are knocked over by wind, rain, and snow, leaving the grass plants at their feet with more available sunlight. In the spring, we will be able to put sheep and cows or yearling calves on the same land with great benefit both to animals and the forage.

Figure 4.15. Widely present in poor pastures, pokeweed has poisonous roots and berries but edible leaves. It is one of many "poisonous" plants whose presence is harmless in our intensively grazed pastures.

Not that we *never* have recourse to mechanical means of weed control; it is a tool we keep in our toolbox, especially useful for the invasions of cane around the wooded edges of pastures in the early stages of reclamation. A brush hog can be invaluable for cutting back thorny growth until grazing can bring it under control. Second-growth blackberry and raspberry canes may be lightly grazed by our dairy cows in the spring, but later the tall, thorny briars, snatching at low-hanging udders and scratching faces, will discourage even the attempt to graze. To address this situation, we immediately follow the cows' second and perhaps third pass over the pastures with high-level mowing or clipping, cutting off weedy top growth but leaving the main body of grass plants untouched.

Most of the shade in our pastures is to be found at the edge of the woods, but our ideal is to have a variety of trees spread around in the field, savannahwise, so that no matter where we want to set up a paddock, it is a simple thing to include some shade. Toward that end, each year we plant a few fruit trees in our pastures, hoping that in the future they'll serve us for shade, fruit, and windfall. If we make it a priority to protect these saplings with woven wire or temporary fence, we can build paddocks around them without worrying much about the few branches the animals can reach through the wire; if we forget, we generally lose the tree, which is good incentive not to forget next time. When we find volunteer saplings of a species we want to encourage in a pasture, we mark them so they won't get mowed next time we clip that area. Black locust are one of our favorites, although the shade they provide may be a little thin: locust trees are legumes, so they set nitrogen.

Seasonally wet areas should be avoided when dampness increases the chance of pugging; and since wet land is also a vector for parasite species like intestinal worms and liver flukes, making a detour around these spots is a good idea anyway. Alternating grazing species is another tool for management of damp spots; the smaller animals, grazing while the ground is soft, improve the forage without tearing up turf, while any parasites they deposit are specific to their species. In the dry season, the passage of cattle over the same land adds a heavy spread of manure without offering a vector for parasite infection.

Steep slopes subject to erosion may be remediated by grazing with small herbivores like goats, sheep, and calves, whose agility, smaller hooves, and lighter weight give them access to areas where heavy cattle cannot, or will not, venture, or where their greater weight increases the danger of trailing. The upper slopes of our home pasture, once an ankle-turning mess of loose rock and sumac stumps where only coarse weeds would grow, are, after a couple of years of mixed grazing by our Katahdins and yearling Jerseys, unrecognizable under their

Figure 4.16. This rock pile on a steep slope in our home pasture has almost disappeared under topsoil and perennial grasses.

new cover of tall bluegrass, orchard grass, and mixed forbs. Where secure footing used to be impossible, now the hillside is held in place by a thick mat of fibrous roots, and the cairns of rocks we have piled up, intending to haul them to the barnyard for pavement, are disappearing in the dense, lush growth.

It helps to remember that the fewer animals we are confining, the less they will influence one another's movements in the desirable surface disturbance we call *herd effect*: the trampling and breaking of the soil surface that is one of the major benefits of rotational grazing. Graziers with a single animal or a very small herd will want to bear this in mind as they plan paddocks. Where thick, persistent weeds or briars need vigorous trampling to suppress their growth and allow light to reach the grass plants underneath, we encourage impact by building long, narrow paddocks, with water at one end and minerals, or a treat, offered at the other. Going back and forth between the two, a few animals will have greater impact than their mere numbers would lead us to expect. We read that some graziers spray dilute molasses solution in areas where they want to see heavy impact; for safety, they must first closely inspect the area for toxic weeds to make sure they aren't baiting a poisonous forage. And if bare areas are included in a paddock, a few flakes of local hay fed in this spot will increase impact while leaving grass seed and surface litter. More tricks of this sort occur to us as we become more intimate with our animals' grazing habits.

Figure 4.17. Young nut tree seedlings can be encouraged for future pasture shade.

Pasture Grasses

Something we heard and read a lot early in our grazing education was the assurance, "There is no such thing as bad rotational grazing; only good, and better." Hearing it gave us a lot of encouragement, something we sorely needed when almost overwhelmed, as we may still sometimes be, by a weighty sense of our own ignorance. Like about pasture composition: warm-season grasses, cool-season, legumes, and forbs, we didn't know one from the other in the beginning, but "there is no such thing as bad rotational grazing," and the pastures improved despite our ignorance. It seemed that whether we had lessons in botany or not, our animals were born knowing what dinner looked like.

Over time, we have begun to understand something about our different forage plants, knowledge that is helpful when we are planning a grazing season. For example, different forage species do most of their growing in different seasons; that is, there are cool-season grasses and forbs, those that prefer spring and fall conditions, warm-season plants that thrive in the summer, legumes high in protein, and alliums for spring tonic. At any given time, different species, even different animals within a species, will favor some forages over others, and when we pay attention, observing our animals' behavior, we learn more about the community we are trying to serve.

Turned onto a paddock of mostly orchard grass and flowering red clover, our dairy cows may graze only the clover to begin with, coming back for the orchard grass in a second pass a few hours later; the sheep, on the same pasture, may ignore both and concentrate on defoliating the forbs. Moved from these conventional pastures to a rough, weedy patch needing reclamation, both species are liable to graze first and voraciously on the supposedly nonforage species like bindweed (morning glory), milkweed, locust seedlings, and hemp dogbane, causing us to speculate whether they are instinctively balancing their minerals. Fescue, on the other hand, is a forage grass that, despite the fact that it grows vigorously all summer long, is only attractive to our animals after the first frosts have sweetened it, or maybe when it is drenched with rain. We begin to have a sense of the growth patterns in our pasture, clover greening up early in the spring, perennial rye lingering into the fall, timothy maturing and setting seed in midsummer, and so on, enabling us more accurately to plan and lay out paddocks. Knowing this sort of detail increases the efficiency of our grazing plans and can save us the frustration of building a pretty paddock and having the animals turn up their noses at it—or worse, trample and defecate on it, overgrazing the few patches they like the look of, and giving less milk that night.

We don't get uptight about it, however. Reams of information are available to the grazier on the nutritional values of forage species, and many of the pasture walks, grazing council meetings, and graziers' conferences we attend devote lots of time to discussing the ideal percentage of legumes in a pasture, carbon/nitrogen ratios, percent available protein, and Brix (a measure of sugar content). Valuable as this information no doubt is, we think for the beginner it is a case of too much, too soon. All that science seems to fall under the heading of "interesting but unnecessary" for the small-scale regenerative landowner, whose goals are as simple as food on the table, the increase of soil fertility, animal health, food security, and the efficient utilization of available resources: sunshine, rainfall, native or naturalized plant species. Unless money is no object, our purposes are best going to be met with a less-is-more philosophy. Nature often does a better job of educating us if we don't forestall her lessons by implementing artificial solutions. More times than we can begin to remember we have encountered a need, or a perceived need, for which we developed a plan that required fairly large amounts of labor or money (and, therefore, had to be put off until we had time or funds) only to discover in the interval that 1) the problem solved itself, or 2) it turned out not to be a problem at all, but a blessing, or 3) some other situation occurred that encompassed the first and put the whole thing in a new light.

Not that we can put our animals on a small paddock of burdock and goldenrod and dust off our hands, thinking we've supplied all the calories they need; common sense is necessary in every endeavor, and that goes doubly for farming. But the art and science of rotational grazing, as a hands-on, gotta-be-there undertaking, is especially calculated to provide an education to the grazier from day one. Put your cows/sheep/goats out in a small paddock of mixed grass, broadleaf, and browse and start moving them daily, and before a week is over you'll begin to have a notion of their grazing patterns. Because moves are frequent, they'll pick up what they need to survive on any given day, and you can plan your next paddock to avoid the shortcomings

Supplementation

This may be too obvious to mention, but it's important to make sure we are feeding our animals adequately while we are partnering with them in the transition to intensive grazing. Not every piece of land is going to grow adequate forage from day one, and where there is lots of bare soil or a superabundance of thorny or brushy weeds, we may find it desirable at first to supplement our animals with some hay fed on the paddock. Not only does this allow us to encourage trampling of coarse growth, it puts grass seed, manure, and urine where it will do the most good. In addition, the minerals that pass through livestock and are added to the soil when we feed hay on pasture are value for money.

Keeping our animals well fed and happy is important on another count as well: community relations. Enough people—usually people with no background in animal husbandry—are going to see our small paddocks and well-grazed forage and assume that our animals are being starved; we don't need to give them substance for the assumption.

of the last. If you are at all attentive, you'll have begun a lifetime education in animal nutrition, with your own livestock as tutors.

Record Keeping

Finally, grass management is one of the many farming endeavors in which record keeping is invaluable. Get a topographical map of your land from the geological survey material available online, or hire a pilot and a Piper Cub and take aerial photographs of your farm, or draw your farm freehand and put down the distances in paces. Make copies; then draw, write, or otherwise record dates, paddock sizes, number, age and species of livestock, grass condition, forage species, duration of grazing—everything you observe about your land, as often as you can manage, to develop and refine a working knowledge of it. Then use this information to help you plan future grazing. Without knowledge of the mistakes of the past, we are doomed to repeat them; without record of the successes of the past, we are doomed to forget them. With detailed notes on each year's grazing patterns, gradually we accumulate the information that will see us through a long dry spell or an extended wet season with a full milk bucket and fat steers; information and knowledge that, over time, will mean more fertility and greater resilience for the homestead, and more and better food on the table for the homesteaders.

CHAPTER FIVE

Fence

Good enough is perfect.

—*Joel Salatin*

How may rotational grazing be instituted, if at all, on only a few acres, perhaps even less than an acre? Are the principles that can be used on a large farm applicable on the microscale? The answer to this question is an urgent "yes." Certainly, there are parameters for intensive grazing that may require some imagination to contrive on the minihomestead, but the principles of short-duration, high-impact grazing followed by extensive rest and regrowth are as well within the reach of the small landholder as that of the big rancher—with the additional ease afforded by very small scale. The owner of even ½ acre of grass, weeds, or brushy land can intensively graze one, or a few, goats or sheep, for much or all of the year, supplementing with hay in periods when the forage is dormant. The tiny field that has been groomed heretofore by a John Deere can, with a small investment in portable electric fencing, become home and dinner to a young steer, two or three dairy goats, or a small herd of sheep. Tractor repairs and fuel bills can be replaced by dividends in meat, milk, and fiber; noisy afternoons mowing in the summer heat, by a quiet daily stroll out to move the animals to fresh forage. The roar of the tractor engine can give way to the sound of ruminants tearing up mouthfuls of grass and the quiet ratchet of the fence reel as you move them onto a new paddock.

Figure 5.1. Intensively grazed animals are accustomed to the proximity of their human beings.

How small is too small? A single steer moved daily over good grass may, at peak periods in our area, graze as little as a 20-by-20-foot paddock, or about eight paces by eight paces for an average-height woman, in twenty-four hours. On a single acre there would be upward of one hundred such paddocks! Of course, paddock sizes vary substantially according to forage condition and the species of animals pastured, but our imaginary steer conceivably would not return to a given paddock for *three months*—under most conditions a more than ample period for forage rest and regrowth. (In a real rotation, paddock sizes would vary considerably over a season, and in any case paddocks should always allow some freedom of movement; these numbers are merely to illustrate the more-than-adequacy of a small pasture.) Erecting and taking down small paddocks requires only a few minutes a day, minutes spent interacting with and observing animals with an exactness and attention that the farmer of large numbers cannot hope to achieve. Frequent and intimate livestock handling results in quiet, tractable animals, easily fenced and controlled under most circumstances. Time spent out in the pasture means time learning its secrets: its plant and animal species, its microecosystems, its areas of prime grazing for each season. It means daily information about forage conditions as they are affected by weather, grazing, and soil temperatures.

It means, as we commit ourselves to the management of the plants and animals under our stewardship, that we ourselves become natives, people of *this* place, people belonging to a particular interspecies community of the globe.

Intensive Grazing

The tenets of intensive grazing are not complicated. Livestock are confined to very small areas where they must graze less selectively and more thoroughly than in a large paddock—where, ideally, in a short time every square foot of the paddock will be either grazed, trampled, manured, or urinated on. Impact will be heavy, but duration brief: after this time, the animals are moved onto fresh forage, and the grazed paddock is given a long period of recovery before being grazed again. Duration of rest varies with the weather and seasons being less a function of time than of regrowth. Photosynthesizing plants pump carbon compounds to the soil biota, which in turn chelate minerals to make them available to plant roots; microorganisms decay root mass sympathetically "pruned" as a result of the removal of aerial parts of the plant; the added organic matter increases fertility and the soil's ability to absorb and store water; root regrowth holds soil against erosion. Plant recovery is vigorous; more forage is made available for the next grazing period, and so on.

Are the results of intensive rotation really so radically different from those of conventional grazing? Absolutely. It is the contrast between periodic maintenance in the form of mechanically applied fuel, fertilizers, pesticides, and herbicides, and periodic maintenance in the form of walking the land, shifting fences, and observing animal and plant life. It is the difference, on the one hand, between

Figure 5.2. Daisies make an edible part of this midsummer pasture.

applying large-scale effort and inputs to establish and give advantage to a small number of select plant forms over others which are, perhaps, better adapted—because naturally so—to the place and climate on the one hand, or the daily observation of the natural succession of diverse native plant species in a tidal flow of response to periodic grazing and rest, and all the variations of meteorological event, on the other. *It is the difference between the tension inherent in a system where every parameter is manipulated by human intervention and the dynamic energy flowing from biological communities that are husbanded with reservation and respect.*

Electrical Fences

Although farmers have practiced some level of rotational grazing for millennia, the intensive management that is the hallmark of the modern rotational grazing movement has been made possible, or at least practical, by the invention of lightweight portable electrical fencing. Traditional fences—stone, brush, hedge, rail, board, or barbed wire—are physical barriers; that is, they work because their mass, tension, and physical integrity mean the animals confined cannot easily break through them. Electrical fences, which may be lightweight or even flimsy, act primarily as psychological barriers, keeping animals confined through intimidation: in other words, the unpleasant experience of grounding an electrical current teaches animals to avoid contact with the source of the charge. It is not the strength of the fence but the unpleasantness of the jolt that confines the animals.

The original "hot wire" fence of the last century—usually a single strand, or maybe two, of permanently installed electrified aluminum or steel wire—is now giving way either to strong fences of taut steel wire on permanent posts, also known as "high-tensile" fence, or to lightweight portable temporary electrical fences of polymer/stainless steel twine or netting. It is these latter that are so useful—indispensable, even—to the rotational grazier. They are built of relatively few simple elements. A fencing system for small-scale rotational grazing requires only a fence charger, ground rod, and power source; a reel wound with polytwine or polytape; and posts on which to hang the wire. It is a simple matter requiring only a little exertion to erect, take down, and move even large paddocks on a daily basis. In addition, because the fences are set up for only short periods, they can be installed in places where a permanent fence would be impossible or impractical, giving portable electrical fencing the further advantage of great flexibility. A line can be set up around trees, bordering a small pond or a creek, even across a lane or driveway, making it possible to paddock animals where they can groom orchard floors, stream banks, and road edges, and yet be moved before trees are girdled, banks eroded, or it is necessary to reopen road access.

Fencing Equipment—Posts, Reels, Chargers, and More

Portable electric fence posts come in several different types, each with its uses, advantages, and disadvantages.

Most useful in our small operation are the slim fiberglass posts with integral step-in braces having one or more permanent hooks or clips on which to hang polytwine. These are pushed easily with foot pressure into all but the driest or most frozen soils,

Figure 5.3. Plastic posts with integral hooks and fiberglass posts with one or several permanent clips are indispensable to the small-scale grazier.

and wire can be fastened conveniently at regular heights. An armload of these posts is light to carry, although there is the minor annoyance that occasionally they get hooked together and require a moment for freeing them. A more important disadvantage is in the clips and step-in braces themselves. The rigid plastic of which these are made grows brittle in cold weather, when they have an unfortunate habit of snapping off; the glue that fastens them to the fiberglass shaft also has a tendency, in some brands at least, to fail under the pressure of regular use. One manufacturer recommends laying new posts out in the weather for a week or so after purchase, allowing the adhesive to cure before use. In our neighborhood, the usual advice farmers give is to buy high-end posts from the manufacturer, avoiding cheaper versions such as are found at the local farm store.

Of course, fiberglass is subject to degradation over time and from exposure to sunlight, and broken clips and foot braces render fence posts all but useless. Plastic or wire replacement clips are available, but there's no way we know of to replace foot braces. An alternative option is the molded plastic post with integral hooks and clips, a style we had overlooked until we heard Ohio grazing scientist Bob Hendershot speaking at the North Central Ohio Dairy Grazing Conference one January.

"You all know what these are like," he said, holding up a sunburned, splintery fiberglass post with broken clips and foot brace. The audience nodded; we had been handling posts like this one daily since the previous November. "Now look here." He held up a white O'Brien post and bent it into the shape of a *U*. When he released it, it sprang back to its original form. He plucked at the hooks, bending them far out to show us how much punishment they could take. "Give me plastic posts every time," he said.

We appreciate his point. Molded plastic posts are lightweight, strong, and resilient, and their metal spikes and broad foot braces make them easy to drive into all but the hardest of dry or frozen ground. On the other side of the ledger, although light, they are bulkier than fiberglass posts, and fewer can be carried in an armload; in addition, their flexibility, while it means that they stand up to more punishment, makes them less appropriate for fence corners and curved fence lines, where rigidity is needed. But for the purposes of the small operation, where fences are usually very short, these disadvantages are not overly significant.

Yet another basic electrical fence post is used primarily in semipermanent situations, or for confining large numbers of stock in bigger paddocks. The "pigtail" post is of coated metal construction, with an open, 400-degree spiral at the upper end. Much stronger and heavier than fiberglass or plastic posts, these may be useful to the small-scale grazier in situations where a perimeter fence (outlining an entire field, pasture, or clearing) or power line (length of wire carrying current to a paddock) will be erected for a longer period of time. And where a large paddock, or softer ground, makes extrastrong corner posts necessary, a few pigtails can be used to anchor corners where a lightweight post might be pulled out of the ground.

Most versatile are simple fiberglass posts without clips or foot braces. These are driven with a mallet, and the polywire or polytape attached at any height convenient with slip-on metal clips or springs. They may be bought relatively inexpensively and are easy to carry in a bundle because, having no permanent clips, they cannot tangle together; however, the necessity of carrying a mallet as well as fence posts and reel is an inconvenience, and the pocket full of wire clips necessary for fastening wire to the posts is another nuisance. At the Sow's Ear we mostly find they are great for herding cows with—just grab one and tap the nearest cow on the spine to drive the animals in for milking.

There are other styles of post available for electrical fencing systems; we have only named the ones we know best. A quick look online at stock fencing sources will acquaint you with a wider range of options.

Figure 5.4. This electric netting foils predators while concentrating poultry impact where it is most needed.

Tape, Twine, and Netting

Once posts are set to outline a paddock, there is the matter of what sort of wire to string on them. Polymer/wire fencing for small-scale graziers takes three basic forms: tape, twine, or netting. Each of these has its variations.

Many people whose first exposure to portable electric fence is with horse fence are at least superficially familiar with polytape. This is the white or colored mesh tape (it looks from a distance a bit like plastic caution tape) often hung on permanent posts around horse corrals. Its high visibility makes it a good choice in certain situations: people, as well as livestock, are less likely to touch it inadvertently. Where fence must be erected across a deer trail, polytape is more likely to be jumped than charged and dragged. For the grazier with a single animal or a small flock or herd, however, we know of no advantage to polytape. On the other hand, we know of a few disadvantages. Polytape is usually more expensive than twine, and less of it fits on a reel; wind and precipitation are also more likely to cause sagging with a tape fence. We do keep one reel of orange polytape, however, for a specific use: cutting and driving livestock (see Reels on page 102).

Because of its greater visibility, it is more effective than twine in this capacity.

In addition to polytape there are the various weights of polytwine (often just called "polywire")—twisted or braided polymer filament with strands of fine stainless steel wire. These generally come in large spools (600 feet or longer) and differ in weight, durability, color, and the number of charge-carrying stainless steel wires they contain, the standard being six or nine. Obviously, the greater number of conductive wires means more contact with the animal being fenced; it also means that when, over long-term use of the fence, repeated flexing breaks a wire here or there, there are more remaining wires to carry the charge. With lighter-weight twine, on the other hand, more twine can be loaded on a reel, making longer single-reel fences possible—an advantage to the grazier with more acres, but for grazing on a very small scale, not necessarily a perk. The various colors are for the purpose of making the fence more visible to different species of livestock (and wildlife), or less obtrusive to human beings.

Polynetting is the portable electric equivalent of woven wire. It comes in standard panel lengths mounted on permanent posts about 12 feet apart and can be found in a number of different sizes of mesh and with varying numbers of charged and uncharged strands. Often polynetting is marketed as a barrier for raccoons, coyotes, or even bears; it is the only form of electric fence we know of that is of any use at keeping poultry in as well as predators out. But while it is a useful tool, we minimize our use of it, because of the awkwardness of setup and takedown. You can't reel it up, but must fold or roll the fence panel into a bundle, neat enough to be unrolled or unfolded again without tangling but small enough to be carried with at least some level of convenience. Because its length cannot be adjusted, making a paddock is awkward: there is a certain amount of guesswork and trial and error as you attempt to outline a predetermined area and bring the ends of the panel together. And because the posts are permanently attached to the netting, the step-in spikes inevitably get tangled in the mesh, complicating setup and takedown. That said, polynetting is, as we said, the only electric fence we know of for poultry; we also find that it holds a litter of young pigs more securely than do two strands of polywire.

Reels

When using polytwine and polytape for temporary fencing, fence reels are indispensable. There are a number of types of fence reel on the market, as well as some reels not intended for fencing that can be very useful in this application.

Of late the fencing industry has provided a number of useful and lightweight fence reels. These may be either simple or geared: simple reels turn in a 1:1 ratio to the crank handle, geared reels revolve at a higher rate (usually in a 3:1 ratio). Because of this, geared reels are faster to use and require less effort, both in rolling up and in unrolling a fence. Indeed, some geared reels unroll so fast that some of us avoid using them: when the grazier makes a sudden stop—to hook line on a post, for example—the reel may keep turning and throw a loop of twine around her ankles. The largest reels—those capable of carrying the most twine—are going to be geared reels.

For the purposes of the smallholder the simple reel is usually adequate and, being lighter to carry, may be more convenient on that score as well. They are usually somewhat smaller than a geared reel and easily load with more than enough twine for a large cell or paddock. For very short lengths of

Figure 5.5. Simple reel hung on one post.

Figure 5.6. Geared reel hung on two posts.

twine—such as might be used for back fences, gates, or temporary barriers—there are even smaller, very economically priced simple reels of durable plastic made just for this purpose. Or, many people use the sort of reel made for winding up extension cords to carry shorter lengths of tape or twine. We use one of these when we need to cut a few animals from a group, drive them into the corral, or load them on a trailer; three or four people strung out along 20 or 30 yards of polytape can cut and drive our fairly docile animals with very little bother.

Two reels of polywire are a minimum for most smallholdings, and three are even better: one for today's paddock, one for the next, and a spare one for a power line or an emergency. Both geared and simple reels should be equipped with a line guide to prevent the twine slipping off the reel and wrapping itself around the axle or handle when the fence is being reeled up. Having to stop in the middle of building or taking down a paddock to unwrap yards of twine from around a reel axle is enough to convince anyone of the value of having a line guide. Most reels now come with the line guide included, but if you don't see one on a unit, ask for one and order it separately, if necessary; it will save you untold time and headache.

Then there are multireel systems, arrangements of multiple posts and lines permanently attached so that they may be set up and taken down as a unit. Of the earlier versions of this we have heard no good at all, the overriding objection to them being that the lines, being on separate spindles but controlled by a single crank handle, could not be kept equally taut, some always being slack while others were pulled overtight. More recently we have seen versions that address this failing, units of several reels of twine permanently attached to about ten posts, all controlled by a single slip-geared winding system. The Gallagher Smart Fence does seem to have overcome the objections to earlier multireel systems, and individual wires can be tightened, or loosened, then restored to equal tension by turning the crank.

At first glance this one-unit-does-it-all arrangement looks like it might offer advantages to the small-scale grazier; we have three reasons, however, for leaving it off our Christmas wish list. The first is the nature of our pastures, which are hilly, even very steep. While these systems may be able to adjust for different elevations, they are inevitably more cumbersome to carry. And the two processes of setting posts and reeling out twine—which, especially on steep slopes, are more easily accomplished separately—are, with a multireel system, necessarily combined.

The second reason has to do with the size of our operation. Paddocks as small as we generally use take only a few minutes to erect anyway, so any time savings afforded by a multireel system would be negligible. In fact, a neighbor who grazes some thirty cow-calf pairs tells us that even in his operation—where the daily erection of fence takes considerably longer than on our small farm—the time savings is negligible. In addition, our need for three- or four-strand fencing is seasonal and of short duration, being only a few weeks in spring when the lambs are small enough to slip under a two-strand fence.

Finally, there is the matter of cost. Right now, 320 feet of smart fence costs about $250, while twenty posts, a good reel, and 1,600 feet of twine can be had for about half that. Nearly always, we find that the simplest tool is the best in the long run—easiest to use, simplest to troubleshoot, least expensive to buy or replace.

Fence Chargers, Ground Rods, and Power Sources

Of course, any electrical fence system is only as strong as its charger, and only as reliable as its power source. Chargers vary as to both how much power they send through the line, and how frequent the pulse. Their most common sources of electricity are grid current, dry cell battery, and solar panel. In addition, any fence charger has to be attached to at least one ground rod—a conductive rod driven several feet into the ground—to close the electrical circuit and allow power to flow through the line. These three elements—charger, source, and ground—are necessary for any powered fence system.

Figure 5.7. This Katahdin ewe's paddock is weedy but contains plenty of palatable forage.

Fence chargers are generally categorized by their power output, rated in "joules" ("jolts," local farmers call them). We have a 1-joule charger on the fence in our small home pasture; up at the monastery farm, the fences are longer and we use a 2-joule charger. Missouri stockman Greg Judy records in *Comeback Farms* that his high-joule unit will knock a hog several feet, and give him, Mr. Judy, a charley horse that will linger for hours. This is a good charger for someone running, as Mr. Judy does, a large number of mixed-species livestock behind long electric fences. For us it would be overkill, and an occupational hazard as well; anyone who has ever touched a charged fence can attest to the unpleasantness of intercepting even a relatively small charge.

Generally, the shorter your fences, the fewer joules necessary. The exception is netting fence, which requires more power per linear foot; netting usually comes with a fence power recommendation. There is no advantage to carrying more charge than your fencing system really needs, and, of course, the lower-power chargers are less expensive. All fence chargers are not created equal, and it is probably worth the money to buy a name-brand unit, not the cheapest one at the farm store. As of now, a good 1-joule charger starts between $100 and $150 and goes up from there. Used chargers may be found at farm auctions and can be worth the money. They built them well back in the day; the one in our big barn is a Sears, Roebuck & Co. unit purchased new by Shawn's dad in the 1970s, and it works as well as ever.

A ground rod being necessary to close the fence circuit, you have to set up your charger where it can be wired to a ground—that is to say, over dirt. Although it is possible to put a ground rod in a dirt-floored barn or shed, wet soil makes a better conductor, and for that reason we wouldn't normally set ground rods under a roof. Chargers can be hung anywhere there is power and a ground: on exterior walls, where there is an electrical outlet proximate to the ground rod; or on an inside wall of a barn or shed, with the ground wire run through a partly open window, a hole drilled in the wall for that purpose, or even a crack between siding boards. The ground terminal on the charger is then attached to the ground rod with wire and an alligator clip. We have been told that it is advisable to use wire, clips, and ground rod made of the same metal—all steel, or all copper—to avoid corrosion and increase conductivity. More ground rods mean a better circuit.

When the charger is hung inside a building, the charge wire, as well as the wire to the ground rod, also has to be externalized (passed through the wall)—a partially open window will do, or drill a hole (it should have a nonconductive liner, such as a short piece of plastic or ceramic tubing—tubes from old knob-and-tube house wiring are ideal). Once the charge wire is outside, it is hung on insulators, plastic or ceramic, which can be attached to the outside of the building, or to a post. You can start your temporary fences from the barn wall, tying the polytwine to the charged wire and proceeding from there to erect fence, or you can extend the metal line to a gate or corral post, or any other fixed point from which it will be convenient to start your paddock fences. Where pastures are at a distance from the charger it can be helpful to string the power line on high posts, or from roof peaks or high walls, out to a fixed point in the pasture where you can access it for fences; this keeps it out of the way of human and animal traffic in places where it is not wanted.

The intensive grazier, who has pastures and paddocks wherever there is forage to be grazed, may find that he needs more than one station for a

Figure 5.8. Fence charger on roofed post.

charger. Some paddocks could otherwise require an inconveniently long charged line back to the central power point. We have several locations where a permanently installed ground rod stands next to a small, roofed fence post; where the roof pitch has been thoughtfully planned, it can also be used as a bracket for a solar cell. Otherwise, extension cords rated for outdoor use are run from the nearest building—house, barn, or shed—to provide power. The charger is installed under the post roof, with the juncture of charger cord and extension cord being protected by this shelter as well. As long as all the elements—charger, power source, and ground rod—are present, you can put a charger practically anywhere.

We charge our fences with either grid current or solar panels. Grid is, of course, the simplest to use; that is, it requires the least effort to set up. Solar is a good option where there is no laid-on electricity and can be very portable, but is subject to brownout. Rechargeable dry cells charged from the grid or by a generator and used in rotation, are cumbersome and require regular maintenance, but constitute a viable third option.

Solar Chargers

Solar fence chargers are available in pricey but convenient self-contained units, or you can buy the parts separately and assemble your own. Fence charger kits that are powered by a small—as small as 5 watt—solar cell are readily available. A very small unit may require no mounting—the case simply rests on the ground—while others may be easily mounted on a permanent post in the pasture, or on a small cart or dolly to be moved wherever setup is desired.

Solar fence can go where it is needed—within certain limitations. The necessity of having a ground rod in proximity to the charger means you can't set it up just *anywhere*. Even the short ground rods that generally come with low-watt kits require some effort to drive into the ground, and to pull up when the fence needs to be moved. In addition, the batteries these units use for storage lose capacity in cold weather, just when natural sunlight is least direct and day length shortest, hence energy storage is consequently most important. Nevertheless, solar is a good way to power a fence when the distance to grid electricity makes running an extension cord inconvenient and carrying a dry cell battery onerous. Our 10-watt panel and 2-joule charger make a good pairing mounted on a dolly and set up in any of the places where we have permanently installed ground rods. We find that even in winter the cows are sufficiently respectful of a solar-generated jolt; the sheep, with their thick fleece, on the other hand, don't stay in without two strands of fence, and not always, even then. Nevertheless, without solar as an

Figure 5.9. A solar panel with dry cell battery and fence charger can be mounted on a two-wheel dolly and set up wherever there is a ground rod.

option we would have found it difficult, in the beginning, to charge fence in some of our more out-of-the-way pastures.

Rechargeable Dry Cell Batteries

Although we have not used house-charged dry cell batteries in our own operation, we have more than one neighbor who has relied on them for years to confine cattle or horses, with acceptable results. Like solar, their usefulness is in the fact that they can supply power where there is no grid current. Since these batteries have to be recharged on a regular basis, it is necessary to have at least two for uninterrupted service, one battery charging while the other is in the field. The systems we have seen charge by means of a simple trickle charger plugged into wall current. Creative nongrid people might come up with other ways to recharge them: bicycle, wind, or water power, for example. Batteries must be protected from precipitation, but we have seen this accomplished with a couple of bricks under the battery and a bucket over it. Keep in mind, though, that marine dry cell batteries weigh a good deal, besides being awkward to carry safely, so this is not a chore to be passed off to the children. And while for the nonresident grazier, battery and solar-charged units can be good options, with no one on-site, such portable units do present an invitation to the dishonest.

Some Like It Hot

Finally, any discussion of charged fences would be incomplete if it did not include instructions on how to test a fence for power. Fence testers are available for a few dollars from any electrical fence supplier or your farm store, and are simple to use; a readout unit, usually with a dial or lights indicating the level of charge, is attached by a length of insulated wire to a short (3 inches or so) ground rod. This latter is pushed into the soil near the fence to be tested, and when the metal contact point on the indicator touches the fence wire it lights up, or the needle jumps, or whatever, to show the level of charge being carried by the fence. If the charge your fence is carrying is adequate, you're good to go. If the tester indicates that the charge is low, it is time to start looking for the problem.

The first thing to check, of course, is whether your fence charger is actually turned on. Most units have a flashing light to indicate that the charger is on; they also make a quiet click with each electrical pulse. Make sure the charger is plugged in; some units can only be turned off by unplugging, and the last person to move the livestock may not have plugged it back in. If the problem isn't the charger, check your connection to the fence; is it making good contact? Is your charger hooked up to the

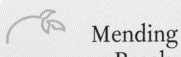

Mending a Break

The life of a reel of polywire depends on the integrity of its conductive parts. Since most polytwine has at least six strands of fine wire, one or several wires can be broken without interrupting the flow of current. Twine is not indestructible, however, and since polymer twine or tape is twisted or braided out of strands of metal wire and polymer filament, it can be difficult to detect where, or whether, the conductive strands are compromised. One way to find broken wires in your polytwine is to walk your fence at night, while the fence is plugged in. In places where there is a broken wire, the electrical charge jumping across broken ends will make a spark plainly visible; at this point, the twine can be knotted over the break (turn off the fence first) to renew contact between the wires.

ground rod or rods? Use your fence tester to determine the area of the fence where the charge is dropping; with the short fences of a smallholding this is seldom a lengthy job. Do a visual check: Is the fence touching the ground anywhere, or a metal post or fence wire—like an uncharged perimeter fence, for example? Is the twine or tape broken or frayed? Is the line touching a lot of plant material, especially wet plants in soggy ground? This is especially easy to spot at night, when the charge jumping from a low strand of polywire to dew-saturated weeds can look like a line of tiny fireflies. Any of these things can pull down the power in a fence, and with simple portable fence systems it is usually easy to spot the problem. Trim the interfering grass or weeds, move the grounded portion of fence, and the problem is solved.

Bringing the Charge to the Fence

There are two ways to get power to a paddock fence. The easiest is to set up semipermanent perimeter fences of light aluminum electrical wire—or, if you have the money to invest and are sure you know where you want them, permanent, heavy-duty, high-tensile stainless steel fence—around your entire pasture. Once your forage is outlined in charged fence, it is a simple matter to tie polywire to this and electrify temporary paddock fences. But if you don't have the money to put up permanent charged fences, or don't want to wait until you have charged perimeter fences to begin improving your pastures with grazing animals, you can still get started grazing right away.

One way is simply to run a charge line—a single strand of polywire on widely spaced posts—from the charger in the barn, or wherever it is, out to the pasture. We usually hang, clip, or tie the end of the polywire to a short lead from the fence charger and run this line out to the area designated for the new paddock. Sometimes it may be very long and require a reel just for the purpose, but on a small acreage this is often unnecessary. Generally the first paddock is made by encircling the desired area with the same reel used for the charge line, closing the paddock by looping back to the first paddock post, and hanging the reel on the next-to-bottom hook or on a double post. Some people are comfortable just setting the reel on the ground, but this usually means that the line is in contact with grass or weeds, partially

Alternate Reels

For a line back to the charger, a full-size reel is not necessarily essential. If the distance back to the barn is not far you can wind a short length of polytwine onto an inexpensive reel of the kind used for extension cords and make this your charge line. We also use these lightweight reels whenever we need a very short fence line, as when we divide a paddock into two parts. Don't think to save money by using only these flimsy, hardware-store reels, however; such lightweight tools are not sufficiently durable or convenient to be used in place of genuine fence reels.

grounding the fence—even, at three o'clock on a dewy night, significantly reducing the fence's jolt. We sleep better knowing there is an unimpeded electrical charge between our animals and the county road, so we hang our fence reels. Successive paddocks need not begin at the charger; a second reel can be used to enclose a new area next to the first paddock, after which the first paddock is partially reeled up, leaving the charge line in place to electrify the new fence.

A second way to carry a charge out to the pasture is to run aluminum fence wire overhead, on insulators attached to tall posts or high on the outside walls of outbuildings. These should be hung judiciously, taking into consideration what other traffic will be using the area. More than once we have strung lightweight fence wire on high posts to pass over a gate or lane, only to find that a loaded hay wagon was just too high to go under. And a charge line that is out of the way when strung through the orchard in the spring is going to be in exactly the wrong place when you head out with your aluminum orchard ladder in August to pick apples.

Setting Up Fence

It may seem that setting up portable fence paddocks would require little thought and no explanation. After all, this is a matter of pushing in posts with your feet and unreeling, right? How hard could it be? But once we get out in the pasture, there are still some details to work out.

Distance between posts depends on the species, age, size, number, and tractability of the animals you are confining, as well as your own convenience. In our hilly pastures, we find that ten paces is about as far as we want between posts in a sheep fence, while our mild-tempered dairy cows can be contained with a fence on posts fifteen or even twenty paces apart. Keep in mind that your pace is approximately half your height; this will help to estimate how many square feet are being grazed in a given paddock, and makes it possible to establish equivalencies when more than one person is setting up fences. (For most adults the difference will be small enough that an extra step more or less will make it up.)

Stepping in a post is usually easy, but under certain circumstances there may be difficulties. In midsummer, very dry soil, particularly in a pasture not yet reclaimed to full ground cover by grazing, may become so hard that it is difficult to drive in a post. This is especially true with fiberglass posts with plastic foot braces, which may even break off if the user is too rough with them. Molded plastic posts, on the other hand, with integral braces and metal foot spikes are generally able to penetrate all but the hardest soil. In frozen ground the same rules

Figure 5.10. Feeding out hay in the pasture keeps manure and waste hay out of the barn and on the field.

Figure 5.11. Walking down the fence line: treading down forage under the polywire to avoid grounding the fence charge.

apply unless the temperatures drop below zero for an extended period—then it may be necessary to hammer a spike into the ground to make each hole; when our temperatures fall below zero for weeks, we use a fire poker and a sledgehammer to set posts. A neighboring grazier carries a cordless drill with a large masonry bit when he sets up fences in frozen ground. Fortunately, such cold temperatures usually coincide with heavy snow and ice, so we feed hay at those times and don't move paddock fences much.

At the other end of the scale is ground that is too soft or wet to hold a post upright. For the most part, if your ground is this soft you don't want to put the animals on it; they will pug it, driving air and water pockets out of the soil, resulting in soil that will dry hard and packed. There are cases, nevertheless, when you will want to paddock animals near a boggy area, or at the edge of the woods, where moisture or deep leaf mast mean fence posts do not set well enough to hold up against cross tension. This is a case where one or two pigtail posts can come in handy for bracing corners, or you can use two fiberglass or plastic posts at a corner, the second one set at an angle with a backstay, and brace the fence that way. With the short fences of the smallholder there isn't a great deal of line tension, so it's not too difficult to brace a corner.

Any electrical fence can be compromised by weed pressure; that is, excessive contact with pasture plants. On a damp night, our fence line may look like a string of tiny Christmas lights, with each stem or blade of grass making its own point of contact for an arc to the ground. Setting up a new paddock in tall grass, we often find it a good idea to "walk the fence line down"; that is, tread down the plants between fence posts as we set them up. To do this, we just set a post, then walk a straight line directly to the point where we intend to set the next post, pushing down tall grass and weeds with our feet. In high summer, in pasture we are reclaiming from tough stuff like overgrowths of cane, briar, and seedling trees, it is sometimes helpful to get out the weed whacker and mow a lane for a fence line, but we seldom find this necessary.

Which End Goes Where?

When unreeling line for a fence, save yourself lots of headache and consider ahead of time whether your reel needs to end up in the field or at the charger. In general it will be most expeditious to tie or wrap the handle end of the twine or tape to the line from the charger and unroll *away* from the charger; that is, in the direction of the paddock. This will put the reel out in the pasture; once we have outlined the paddock, we can hook it to the fence line at the point of closure. This gives us a convenient way to close the paddock, and also makes it easy, when it is time to move the animals onto fresh forage, to reel up only partway and then outline a new paddock. Or if we are leap-frogging paddocks (see Moving Livestock on page 117), we can reel up the old paddock and release the livestock into the new one while leaving the power line still in place. With the reel at the charger end of the fence line, on the other hand, no adjustment can be made to the amount of line available in the field.

No Spiral Fences

Some fences will require more than one strand, as perhaps for sheep or pigs. It is perfectly feasible to set up a fence of multiple strands with a single reel, but if you do, you want to be sure to *reverse your direction at the point of closure*, rather than unreeling fence line in a spiral. Reversing direction whenever you reach your starting point means that you maintain an opening in the paddock fence, by which animals can be let in or out while the fence still stands. With a fence laid down spiral fashion, animals can only be put in or out by reeling back to a single strand of fence. For animals requiring more than one strand in the first place, such a reduction of their confining barrier extends a temptation to charge the fence, especially when the fresh green grass of their next paddock is waiting for them on the other side.

Gates

What constitutes a "gate" when one is dealing with temporary, portable fence? Anywhere you can open the fence and allow animals to move into or out of a paddock without stepping over or under a fence line, whether that be by reeling up a bit of twine, unhooking a plastic gate handle, or pulling up an end post and folding back a section of fence, is a gate. Planning where gates are to be located is one of the primary considerations when you are setting up a paddock, an issue made more urgent by the temporary nature of the fence. Animals impatient to get to their new grass or to the dairy to be milked should not be required to wait while you fiddle with the fence, trying to compensate for a poorly planned

Figure 5.12. Calves moving through an open fence (*right*) onto a new paddock.

paddock. A gate can be at the end of a temporary loop or dead-ended into a permanent fence; it can open either to the left or to the right. And it can be erected for ease or difficulty of operation: for example, don't loop your twine for a multiple-strand fence *around* a strand of permanent hotwire, and then expect to be able to open the fence at that spot.

Points of Paddock Design

There are a number of considerations when you are determining where to erect a paddock fence; one of the first is the location and type of water source.

Water in the Paddock

Generally, by far the simplest and best water system for the intensive grazier is a portable tank with an automatic fill valve. Whether the water comes from your well or your rain barrel, a spring or an elevated culvert, the forage and soil will best be served by a water system that moves with the animals. With a tank of manageable size, paddocks can be built wherever there is forage that needs to be grazed, and the water put inside the fence. We use this system nine times out of ten, following our animals with half-barrels plumbed with a Jobe float or Hudson valve and filled through ordinary garden hose by elevated rain barrels or cisterns. This

Figure 5.13. Using fence to split access to a permanent water tank. Courtesy of Marcus Grodi.

method keeps water (and urine, manure, and impact) on the pasture and eliminates the trailing inevitable around stationary water sources.

That said, there will sometimes be reasons for watering the animals at a pond, stream, or permanent tank, and this can be done in such a way as to minimize trailing and puddling.

Where a stationary tank is to be used over a number of days and for multiple paddocks, it is not necessary or desirable to build each paddock with 360-degree access to the tank. You can erect temporary fence so that only one side of the tank, even a part of one side, can be approached from each successive grazing area, the paddocks radiating outward from the tank pie-fashion, so that the trailing that occurs when animals use the same route to water time after time is made negligible by the short period they are on that section of pasture. Each time they are moved, a different side of the tank is fenced and the trampled area in the former paddock is given time to recover. Root dieback, grass regrowth, and the activities of insects and worms drawn to the area by its increased fertility quickly repair any damage done by the hooves of our livestock. More than the number of animals in an area, the amount of time spent there is responsible for negative effects like trailing and pugging, and in an intensive grazing system, impact, even around water holes, is kept brief, while rest time is extended.

The same principles should govern use of pond and creek banks for livestock watering. While impact on the verge of the waterway can be beneficial, rounding the profile of steep banks and driving in grass seed to encourage plants to clothe bare soil, it is of great importance that it be of short duration—disturbing, but not tearing up, the ground cover. When grazing on the side of a stream or pond where livestock are given access to the waterway, we move the point of access with each change of paddock, and keep paddock durations very short, to get the most benefit from our intensive grazing practices.

Managing Impact

Because intensive rotational grazing is a system designed to spread impact, manure, and urine more evenly over pasture and paddock, it helps to separate amenities—water, minerals, shade, supplemental hay, treats—as far apart as may be within the grazing area. Movement between points makes sure animals visit every part of the paddock and trample what they don't eat. This practice can even be used to increase traffic in a spot that would otherwise be avoided; placing the mineral pan in a patch of thistles, for example, means these less-favored plants will get a thorough stomping; feeding winter hay where woody perennials are a problem cuts down standing plants while spreading forage seed and pressing it into the soil.

Shelter

What sort of shelter you build into a paddock is going to be determined on a day-by-day basis; that is, it will be a function of the weather, of what you are sheltering the animal or animals *from*. For animals with minimal shelter requirements—cows or sheep, for example—there will be many days when no extremes of temperature, wind, or precipitation are expected, during which nothing at all is necessary. Heartlessness? Even with access to a shed, in wet, cold, or snowy weather our cows and sheep usually choose to rest in the field. After all, they are animals, and while

Figure 5.14. In their thick fleeces, sheep don't seem to mind even the coldest temperatures.

many, given free access to a snug barn, will no doubt enjoy ruminating there, they seldom need a roof for comfort. Animals that stay on the pasture promote their own health and that of the farm.

But on days when temperatures are expected to be very high, or days in early spring when, despite moderate temperatures, intense sunlight can overheat animals still adjusting from winter, a patch of shade may be desirable or even necessary. Often, throwing a loop of fence around a tree is all that is necessary. But when erecting a paddock with a shade tree, remember to consider the sun's east–west path and the track of the tree's shadow over the course of the day. A tree on the east side of a paddock may provide adequate shade for the morning, but in the afternoon, during the period of most intense heat, that shadow will fall on the east side of the tree and *outside* of the paddock fence, where it is inaccessible to livestock. And in high summer, in the Northern Hemisphere, shadows fall largely to the south of an object, so a tree just within that side of a paddock will provide little shelter to animals whose paddock is to the north. We keep one eye on the weather map all the year, but especially in summer, and when high temperatures are expected we make sure to include generous time-appropriate shade in our paddocks, even if it means moving the animals from one side of the farm to another.

Rain or snow, even a heavy fall, is seldom a threat to cattle or sheep in our climate, but when high winds and long periods of heavy precipitation are expected, our animals are more comfortable—and so are we—if they have some way of getting out of the worst of it. A windbreak can go a long way to mitigating the inconvenience of a period of inclement weather, and animals will seek the protection of even a leafless tree line when the wind gets up; a rise in the ground is also sometimes adequate to provide a place out of the wind, if the rise is situated across the direction from which the wind is coming. In heavy winter weather we make sure to include some forest edge on the north or west of a paddock, since those are the directions from which our prevailing winds come; when we build paddocks on the east side of the farm, a ridge to the west cuts off the worst of the blast. And in a genuine blizzard or storm, we may give at least the young livestock access to a run-in shed for the duration.

Goats, on the other hand, are not as hardy as their bovine and ovine cousins, and need more elaborate shelter, more of the time. Ours are mostly tethered at the edge of the woods, where it is shady, and do fine with just a run-in shed or the open stall under the bank barn for shelter from heavy rain or snow. In winter, when there is little for them to eat on our rocky hillsides, we pen them in the nursery pasture with access to the barn and feed them hay. On another farm in our neighborhood, large permanent paddocks are each equipped with a small, three-sided pole shed covered with roofing metal; when shelter is necessary, smaller, daily paddocks can be built to include access to a shed.

Shape

In general, grazing is done most economically in compact rectangular or square paddocks, where animals may with the fewest steps reach the greatest amount of forage. This reduces unnecessary trampling of palatable plant material, as well as keeping in closest proximity animals whose natural inclination is for bunching and herding. Additionally, in a square or compact paddock sight lines usually permit grazing animals to keep in comfortable visual contact with their pasture mates without losing grazing time. We usually avoid long, narrow paddocks (except in situations where we use their disadvantages to our advantage, as in trampling weeds) and paddocks that include substantial visual or physical barriers (paddocks straddling the brow of a hill, for example, or with a long lane back to water); under these circumstances, animals lower in the pecking order will follow dominant animals to water, forage, or shade but will not necessarily get the time to meet their own needs before the herd shifts ground.

Figure 5.15. Hemp dogbane. This perennial plant is considered a noxious weed by herbicide manufacturers; intensively grazed livestock devour it without harm. Maybe they know something we don't.

That said, there is nothing wrong with an oddly shaped paddock where it serves our needs, especially in the very small grazing areas appropriate to low stock numbers. Detours or keyholes around sensitive points like young fruit trees don't significantly decrease the effectiveness of grazing, while allowing us to enjoy the great benefits of using animals for orchard cleanup. A muddy spot will usually be fenced out of a paddock, either completely, or by throwing a short line around it and linking back to the fence. We have to keep in mind the turn radius of our animals, and the size of their personal space, so that paddocks are designed to avoid narrow places where small animals, like lambs or calves, can be shoved through or under the fence as livestock jostle for position at the water tank or in a lane.

Straight or Wavy?

Curvy fences are sometimes held up as a sign of enlightened paddock building, evidence that the grazier is sensitive to every undulation of his farm. Intensive grazing works best when variations in terrain and forage type are taken into consideration and paddocks are built with a degree of uniformity in their composition, so our fences will often vary from the straight, as we skirt around a rise of ground or tailor a paddock for efficient trampling of a patch of coarse weeds or briars. But there is no reason to waggle your fence line for every little rise or hummock; straight fences, with a single line of tension, are sturdier than roving fences, and it is far easier to calculate the size of an area of forage when it is built with straight lines and more-or-less 90-degree corners.

Permanent Perimeter Fences

Paddock design is largely affected by whether the grazing area has a perimeter fence, and, if so, what kind. Pasture with an electrified perimeter fence is the easiest sort in which to build temporary paddocks, since in this case the perimeter fence itself can often function as at least one side of the paddock fence, and the temporary fence will be electrified by contact with the permanent fence. Under these conditions we can simply outline successive areas of fresh forage, the permanent fence making one or more sides, and in this way move our animals gradually across the pasture.

Where the perimeter fence is not electrified we may still be able to use it as part of our paddock fence, if it is of a sort that will contain the species of animal we are confining. Barbed wire won't hold sheep or goats but will likely be perfectly adequate for bovines; woven wire should stop the sheep, but goats and pigs are another matter. If the perimeter fence is not of a sort that will contain our livestock reliably, we will need to encircle each paddock completely with the necessary electrified fencing. In either case—whether we use the permanent fence as one side of our paddock fence or not—it is important to make sure that we don't let the temporary fence ground (that is, make a conductive contact) on the perimeter fence; anything that compromises the charge in a fence decreases its psychological impact and therefore its effectiveness.

Access—Location in the Field

Perhaps it goes without saying, in a system designed around timed grazing, that where it is possible we limit or prevent animals—and people!—from walking over ungrazed areas on their way to and from their paddocks. In other words, we design our grazing patterns with successive paddocks moving *away* from any point to which the animals will be returning often—for example, a stationary water tank or (for dairy animals) the barn; we don't, if it can be avoided, start at the farthest point and then work our way back. This won't always be possible, as when the condition of the forage—more mature grasses at the back of the field needing to be grazed before the nearer portions, for example—make it desirable to graze the more distant areas first. In this case, we sacrifice a small area of forage as a temporary lane for livestock movement, so that the premature grazing and back-and-forth impact can be limited. We build the lane first and use it for one or more days' paddocks, working our way out to the distant point, so that none of the trampling of animal movement is happening on forage we are saving for a later time. Driving animals over ungrazed forage not only damages the plant material, it is a pain in the neck, since all that nice, lush growth is a positive invitation to Bossy to settle down and get her dinner.

Access—Activity Patterns

Good paddock design reflects the activity patterns of the animals being enclosed. Livestock that leave the paddock on a regular basis—such as a dairy animal that comes up to the barn once or twice a day for milking—have different fence requirements from those animals that don't leave the pasture. For one thing the former need gates—fence openings—that are easy to operate. In addition, where there is a power line going back to the charger, those openings need to be on *the same side of the line as the barn door.*

Think about this for a minute. Imagine that your paddock is at the extreme end of a line fence, the other end of which is attached to the fence charger, wherever that may be, as, for example, on the side of the barn. Supposing the dairy door is on the front of the barn, then, if your paddock opening is on the *back* side of the power line, that line is between you and the dairy door. Your cow, who knows where the dairy is as well as you do, will want to make directly for the dairy door. You, on the other hand, *if* you remember that the line fence is in the way, will want to lead her away, around to the other side of the line fence, which she will resist. But it is more likely that it won't occur to you which side of the fence you need to be on until you are halfway back to the barn. Now your cow *really* won't want to walk back with you. So what, say you—I've turned the fence off, I'll just drop a post and lead her over it right here, or I'll lift it and let her pass underneath. There are two objections to this solution: First, since your fence is only a psychological barrier, you don't want to teach your livestock that under certain conditions it is to be ignored. Secondly, you probably won't be able to do it anyway, because your animal won't want to risk the shock.

Moving Livestock—Making Paddock Shifts

When animals are confined by a portable temporary fence, a primary challenge has to do with making the moves between paddocks. It will be obvious that, for the sake of convenience, the grazier will probably want to have at least two reels of twine or tape for fencing: one for today's paddock, one for tomorrow's. In the simplest scenario, the animal being moved to a new paddock is left in the existing paddock while the next grazing area is outlined, then calmly led or driven into the new area, enclosed, and the old paddock reeled up. In the simplest of all possible situations, this is an excellent plan—and therefore it is frequently impossible.

There are a host of reasons why this vision may be oversimplified. About half the time the difficulty will be simply logistic. There may be more animals in the paddock than the grazier can easily lead or drive at once; herding animals will resist being separated, and you can't make them understand the concept of "just a minute." Or the single animal being grazed—a yearling beef steer, for example—may not be well trained to the lead. Circumstances may prevent placing the gates of subsequent paddocks in proximity. It is good, therefore, to start your grazing career with a few simple tricks in your bag.

Imagine a paddock—we'll call it "paddock X"—as a loop at the end of a long line: the loop, a closed curve of fence line, is the paddock itself, and must turn either to the right or the left; that is, be either a clockwise or a counterclockwise loop. When the new paddock is to be on the side to which the old one opens, you can build three sides of the new paddock, omitting the side that would be shared with the existing paddock and extending the third side as far as necessary to contain the animals while you reel up the old paddock. Doing so will eliminate the intervening side and let the animals move onto the new forage outlined for them; then the fourth side of the paddock is set up. For the next paddock, the original reel will be made to bypass the second paddock, a new paddock will be outlined, and the process repeated.

If, however, the new paddock is to be on an off-side of the old one, the intervening section of fence would be the last section to be reeled up—that is, before the animals could be released into the new paddock, the old one would have to be taken down entirely, leaving them unconfined for a short time. When this happens, we "leapfrog" paddocks: when

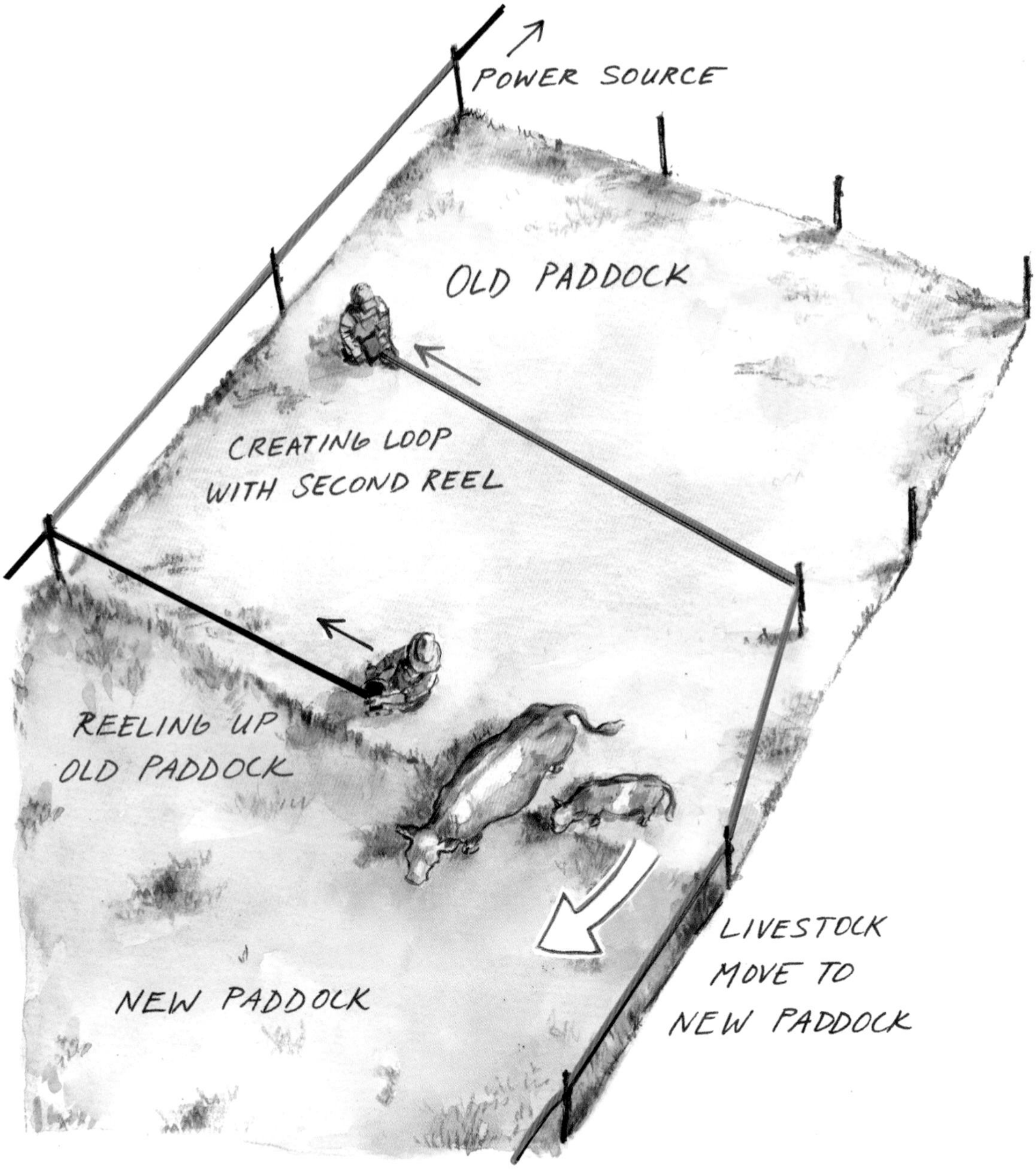

Figure 5.16. By "leapfrogging," one or two people can shift a paddock without releasing the animals. Illustration by Elara Tanguy.

a subsequent paddock is unreeled to the back of the current one, the fourth, tangential side of the new paddock is temporarily looped around a small area of the old paddock *so as to include the animal or animals inside the shared section*. The new line is closed, confining them until the old one is reeled up and the animals are released onto the new forage. Then the new fence line can be tightened up to exclude the already-grazed area. This method has the virtue of letting you change paddocks while the animals remain constantly fenced, a real advantage on land with poor or nonexistent perimeter fences.

The leapfrog method, since it means temporarily confining animals in a very small area, works best when these are trained to tolerate your near approach, even touch. There are circumstances, however, when such is not the case. In moving livestock that are skittish when closely confined—sheep come to mind—it may not be possible to shift paddocks by leapfrogging. The enclosure of chancy animals in the small area of overlap between old paddock and new may be undesirable or impractical: either they may bolt out of the area intended for the overlap before you can close the line, or, in their nervousness, they may charge the fence. In such a case it is possible, so long as the new paddock is on the open side of the old one, to build the second paddock ending in a line thrown out around the first paddock gate, forming a temporary chute. The gate of the old paddock is then opened, allowing you to drive the livestock through the chute and into their new paddock without encroaching on their personal zone. Once the animals have been moved onto new forage, the chute can be reeled back, the new gate closed, and the first paddock reeled up.

Figure 5.17. A mixed herd on fall forage moves into their new twenty-four-hour paddock. Courtesy of the Franciscan Sisters T.O.R. of Penance of the Sorrowful Mother.

Long Moves

Sometimes you may have to move multiple animals over a long distance between paddocks, perhaps to avoid an obstacle—a garden, for example. We often make seasonal moves from one end of the monastery farm to the other, with no permanent interior fences to help us direct the animals. While for shorter distances it is possible to erect a long, temporary lane or chute through which to drive the animals, our long moves are over several hundred yards, so we have developed less labor-intensive means of directing livestock. One that works well for us requires at least three and preferably four people (any ambulatory person will do—our five-year-old makes a hand). Two or three people string out carrying a short section of line and make a walking fence, pushing the livestock ahead of them; with three or more people, this fence can be angled in to confine the animals from the side as well. Moving slowly and avoiding excitement, we can usually drive even large groups of animals across the farm to a new, previously erected paddock without incident.

Sometimes, though, if we are short-handed, we'll use another method that can be accomplished with as few as two people. We start by erecting three sides of the new paddock, leaving it open on the side from which the animals will be coming. The third side strung is extended out a good distance to make a weir or funnel directing animals into the paddock. Then we open the old paddock at the reel, and one person walks away toward the distant paddock *carrying the reel with her*. This acts as one side of a lane fence, so that a second person walking behind the animals can push them slowly along the line toward the new paddock and into the weir. Once the livestock are within the arm of the weir, that line can be swung out and around, driving them into the paddock and closing behind them.

Cell/Paddock Divisions

Building small paddocks is not time consuming unless the terrain is very difficult. Nevertheless, after intensive rotations became part of our regular schedule, we were glad to develop methods to streamline the operation. One of these is the use of cell/paddock divisions. Essentially the construction of a paddock within a paddock, this method also builds greater security into a temporary fencing system, a benefit for the grazier with compromised or nonexistent perimeter fences. In this terminology, "cell" refers to a grazing area delineated for several days' paddocks, and broken into sections by interior lines. This simplifies daily paddock shifts by reducing the number of fence lines being moved at a time; in addition, when the water tank is placed conveniently it can eliminate the need to move the tank on a daily basis. Further, a paddock inside a cell has a secondary fence line on at least one side, with the additional security this creates.

Simply, in a cell/paddock arrangement, a perimeter fence is erected and a portion of the area enclosed is then demarcated with only the necessary interior lines. For each paddock shift the interior lines are moved, but the exterior lines remain the same. As with simple paddocks, some thought must be given to the direction in which the fence is laid out—clockwise or counterclockwise—if the animals are going to be taken out of the paddock frequently, as when being milked. Where the water tank is not centrally located, shifting paddock can also mean extending the interior lines to include a lane back to the tank.

Figure 5.18. Asters—another edible "weed." Courtesy of the Franciscan Sisters T.O.R. of Penance of the Sorrowful Mother.

Line Grazing

One way to use cell/paddock divisions to advantage during seasons of slow forage growth is line grazing. In this case the cell is erected enclosing several days' forage, with the water tank, shade or shelter (when necessary), and minerals being located at one end of the cell. The first paddock will be at this end, the animals being held off the remainder of the forage by a simple line fence set up from side to side of the cell. Subsequent paddock shifts will be accomplished merely by moving this fence line gradually down the cell to deliver more forage. No back fence or lane is built; the livestock are left with access to a larger and larger area as the cell is opened up. So long as this is practiced for only short times—three or four days is our summer maximum, longer in winter—during periods of slow forage growth, there should be no back grazing, that taking of the "second bite" that is so detrimental to regrowth. Herd effect—trampling

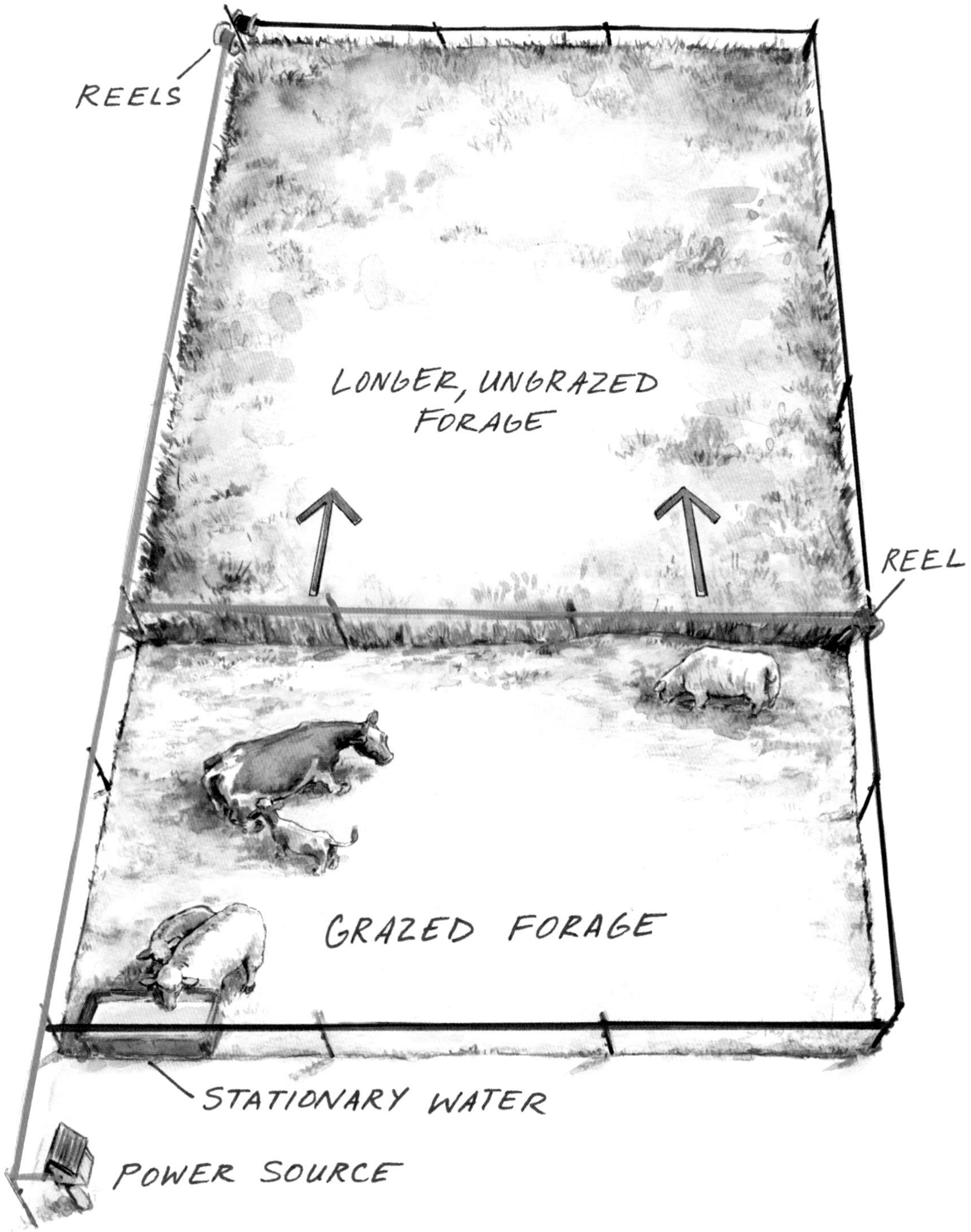

Figure 5.19. On our farm, winter-dormant pastures and stationary frost-free water mean we line graze long cells or "lanes," allowing access back to water without trampling ungrazed forage. Illustration by Elara Tanguy.

and manuring in a severely limited area—is, it is true, somewhat compromised by line grazing, since the animals must go back over previously grazed forage to reach their water tank or shelter. No system is perfect. Observation is necessary in this as in all aspects of intensive grazing; it is up to the grazier, on a daily basis, to evaluate conditions of weather, forage, and animal needs and behavior, to determine the best practices for the day's grazing.

Winter Grazing on an Extended Lane

In this part of the country, winter's biggest challenge is the keeping of stock water before the animals. Fortunately, with the ground frozen and the forage all but dormant, there is less risk of pugging and trailing; so when the temperatures drop below freezing and we have to drain the rainwater cisterns and roll up most of the hoses for the winter, we can still graze our far back acres, although it means a long walk for the animals coming up to water. From a frost-free tank by the barn we can create a long lane with two reels dead-ending in the field, and extend it every twenty-four hours to include a new day's forage. To avoid leaving the end of the paddock open while we make the extension, one reel only is used each time: posts are erected to outline the new area, and a single reel is used to enclose it; then the other reel, the one that had been in place as the end of the last paddock, is reeled back, releasing the animals onto fresh grass. For the subsequent paddock the arrangement is reversed: the second reel outlines the new area, after which the first is reeled back to let the animals through. Because with this method we have an opening at both ends of the lane, it is an easy matter when we need to take an animal up to the barn for milking.

CHAPTER SIX

Ruminants

Whatever the dictates of ignorant idealism, it remains true that good farming has always involved pasture, grazing animals, and animal manure for fertilization.

—*Wendell Berry*

We call our farm the Sow's Ear, which tends to make people think pigs are our primary livestock, but actually, on our farm, dairy animals are Number One. Our road to local food independence wandered without clear direction until our first dairy cow, Isabel, came on the scene. It was with the introduction of fresh milk, twice daily, in large quantities, that we began to see that grass, and the conversion of grass, was the key to feeding our farm without off-farm inputs. Ruminants, in the course of converting grass into other forms—milk, meat, fiber, and manure, not to mention baby animals—ensure the health and continuance of grasslands, the earth's primary means of solar capture, while feeding and clothing human beings at the same time. In the form of milk, especially, we receive generous quantities of the highest quality proteins and fats, abundant and versatile, in a harvest of sunlight-to-food that, renewed fresh daily, means calories are always available, without need of further conversion or storage. Ruminants are truly humanity's most valuable domestic livestock, as history has demonstrated since the dawn of agriculture. The importance of ruminants as a class

Figure 6.1. Nubians can be gentle livestock and good milkers.

on the regenerative farm reflects, not surprisingly, their most-valuable-player status.

While a ruminant of some sort—or several—is a must on the independent farmstead, still, given the numerous species available, choosing which one can be a challenge. The appeal of any number of breeds, their beauty, history, and usefulness, inclines us now this way, now that. There are considerations, however, that can cut through most of the issues.

Choosing a Ruminant

What is needed first of all, when beginning a regenerative grass operation, is a ruminant to fit the conditions of the land *right now*. In other words, the question is not "What is my favorite farm animal?" nor is it "What do I want to be raising in five years, or ten?" but "What species and breed of ruminant best fits the land *as it is*, at the moment?" The farmer who wants to raise Scottish Highland cattle may set as her long-term goal a herd of ten brood cows and a bull of that breed, but if her land is presently choked with raspberry cane and greenbrier, her first livestock should probably be goats. She need not abandon her original dream, however; just factor in an extra step or two.

What is *your* land like? What is its condition—stripped of topsoil; littered with stone, brick, or concrete; covered with trash? What is growing there—ailanthus and sumac, goldenrod and Queen Anne's lace, pokeweed and ragweed? We want to consider the style of grazing or browsing specific to each species of ruminant. Where cows will do best with grass and low ground cover for grazing, sheep—ours, at least—prefer to nibble the leaves from tall broadleaf weeds and woody shrubs, and impact grasses more lightly. And goats, those masters of the art of escape, love briars, berry canes, tree bark, and thorny shrubs, including your new apple trees and your neighbor's prize roses; under proper control they are unsurpassed brush clearers, but left to their own devices they can be ruminant ruin. What else needs remediation: sour soil, eroding gullies, trash-filled stream beds? Smaller animals are easier to work around when there is earthmoving to be done, and less expensive to purchase and keep while they are doing their best for you in the brush-clearing, woodland-edge-opening departments.

Then there are qualities less susceptible to change. What is the land's topography like? Is it steep or level, dry or soggy, sandy or rocky? We need to consider the characteristics of various ruminant species, with their relative weight, agility, and hoof size. A cow will mire in soft soil that might bear the traffic of sheep or goat, and may lack the agility necessary to graze on steep, rocky slopes. Woodland grazing may be adequate for the nutritional needs of a few sheep but wholly inadequate, at least initially, for a couple of Jersey cows. And the farmer who is so fortunate as to purchase land boasting a young orchard will want to give serious consideration to the damage a goat can wreak in even a couple of unsupervised hours.

And what, if anything, was being raised on the land in the recent past? This can be a serious issue, since commercial livestock operations today are frequently the source of considerable and lasting soil and groundwater contamination. We may want, initially, to avoid livestock species with a recent history on the place; that is, we'd think twice about raising pastured hogs where confinement hog manure has been sprayed—as is frequently done on or in the neighborhood of hog confined animal feeding operations (CAFOs)—or about putting sheep on grass that has been used until recently in a conventional sheep operation. Possible infection, infestation, or contamination from manure-borne parasites and pathogens can be avoided if time is allowed to pass before these species are reintroduced to a piece of pasture, and nutrient buildups specific to a species may have a chance to dissipate. Better, maybe, to start our homestead raising animals with no recent history on the land.

Water, its source and availability, must be a consideration when you are choosing a ruminant. How much—or little—is presently available can tip the balance in favor of smaller species requiring less water. And land without a source of frost-free water—be that source a frost-free spigot, a stream, deep pond, or nonseasonal spring—may make it desirable, for the first year or two, or until other sources can be developed, that no livestock be carried over the winter; in which case, a good option is to pasture meat animals (rather than dairy) and butcher them when subfreezing temperatures set in.

To Fit Your Plans

Not that the present state of the land is going to dictate what is possible forever. What you choose to graze now will be determined as much by your long-term goals for your grass-based farm as by the land's condition at time of purchase. What are your intentions for the future? Now is a good time to draw up a comprehensive plan for the farm of your

Figure 6.2. Nigerian dwarf goat and kids. Courtesy of Sara Hildebrand.

dreams. Don't depart completely from reality; don't, perhaps, envision a grove of avocados in Ohio (but then again, with movable high tunnels making it possible to grow winter crops from two garden zones south, maybe you could). Consider the land, and what you hope to be doing here in ten years, twenty, fifty. Do you intend to be certified organic? You'll want to decide that as soon as possible, so you don't make choices that will set back that goal. Are you hoping to raise all of your food here? If so, think about where the permanent improvements will be, the orchards, barns, raised beds, winter tunnels. Does the land need a pond, do you think? Where, and how large? And what complementary enterprises can you envision that, while not strictly agricultural, may be affected by, or may affect, your choice of grazing strategy for the present? The owner of a herd of Nubians, for example, will want to control their access to brushy land intended for a ground-bird sanctuary.

After that, we come up with a tentative (and probably unrealistically optimistic) timeline for our goals. Consider what needs to be started right away, and what you want livestock to do for you toward those ends. Are you clearing trees for a future grazing area? Goats, sheep, or pigs may be helpful in this endeavor. Do you want to open up an area for a large garden? In that case, putting goats, sheep, cows, pigs, and chickens over a piece of land, together or in succession, may be just the preparation the land wants before you sow your first green manure crop. Succession grazing can be a powerful tool in pasture improvement: browsing animals impact coarse overgrowth to open up access, followed by grazing animals to promote increased levels of ground cover. Then possibly a quick, less-than-20-percent rooting-over by the farm hog to eliminate some perennial weeds and small stumps and turn up the soil surface, followed by two or three light hand-broadcast sowings of, say, clover seed, and our pasture improvement program is already well underway.

To Fit Your Situation

Of course, we choose our ruminants not only for the landscaping improvement services they provide, but for other qualities as well. Supposing we have already determined that our rocky, briary future pasture will best be brought under control by an application of goat power, we are in a position to decide which breed of goat we want to start with. There are goats favored for meat production, for dairy, for pets, for fiber, or a combination of these; if we acquire both a mama and a papa we are going to have goats to sell, sooner or later, so we'll probably look into comparative kid prices. Sheep usually (but not always) produce valuable fiber, and there is presently a good market for meat lambs if they are undocked and sexually intact. In addition, there are many artisan cheeses made exclusively from sheep's milk, so if our future plans include dairying and cheese making, we might want to start looking for a local herd of grass-genetics East Friesians or Lacaunes from which to buy foundation stock. And if cows are our choice, do we want beef, dairy, or draft, or a tripurpose breed like Dexter, Devon, or Milking Shorthorn?

And then there are considerations of time and other assets: money, existing outbuildings, and fences. The species, breed, and number we choose should take into consideration these things as well. How often will we be able to move the fence and water tank for our new ruminant(s), or shift and untangle tether lines? Too many animals at the outset make for overextension and discouragement. If the land has fencing of any sort already, we may be planning to use that fence as part of our grazing pattern. If so, its location and condition, as well as the need possibly to extend or repair it, will affect our choice of species to begin improving pastures. Existing barns or other structures will widen the choices available, as they make it possible to shelter goats in inclement weather; store hay for winter and dry season; confine and protect sick animals and animals giving birth; or restrain an animal for milking, shearing, hoof-trimming, or veterinary care.

And we will want to consider the compatibility of a new ruminant with any species we already have or intend to have on the farm. The woman who raises Tibetan mastiffs will want to think hard before adding a flock of sheep to the family; maybe water buffalo would be a better companion species. Conflicts may arise: some merely annoying, others more serious. Goats, the consummate escape artists, are also good safe breakers and will find a way, no matter how, to get into your chicken tractors and gorge themselves on poultry feed. On the other hand, sheep and chickens are good companions; some people report that their chickens are a significant factor in protecting sheep from fly strike. Cows and sheep, on our farm, are complementary grazers, the sheep adding briar control to the pasture grooming; during the hot season we run our yearling calves and Katahdin ewes in the same paddock, while the ram may spend his summers with the big steers, to isolate him from the ewes until the breeding season begins in November. And we are sure that goats, if they could be confined behind fences adequate to contain cows and sheep, would be beneficial additions to our pasture-management team as well. Running a mixed herd can also provide smaller ruminants with some protection from predators like coyotes and wild dogs since, in our neighborhood at least, these animals are usually reluctant to enter a paddock where cattle are grazing.

Figure 6.3. Bottle feeding an orphan lamb.

To Fit Your Neighborhood

In selecting a species and breed of ruminant, it is important to consider who our neighbors are. What is going on next door? Is it a nursery for rare shrubs, or an orchard or vineyard? An escaped goat could wreak havoc there in just a few hours. Is our land bordering a superhighway? We'll want a good fence between our black cows and an unlit road carrying fast traffic at night. Are there zoning laws that might restrict what kind of, or how many, animals we can keep on our land? It's probably a good idea to investigate. And by all means consider the noise and smell factors associated with our proposed species; if at all possible we will want to be on good terms with our neighbors, and goat smells next to their grill pit, or incessant blatting under their bedroom windows from anxious mother sheep on tether calling their straying lambs, may not be the best way to ingratiate ourselves.

To Fit Your Fences

Finally, we want to make sure we have the means of confining our new animals. Match your animals to your fence, or your fence to your animals! If your new land has existing permanent fence that you intend to use for some part of your livestock control, make sure it is suited to the species of animals you intend to confine. Five strands of barbed wire will—usually—contain cattle, but sheep don't consider it much of a barrier if there is something they want on the other side. Woven wire, on the other hand, will generally restrain sheep, but not an adventurous goat, unless the vertical wires are too close together to admit a climbing hoof—and even then, don't underestimate the creativity of the goat. And rail fence, while attractive, isn't going to slow down anything that can slide under the bottom bar. Beginning our grazing career with multiple daily forays to catch escaped animals is a recipe for frustration, so it is important to make sure the fences we intend to use are adequate to contain our chosen livestock species.

Goats—Caprine Grazers

Goats are on the small side as ruminants, weighing from about 50 pounds, for a mature Pygmy doe, to as much as 300 pounds for a full-grown Boer buck. Usually less expensive than cows, goats breed easily, and after a five-month gestation generally produce two or more offspring. A single goat may tether well, and they are great first clearers of underbrush, eating the thorniest and most coarse shrubs by preference. Their light weight makes them less quick to cause trailing, particularly if rotational grazing practices are followed. Somewhat less hardy than bovines or sheep, they need a shelter in which to keep dry and out of drafts in wet or cold weather, but this need not be fancy, and our goat herd did well in a three-sided shed of recycled materials, even in our sometimes-bitter Ohio winters.

Goats can be a great introduction to dairying, since their smaller size makes them less intimidating, and their modest milk production is not as overwhelming as the multiple gallons of milk a freshening cow will produce. Goat milk makes great cheese, too, and early cheese making efforts are quickly rewarded with a tart feta or a creamy chèvre. Goats are not necessarily expensive, and are small enough to be butchered easily, which is a serious consideration if you are raising goats, since a single year can triple your herd size. Their smaller frame also makes transporting them a little easier—we know homesteaders who move their goats in a minivan—and, being on the small side, they are less likely to hurt you accidentally, as when swinging a head around or backing unexpectedly. A well-trained goat can be managed easily even by a child.

Niche

Goats, as we have said, are excellent brush clearers. Ours have been responsible for opening up the understory of our mixed-hardwood forest and clearing the impenetrable cane from the wood's edge; the spot that is now our primary kitchen garden, once a dense cane break, was cleared by goats. That said, if you purchase an adult goat that has spent its whole life in a barnyard eating hay and grain, don't be surprised if it isn't able to make a happy transition to black locust seedlings and raspberry cane on a rough hillside. But a goat or two, or three, judiciously tethered, are the best tool we know of for initial brush clearing. Bear in mind that goats primarily browse, that is, nibble higher, woody plants and plant leaves, so, for grass pasture improvement, goats alone will be less effective than in concert with other ruminant species.

Goats don't have their Houdini reputation for nothing. People who raise goats generally agree that the question is not *if* they will get out, but *when*; and whether it will be your fruit trees, or your neighbors', that they devastate. That said, we would not want to forego the land-reclamation skills of the goatkind, and find that good stout hog wire, charged polynetting, or a strong tether with a swivel at each end gives you fairly reliable control of your goat's whereabouts, provided that her needs—food, water, shade, access to her babies—are all met within the confines of fence or tether. Goats are rapidly reproducing sources of meat, milk, and sometimes fiber, in addition to being brush-clearing dynamos, and they can be a valuable addition to the farm community.

Figure 6.4. Goats require creative fencing strategies; here, a second level of goat fence is being erected above the first. Courtesy of Sara Hildebrand.

Needs

Goats are good at foraging for their own calories; when you have seen your goats standing on their hind legs to nibble choice tree leaves, you'll understand why we say they enjoy self-harvesting. They are usually good mothers, conceiving, bearing, and nursing young with little or no assistance. Although artificial insemination for goats is available, the cost and difficulty of obtaining and storing semen makes it more practical for most people to keep a buck and let him do the job himself. He will be much more adept than you are at detecting when your does are in heat, and knows better than you do what to do about it.

On the other hand, goat bucks really do smell—and we mean *smell*. If your buck has access to the dairy shed, you are likely to find his smell permeating your goats' milk. Actually, in our opinion at least, you are going to find goats' milk rather pungent no matter where your buck hangs out. And while at every workshop we have ever attended or given on dairying, there is at least one goat person who swears that properly handled goat milk doesn't taste like goat, our experience is to the contrary; no matter how meticulous our precautions, our goats' milk always tasted goaty. We speculate that feed, as much as milk-handling procedure, is responsible for the distinctive taste; while alfalfa hay and sweet feed may produce a mild goats' milk, the brush and forbs that are the reason we keep goats make goats' milk taste like, well, goats' milk.

Goats are generally curious and gregarious, so a single goat will want a companion—if not another goat, then some other animal that likes to be near her, or that shares her paddock. Our miniature

Figure 6.5. Although common in poor pasture, poison ivy has never been a problem for our grazing animals.

horse Bridget makes a feisty companion animal, but even a friendly chicken will sometimes bond with a ruminant, or be inclined by the grass seed in its manure to stay nearby. If your area has large predators, like coyotes or wild dogs, some thought should be given to providing goats with a guard animal; a companion chicken will be wholly inadequate for coyote patrol.

Goats benefit from attentive management. They are tender animals, and while they can stand cold, they will be healthier if they have a dry, draft-free shelter in which to escape inclement weather. Predators, as we have noted, must be taken into consideration; if a guard animal is not available, some other form of protection may be necessary: electric netting fence, or a closed barn at night. Goats, of course, require human planning for daily and seasonal forage availability and water access; in addition, if you are keeping breeding animals, males must be removed from the herd until breeding time to prevent too-early or overfrequent pregnancies and the toll they exact from breeding females. Our own experience leads us to advise even the most pharmaceutically opposed to consider tetanus vaccinations for goats, as this is a fatal disease to which they are dismally liable. Finally, for fiber goats, the farmer must arrange for shearing.

Surplus and Services

Goats are prolific! They will almost certainly give you more offspring than you want to eat or milk. Chances are, after a year you will have goats to sell, for meat, dairy, or breeding purposes, unless you are trying to build up a herd for brush clearing—which leads us to another marketable goat product: their land remediation services. Large goat herds, managed by an itinerant grazier, are becoming more common, applying their selective and environmentally friendly pasture grooming skills, for a fee, on lands both public and private. And fiber goats, of course, produce fiber, some of which commands a handsome price.

Shelter and Fence

Goats, as noted previously, need shelter in wet or cold weather. Here in east central Ohio, well within the weather impact of the Great Lakes, we house our goats in a three-sided shed, and this has always been adequate. A roof and a dry floor, backing to the wind, go a long way toward comfort for any species. Tethers work well for controlling one or two goats, but we find shifting more than two tethers a little wearying. Electric net fencing would be our choice for temporary goat enclosures, with two or three ground rods on the charger to make sure there's a good snap. Permanent fences of horse wire have done us good service, too, at least until one or another of our caprine friends wants to go wandering. Goats aren't like cows or sheep, which will stay in a fence so long as the fence is on, with grass and water inside it; as we said before, with goats, it's not a matter of *if*, but *when*.

Lots of people want to go out and pick up an animal before they are really ready—we know, we've done it, more than once—and, admittedly, some of us just need that degree of urgency before we actually buckle down and get a job done. But there is a lot to be said for avoiding the headaches, for you and your neighbors, that are almost guaranteed to come when you get into goat-keeping unprepared; that little bit of preparation—40 feet of airline cable with a swivel at each end, a screw-in stake or a cinder block, a half-barrel of water with maybe a float valve and hose attached, and the corner of a shed with something for the goat to stand on while she is milked—can mean the difference between your enjoying the initial experience, and wondering just why you ever wanted to try.

And while goats are fast and agile, giving the impression that they could keep ahead of almost anything, they are vulnerable to dog-species predators (and any bigger carnivores you might have locally), making it a good idea to use predator-proof fences, barn the goats at night, or employ the services of a guard animal. Here at the Sow's Ear, the farm dogs do guard duty, while up the road at the monastery, the coyotes avoid paddocks with cows in them.

Don't tie an animal, any animal, but especially one that can climb and jump, where she can hang herself up by her cable over a tree branch or fence. "Stuff dies," as the farmers say, but stuff doesn't have to die like *that*.

Food and Water

What a goat should eat is the matter of some debate. An all-grass pasture will be inadequate for goats,

Figure 6.6. This weedy field includes lots of good goat forage.

whose normal grazing/browsing pattern includes lots of high-up things: tree leaves, the growing tips of sassafras bushes, trailing greenbrier, and the little red hips on rosebushes—also black locust seedlings; multiflora rose (that scourge of Ohio pastures); and anything really nasty, tough, and thorny. Our goats browsed in the spring, summer, and fall and ate hay in the winter. We milked the does, ate the little bucks, and everyone seemed to do fine. They need access to salt, or a mineral block formulated for goats, and today we would also probably give them a box of dried kelp.

Like any other ruminant, goats need water constantly available. Not that they would die if you just took them a bucket twice a day, but you'd forget, or you'd send the ten-year-old, who would get distracted and forget why she was there, or it would just be a big nuisance. Better to have a tank in the paddock—it can be small, but not so small that when the goat puts her hoofs on the edge it will tip—and put a float valve and a hose on it.

Milking Goats

Milking a goat is child's play, literally; our children enjoyed milking the goats, except on cold mornings, or when they had to break off a game to go milk. Goats have comparatively small udders, so they milk out quickly, but they are also pretty close to the ground, so you are going to enjoy this task a lot more if you build a raised stanchion where they can stand and enjoy a treat of oats or good hay while you milk them. You'll want a head gate at one end, with a feed crib or hay rack where the animals can reach it, and make the stand high enough that you don't have to squat too low to reach her udder. Some goat stanchions have a built-in milking stool or bench, but we just sat on buckets. We needed two buckets, a 3-gallon for Shawn and a 5-gallon for me—he's about 10 inches taller. Make sure you have a milk pail small enough to go under the goat's udder; goat pails are available that have a half-moon lid, to keep out bits of hay and so forth that might fall in while you are milking, but you can milk just fine into a gallon stainless-steel soup pot or something similar.

Figure 6.7. Although toxic white snakeroot is ubiquitous at our wood's edge in autumn, our ruminants don't graze it. Courtesy of the Franciscan Sisters T.O.R. of Penance of the Sorrowful Mother.

Breeding

If you want your doe to give you milk, she will have to be bred. You may buy her already pregnant, of course, but that only takes care of the first year; after that, you will have to breed her yearly if you want a regular supply of milk. The easiest answer is to buy a buck and let him work out the details. Although artificial insemination (AI) is available for goats, the issue is more complicated than just getting the right stuff in the right place, and timing is essential to success. Frozen semen remains viable for only a short time after it is thawed, and does are able to conceive for about one day in their three-week cycle. If you don't have a professional AI tech for goats in your neighborhood, you can do it yourself, but a nitrogen tank is an expensive piece of equipment and has to be recharged regularly. We don't deny that there are definitely advantages to AI, especially the specialized genetics it makes available, but nature, after all, is hard to beat. If you don't want to own a buck, try to find someone nearby who rents his buck out for stud, and see if you can take Nanny for a short visit.

Health

If you don't like horns on a goat you can disbud while your animals are babies, either with a hot iron made for this purpose, or with caustic paste, both of which should be available at the farm store (see Dehorning Calves on page 177). We used to have the vet disbud them with an iron, then we did it ourselves, then we started using the paste. In the end we decided that horns were good protection for goats and that's why they had them, and when we put up hog wire fence we made sure the spaces in the wire were too big for a goat to get her head in there and get stuck.

People usually castrate baby bucks, partly to prevent them from breeding the young does in the herd (a buck may be able to breed as young as two months of age) and partly because castration is supposed to make the meat taste less like goat. Castrating a little buck is like castrating a little bull and can be done with a knife, castrating band, or burdizzo (see Castration on page 177), but some cultural cuisines require that goats be unaltered and undocked, so you might not want to castrate. If you are going to eat him young, he won't taste too strong yet, and anyway, if you don't really like the taste of goat meat you probably won't like it whether you castrated the buck or not.

Goats look like tough animals, but little things can kill them. Veterinarians recommend vaccines for does in late pregnancy, with boosters for the kids at a few weeks of age, and you can pick up some of these immunizations in the refrigerator section of the farm store. We do not always follow recommendations for vaccinations; lots of the problems people warn you about in ruminants are actually the result of confinement—too many animals in too small a space, poor air quality, wet feet, and so on—or of conventional grazing practices that provide parasites and pathogens with regular access to the host animal. One vaccine we are faithful about giving to our goats, however, is tetanus. Goats are very susceptible to this scourge, and it's bad, very bad. It gets into your soil and you don't know it's there until one of your best little doe kids, just disbudded, or a buck kid recently castrated, gets it. A bit of dirt gets into an open sore—watch them bring up a back leg and delicately scratch the scab over their bud scar—and they contract tetanus. Then they lie down on the ground with their legs straight out and their back arched, and the least thing—a small noise, a touch—causes their muscles to spasm repeatedly, which, since their joints are locked, is extremely painful. They make pathetic little bleats. You read that there is a one-in-ten chance of recovery from tetanus, so you wrap them up good and pour milk down their throats four times a day, and they get pneumonia because they can't swallow very well and the milk gets in their lungs, so after a few days you have to put them down anyway. After that when you see a goat with tetanus you put it down immediately, and then go to the store and get boosters for all the rest. We think it's worthwhile to immunize goats for tetanus.

Sheep—Ovine Grazers

Goats and sheep can be sufficiently similar in appearance that sometimes it's hard to know which you are looking at—our Katahdins are regularly mistaken for goats in early summer, when they have shed their wool jackets—but

Figure 6.8. Small ruminants seem to enjoy steep pastures.

there are some significant differences in care and husbandry. Sheep are small ruminants with a reputation for being stupid, and while we don't want to offend any sheep enthusiasts out there, we have to stand by that epithet. Gospel references to the people of God as His "sheep" give you some idea of how humanity looks from the divine perspective, and it's not complimentary. Sheep are hardy, fiber-producing (usually), and small enough to move in a minivan, or for your eight-year-old to lead by a tether. After a five-month gestation they generally bear twins or triplets; they reproduce well, sell well in food markets specializing in religion- and culture-specific cuisine, and are not so expensive as to make too big an investment for the infant grazier.

Niche

Sheep are more likely to nibble forbs (broad-leaved, nongrass species—read "weeds") than to graze a lawn or eat bushes, but they like all three. Because sheep are small in size, their grazing impact will be less, which means less pugging and trailing, but also that their pasture remediation will be slower unless you have quite a few and run them in very small, short-duration paddocks. We find ours the perfect remedy for giant ragweed, which germinates so efficiently in bare waste ground, like the rough patch where we threw out hay too many times last winter. They also eat the invasive scourge Japanese knotweed (*Polygonum cuspidatum*), which locals around here call river weed, that travels underground and can, seemingly, push its way beneath a two-lane blacktop road to come up on the other side. Like goats, sheep are great on hillsides too steep for cows, and since they prefer a different mix of forage, the two species make great grazing companions.

Needs

As far as what needs they are capable of providing for themselves, well, sheep can eat and find their way to water if you put it in their pen. They can generally breed and birth unassisted and don't need much shelter. And, blessedly, they like to stay together, so when one gets out of the fence he just hangs close looking for a way back in; when we leave the fence charger off and they all get out, they go back in all together instead of piecemeal.

Although they aren't brilliant, sheep are, as cooperative members of the community, on firm ground. We find them an unobtrusive part of the farm ecosystem except for a few weeks during the lambing season, and they seem to give and receive benefits with little fanfare. Their grazing and browsing improve pasture for other ruminant species, while their nonaggressive personalities (rams sometimes excepted) mean we don't have cow-sheep-pig fracases breaking out in the paddock. When we run sheep with cattle we don't have any trouble with predation, with the possible exception of wild dogs. (While well-trained guard dogs are a sheep's best friend, wild dogs—or the family pet when the moon is full—are a real threat to sheep, being generally unintimidated by cattle; when the mood is on him, even the nicest pooch may chase a panting sheep to exhaustion and then tear her insides out.)

Our sheep are on forage twelve months of the year, with hay only when thick ice makes the winter-cured standing grass too hard to get at. We keep water in the paddock, either gravity-fed from our spring tank on the side of the hill, with rainwater from the IBCs that hold roof runoff, or by stream-side access when they graze along Jeddo's Run at the bottom of the south pasture. Only salt and minerals are bought in; we make sure we get a mineral block without copper, since this metal is toxic to sheep.

Planning for forage availability and water access—the basis of rotational grazing—along with timing breeding periods are the responsibility of the farmer, who will probably find that these basic services obviate the necessity of such things as parasite and pathogen management as well.

We hear people comment on the readiness with which sheep get lamed and have noticed as much in our own flock, pastured as they usually are on a steep hillside. When we see a lamb limping, we move it into our hospital pen—a small enclosure of flatish ground with good grass and barn access—or pasture the whole flock down on the level until the lame sheep is sound. So far this has always worked, and we know other farmers who have the same approach.

The other and obvious service for which sheep will rely on the farmer is the yearly shearing, unless they are hair sheep and shed off in the spring.

Surplus and Services

Lambs, of course. Sheep generally produce twins, half of which on average will be ram lambs, mostly unwanted for breeding purposes. Depending on breed, some do presently bring a high price for breeding stock, but this is generally a temporary window; once the demand is met, some other breed takes first place in popularity. An exception right now is for animals with grass genetics—that is, animals that do well on forage alone—and with proven parasite resistance, like our Katahdins.

Figure 6.9. Even in winter, sheep can find something to graze.

These strains retain their value for breeding because they have inherited qualities that don't lose favor in nature or in the market. There is also a demand in certain religious and cultural cuisines for undocked, uncastrated lambs for butchering.

In addition to lambs, most sheep will give you wool to sell. Not Katahdins, though—they are hair sheep, meaning that they shed. No wool to spin, but no need to find a shearer, either.

Shelter and Fence

Sheep can get through a pretty reliable-looking fence, like five strands of tight barbed wire, but usually don't bother unless what they want isn't inside their paddock. So the rotational grazier is by definition all right—she is dedicated to keeping her herd or flock on a small paddock with everything they need, and moving them before they run out. Most of the time, our sheep are held behind two strands of polywire on step-in posts. Winter paddocks might be big enough for a week, since there is no regrowth to tempt them into taking a "second bite," and they remain in the fence until we move them. Summer paddocks are usually shifted every day or two, leaving little chance of the sheep running out of forage before we move them to a fresh area—so they mostly stay in.

Lambs are another matter. A baby lamb is a tiny thing that can easily slip under the two strands we use to contain adult sheep, so when the lambs are young we run a third strand lower down, and that is usually enough to hold them. When a lamb does get out—a bottle lamb, usually, since they learn to associate people with nourishment and want to be where the people are—it generally stays fairly close to the flock. And, since it has let itself out, it can let itself in as well, if there is something it wants inside the paddock, like the company of the other lambs, or a go at the water tank.

Fences are not just needed for keeping sheep in; ideally, they are also for keeping predators out. For this purpose, the new portable electric fences may actually offer a degree of protection not provided by barbed wire or even most woven wire. Some electric netting is sold specifically for predator exclusion, and there are graphic ads showing bears and wild hogs backing away from a panel of electro-net. Anything that can psych a bear or a boar ought to eat coyotes for lunch. Even polywire, if it is strung at dog-nose-level, is said to stop a coyote. We can't vouch for it personally, not having seen the event, but we can say that when coyotes came howling down our hollow last winter, the sheep, behind two strands of hot twine, remained untouched. We still kept the shotgun loaded, though; nature is nature, red in tooth and claw.

We find that weather doesn't bother sheep much. Sheep are naturally protected against all kinds of meteorological events, and don't object to rain, snow, or cold. They do need to be able to get out of the sun on hot days, but winter doesn't seem to disturb them. Occasionally, when the weather drops below -10 degrees or so, we will turn them into the open center bay of the big barn, but even then they seem just as comfortable in the field.

On the other hand, mysterious things happen to them. Those thick wooly pelts hide a good deal, and if a sheep is growing thin under some stress it will be hard for you to tell. As the sheep farmers say, "A sick sheep is a dead sheep," because you often don't know there's a problem until there isn't one anymore. It is not heartlessness that makes farmers philosophical about this; perhaps it helps that a single market sheep at 75 pounds might sell for just about what the vet charges for a farm visit and two meds. Overall, we think, sheep can make a good first ruminant for

Figure 6.10. Forage makes good milk.

the regenerative farmer, since they generally reproduce well, sell well, and aren't too big an investment for the beginner.

Food and Water

Aside from winter forage and minerals, our sheep don't get any supplement. We use about sixty bales for a flock of six in a bad winter; none, if the summer has been good for grass and the winter is clement.

Sheep, like any ruminant, need constant access to drinking water, but not being large animals they can be adequately supplied with a small—30-gallon or so—half-barrel for a paddock of half a dozen sheep or thereabouts. Make sure that lambs can reach in; you may need to put a cinder block or round of log next to the tank while the lambs are tiny. Not too high, though—you don't want anyone falling in. Automatic float valves are the easiest way to be sure your animals won't run out of water before you come back to move them onto fresh forage, but you *can* refill their tank with a bucket if there is no hose. Don't compromise in the area of drinking water; your sheep will be happier, and make better members of the farm community, if they don't get stressed for water. Likewise, make sure they have access to a salt block, or a mineral and salt combination made for sheep.

Milking

Ewe's milk is excellent for cheese making; being nutritionally more dense than either cows' or goats' milk, it gives as much as twice the yield of cheese per gallon. We do not milk our Katahdins, but artisan cheese makers around the country are raising sheep for cheese milk, and the market is supposed to be growing. We advocate hand milking for homesteaders in all but the most unusual circumstances, and that goes especially for sheep, which have tiny udders and teats and should be treated very gently. Although we have seen vacuum milkers made for use with sheep, we have the same reservations about them that apply to all mechanical milkers—they cannot be used gently, and they both cause and spread mastitis. That said, if your arthritis won't permit you to milk by hand, you can look into getting a small vacuum milker (see resources) and be conscientious about cleaning it. Like goats, sheep will be much easier to milk if you put them up on a stand with a head gate and feed crib.

Breeding

As with goats.

Health

Our Katahdins have been beautifully healthy since we started the herd three years ago, and we pray fervently that such will continue to be the case. It helps that we bought our breeding stock from an established herd with a good track record for parasite resistance; even more importantly, we have grazed them on a long rotation, with extended rest before we return to any given paddock. We have had no significant problems with them. That said, some issues have come up. Sheep are generally able to lamb unassisted, their babies usually coming as twins and triplets and not too large for the birth canal, but it is a good thing to check the ewes often during lambing season—several times a day and, for the conscientious, once at night—in case anyone has a bad presentation and needs you to turn a lamb. This is not too hard, but of course there are always problems that can crop up. Wash your hand and arm well, put some O.B. lube on it (see the farm store), and have a gentle feel. We've helped with a tight fit but never had to actually turn a lamb; we *have* turned a calf, and the big difference, we understand, besides that a cow can squeeze your arm a lot harder than a ewe can, is that lambs usually come in pairs and you want to get the right parts out together. Our vet encourages us to try little jobs like this ourselves, fortunately, so we aren't afraid to do our best and only call her if we hit a snag.

Always remember, too, when dealing with sheep injury or health issues that a single farm visit from the local vet, plus two meds, is likely to cost about the same as a 75-pound market lamb will bring. We count on isolation and nursing as our best curatives; those individuals that don't respond to quiet quarters, extra care, and good feeding have genetics we don't want to replicate anyway. We have no intention of being heartless, and if disease appeared to be the issue, with chances of infecting the whole herd, we'd call the veterinarian in and consult, but in our twenty years on this farm we have only called in the vet for a handful of small-animal sicknesses, usually when several were sick at once (some bull calves bought in from a local dairy, complete with pneumonia and scours) or when several animals in succession went down (goats with tetanus).

Cows—Bovine Grazers

Cows, we will frankly admit, and in case you hadn't already noticed, are our favorite ruminants. There are a lot of reasons this is so. One may be the fact that both of us grew up with cows; beef cows, not dairy, but the same slow, patient, grass-smelling bovines that make such efficient use of pasture and turn it into creamy milk, sharp Gouda, stretchy mozzarella, and the best, dry-cured, 1¼-inch-thick rib-eye steaks (and which can and do sometimes explode into sudden action and knock you on your rear or take out a fence). Cows turn grass into people food better than anything else we know—lots of food, gallons and gallons, on a daily basis—and, properly managed, leave the pasture behind them better than they found it: mowed, trampled,

Figure 6.11. Dairy cows gathering the solar harvest.

and covered with half-digested cellulose and billions of soil-enhancing microbes. We love cows.

Cows are primarily grazers, not browsers, eating grass and forbs in preference to bushes and brush, although, as with sheep and goats, this is not a hard-and-fast rule. Cows will sometimes eat cane or tree leaves with relish; these long-rooted plants mine the soil more deeply than most pasture plants, and their leaves contain minerals not found in grass. Bovines' larger size means they have a heavier impact on pasture, so if they are managed well, pasture remediation—increase of plant matter, biodiversity, and soil depth—can be a faster process. A single dairy cow can produce an amazing amount of milk, meat, and manure.

Unless you raise miniature cows, your cow is going to be a lot larger than you are. This may make her more intimidating initially, but in fact we find our cows just as tractable as our smaller ruminants—more tractable than some!—and although their size does make them harder to immobilize (for shots or hoof trimming, for example) they are generally docile, especially when they are handled and moved regularly, as in an intensive grazing system. This applies especially to dairy animals, of course, which become completely familiar with the milker's touch. Although they are seldom aggressive—a ram or buck is much more likely to get belligerent than one of our dairy cows or beef steers—it is important to remember that, as big as they are, accidents can happen. Without intending to hurt you a cow can step on your foot, which is painful, or swing her head around suddenly and bang into you. It is for us, with our greater agility and judgment, to be aware of our bovine's location and condition so that we are not in the way when she moves.

Cows can navigate a good bit of steep ground, but you don't want to ask too much of them, and we don't put them in paddocks that are all steep—they need to be able to stand comfortably along the lowest paddock fence without danger of slipping underneath and getting entangled. If a bovine can't get up it can die quickly of bloat; long ago we lost a calf this way, and it has made us careful. Mostly we put only young calves on the steep paddocks, since they are more agile than the large cows, and we make sure to include some flat ground at the lowest point. Cows' larger size makes moving them over long distances more involved than moving sheep or goats: you can't put a full-grown cow in the back of the van, so if it's too far to lead her, you will have to find a stock trailer, or hire someone who has one.

Cows are more expensive than sheep or goats, so you have a bigger investment; however, their offspring, a single calf born after a nine-and-a-half month gestation, have considerably more cash value than the offspring of most goats or sheep. (That said, if we were in this for money, sheep are said to give better returns. But if *you* are in this for money, you're reading the wrong book.) Calves take longer to reach maturity than do sheep or goats; hence, they grow your herd more slowly. Cows are hardy and generally calve unassisted.

Niche

Cows are excellent at improving a pasture for other species—not only for grazing animals, but all the small birds, mammals, reptiles, insects, arachnids, and so on, that make up the grassland biome. Trampling and grazing reduces forage height, increasing access for small animals, and the microbial life in a cow's gut is just what poor soil needs to start breaking down plant litter. Chickens put on pasture behind cattle benefit greatly from the shortened grass (easier to navigate in and less likely to hide a chicken-hunting predator), and the chicken diet gets a boost from fly larvae and undigested grass seed in cow manure, as well as the increase of worm and insect life, and generally greater biodiversity. Cows' size intimidates some predators that might endanger smaller animals pastured alone; and the great volume of milk they produce means that for much of the year there is plenty of good milk surplus to feed nearly all the animals on the farm.

Needs

On good pasture properly managed, cows can self-harvest their forage all through the year, except when there is a lot of ice on the ground. Snow alone is not a problem, and they will paw through as much as 2 feet of it to get at the grass underneath. Our own dairy cows are on pasture twelve months of the year and have no problem grazing through 12 inches of snow, provided it is still fluffy; when it becomes packed hard, or freezing rain makes a coat of ice over the top, we haul hay to the pasture and spread it for them to eat there. Our cows can usually find all the shelter they need in a belt of trees for protection from sun, wind, rain, or snow, and they generally calve unassisted.

Figure 6.12. A few inches of snow are not an obstacle to grazing.

Cows are good companion animals with complementary grazers like sheep and goats, and benefit greatly when they are followed in rotation by scratching birds like chickens and turkeys. Cows need the farmer to provide pasture and fence management, in order to ensure constant access to fresh forage, seasonal shade, and ample water. Provision and storage of surplus grass food—hay—for icy days and periods of drought are the responsibility of the farmer, and so are supplying minerals, or a natural mineral supplement like kelp, reproductive planning, and (possibly) artificial insemination.

Surplus and Services

A single milk cow of moderate productivity will give you hundreds of gallons of milk a year, more than most families will consume even if they make a lot of cheese. If you don't keep chickens and pigs you may be able to exchange some of this extra milk for money or goods, but better yet, get some chickens and a pig and feed the surplus to them, and enjoy reduced feed bills and better quality eggs and bacon.

Heifer calves are future milk producers, and unless you have lots of land and the desire to go into raw milk production on a large scale, or possibly explore artisan cheese making, you'll soon have more heifers than you need. These can be sold the fall of their first year, if you don't have grass to feed them over the winter—or they can be overwintered, bred, and sold as bred heifers, which bring a higher price. If we have plenty of grass and need more mouths to graze it, we may even keep heifers over a second winter, calve them out, start them off hand-milking, breed them back, and sell them as family milk cows, for which there is a good demand. Bull calves are mostly freezer-bound; depending on the number of farmhands you are feeding, you may want to sell halves or quarters. And grazing itself is a valuable service; if you have friendly neighbors with an empty lot they want kept in order, you may be able to barter grazing for other goods or services.

Shelter and Fence

Cows are weather-hardy in general, and in Oklahoma and Texas we didn't worry about shelter for them, aside from access to a tree or two for shade. Here in eastern Ohio most of our neighboring graziers winter their cows in a wood lot. For young, tender stock we have a three-sided shed backing to the north to use during really nasty weather, like deep snow and negative temperatures at the same time. Fencing is the biggest issue, as in "the most important," but cows are usually easier to contain in paddocks than either goats or sheep, one strand of polywire being enough for full-grown animals, and two for calves.

Food and Water

Contrary to what you will often hear, cows do not require grain feeding. We are contacted regularly by people who say that the farmer next door, or a man at the market, or the local vet, or someone, told them they couldn't raise their beef steers on just grass; even dairy people who believe in an all-grass diet for bovines will urge you with their next breath to feed young animals grain. There are plenty of all-grass beef herds and dairies out there to demonstrate that grain isn't necessary, but grain feeding is a practice firmly established in the general mind, probably because if you pump a steer full of grain—like, 20 pounds a day—he gets *lots* bigger, lots faster, which by itself seems to some people to prove its superiority as a practice. (Do we look at 120-pound nine-year-olds and exclaim over their superior health?) Fine, maybe, if that's what you want, but cows are designed for eating grass and forage, and, started young, will thrive this way, with the possible exception of commercial milk-production breeds, or a cull cow from a grain-based dairy operation. Genetics and long usage may render these animals unfit for a natural diet, and, in fact, an older animal of any breed, accustomed to a high-grain diet, might be unable comfortably to shift from concentrates and grain feeds entirely to forage.

Water, as for any animal, needs to be available on demand, and cows drink a great deal, up to 15 gallons or more per day, the actual amount depending on the weather, whether she is lactating, and what she is eating. Cows on snowy forage or grazing in the rain will require less water; in high summer heat, more. Our national soils now contain fewer of the microbes that make minerals available for plant roots, so our pastures can no longer produce forage with adequate micronutrients, and you will probably want to provide your animals with a mineral supplement: either a mineral block, loose minerals free-choice, or a natural mineral source like dried kelp. And you should always make salt available to livestock.

Buying a Ruminant

Finding a ruminant these days is not as easy as shopping for a pair of shoes; there's not one for sale on every street corner. So where do we start? And how do we know what to look for?

First of all, we want to do some self-educating. Read magazine articles, books, and Internet sites on the breeds you are considering. Get to know something of their qualities and characteristics, both good and bad. Look around at options—just because you've heard that Devons are a three-purpose breed—draft, dairy, and beef—don't build up a dream in which only Devons will do. Learn all you can about your dream breed, but look closely at others that also seem to offer the qualities you think you want.

If it's a heritage breed that interests you, you might check with the Livestock Conservancy for a breeder near you. Even if you don't think your chosen breed has endangered status, this organization may have information for you; there are many breeds not on the endangered list that are still in need of protection, and a good many of these are listed with the conservancy. In any case, what would be most useful to you in your new enterprise is a mentor, someone to whom you can turn for advice and assistance, and you may find one as your research takes you to the places, and people, where your favorite breeds are known. Find out about local 4-H chapters, and ask if there is one that specializes in the species you are interested in; 4-H leaders are very often adults who themselves grew up in 4-H and may well be able to tell you of someone in the area who raises your pet breed, or to direct you to better sources of information. Similar local groups, like the Future Farmers of America (FFA), may offer the same advantage, and a call to the county extension office can give you some immediate contacts.

At any rate, a good way to begin to get to know the local farm community is to attend some meetings, whether of the 4-H or FFA variety, or the spring planning meeting for the town farmers' market. Show up and pay attention. Go more than once. Ask questions and listen to the answers. Help when there is a project, like the annual Clean Water Day at the city park, or the organic farmer who puts out a call for hands for a high-tunnel raising. This not only exposes you to people who may be your future mentors but gives the local farmers, who have seen many wannabes come and go, a chance to warm up to you. All of these things, while seeming scarcely related to the project of finding your first ruminant, are building for you a framework of resources in your community, while at the same time taking baby steps toward making you part of it.

Figure 6.13. Once they are trained to respect hotwire, several animals are as easy to shift as one.

Meanwhile, check the farm papers and read the want ads to get an idea of what is readily available out there. Don't be discouraged if none of the advertisements resembles anything like what you are looking for; if it's a rural paper, most of the ads are probably aimed either at commercial farmers or hobbyists—that is, they are for the sale of an entire dairy herd, or for a single high-priced breeding animal. Once in a while, of course, you may actually find what you are looking for in the newspaper, and in the meantime you can, at any rate, look at the livestock sales column and get some idea of what the type of animal you want is selling for this week at the auction barn.

It will be ideal, when you think you have found an animal of interest, if you can take with you a person with some experience in farm animals and their purchase, both for the advice (and restraint) these people may be able to offer, and so that you can avoid being the lone, wide-eyed greenhorn in the barn lot, staring with little comprehension at the animal on display, wondering, principally, whether the seller will want to "bargain" (something most city folks don't get a lot of practice at) and how, if you buy the animal, you are going to move it. A mentor is a great buffer in such a situation, being able to offer real advice and ask intelligent questions. He or she may spot problems you overlook: a slight limp, bad feet, an awkward teat, a scar that wants explanation.

What does the perfect first cow (sheep, goat) look like? Well, everyone is different, of course, but on the grassfed homestead we can make some generalities. Grassfed, of course; this animal should be receiving little or no grain. Young, but not too young: four or five (three or four, for sheep or goats) years old, so she's already calved (lambed, kidded) two or three times and proved she can do it. Presently lactating, and accustomed to being hand-milked. Bred (pregnant), but not too far along, so you won't have to start your grazing career trying to figure out how to catch her in heat and breed her back right away, and so you'll get several months of milking in before you have to dry her off to give her a rest before calving. Shapely, in excellent health, with straight teats close, but not too close, together, and pointing down or slightly out (not in or back), the length of your pinkie finger or a little longer and not too fat, on a nice full udder of moderate size (she shouldn't have to waddle around it, except for a couple of days right after calving, lambing, or kidding), and not too close to the ground. Good-tempered, of course; and, if possible, get the owner to sit in the target zone, where you'll be sitting to milk, and let you see how calmly she submits to being handled. Trained to respect hot fences.

Get as much in the way of written records on your prospective purchase as you can. Find out where the present owner acquired the animal, and when. If it is registered, you will of course ask to see these papers, which will be yours if you buy the animal. Ask about its lineage and breeding history, its milk production records, and, if the animal is being sold as "bred," ask for a guarantee of pregnancy—a vet certificate to the effect, or written assurance that, should the animal cycle back into heat after you get her home, some part of the selling price will be returned to you.

And whatever the case, be prepared to walk away from any animal you go out to inspect. Don't buy on impulse; better to tell the owner that you'll think

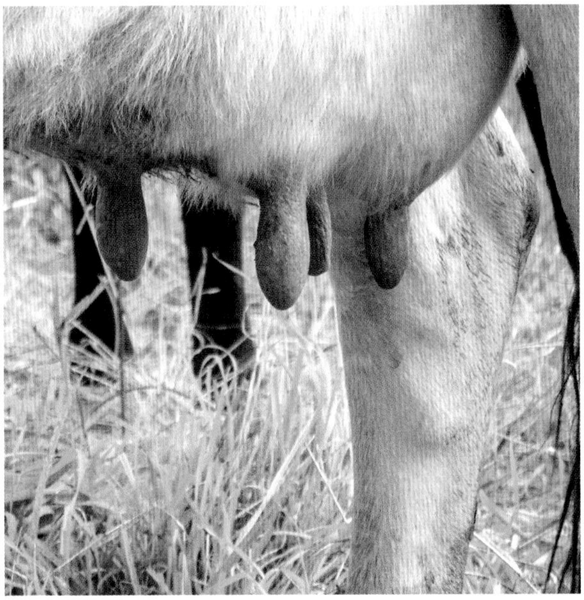

Figure 6.14. Teats of a handy size.

it over and call her back. Even if you are nearly certain you want this animal, give yourself the time it takes to drive home to consider. Not only does this protect you from making an impulse purchase, it is advantageous to you in the bargaining process if you don't seem too eager.

What to Avoid

No culls. After we are more experienced will be soon enough to buy a bargain animal with a flaw we think we can fix; we stack the odds against success when we buy someone else's problem. Also, avoid buying

Has to Be Heritage?

We are setting ourselves a big task when we undertake to recover and reimplement the methods and resources of the preindustrial, precommercial, sun-nourished smallholding, and it might be natural, in our search, to assume that as a part of that goal it is necessary for us to raise only what are officially termed "heritage" breeds; that is, true breeds of livestock that have an extended history in farming and are or have been endangered. To do so is admirable and valuable; the preservation of genetics that pre-date modern breed "improvements" is an important, even a vital, job.

Nevertheless, the question we encourage you to ask yourself when you are selecting livestock is whether it is necessarily *your* job at the moment. While it may be a real advantage to you to start your regenerative farming career with heritage genetics, it may also possibly place you under some added burdens, at a time when you are already taking on a big job: financial burdens, since true heritage stock is usually pricey, but others as well. It may be difficult to find a breeder of the livestock you want anywhere close to you, and traveling across three states to get your new baby is romantic and exciting but rather hard on the baby, not to mention using a lot of gas. If no one in your area has experience with your chosen breed, you are going to be on your own—possibly barring phone advice—when it comes to figuring out what is "normal" for this animal. When it comes time for breeding—after all, breeding is the goal of heritage breeders—finding a stud or semen from a stud may prove difficult. And before long there will be the need to increase diversity; that is, to access genetics sufficiently removed from your animals' bloodlines so that you may further your breeding efforts.

By all means, if your spirit compels you, or if there is a local source, raise heritage animals; just don't make it a hurdle you think you have to get over or your farming venture won't be genuine. Lots of breeds not listed as "heritage" have a long and respectable history; only, having been raised in large numbers right through the advent of commercial farming, they have never been endangered. And while the modern genotype for some of these may have altered since the days of the family cow or cottage pig, they may—probably will—perform just fine for you on natural feeds and in natural settings, *provided you start with young stock whose powers of adjustment have not yet evaporated.* Raising animals of the breeds available in your area means, at any rate, that you are likely to have an expert nearby when you have questions, and ample variety in breeding genetics when you decide to raise some offspring.

Keep in mind that, in our quest for best, we don't want to box ourselves into a set of "musts" that, with their numbers and complications, make it impossible for us even to get under way.

someone's pet, and never buy an animal because you feel sorry for it. Don't buy one for which you are not yet set up; your love affair with farming may take a serious hit when you start off having to pull your new cow out of the neighbor's swimming pool (really happened), or chase your just-purchased sheep down the highway. Have we ourselves ever brought home livestock before we were really ready for it? The answer is yes, and more than once—but we can't recall that it ever turned out better than if we had gone home first and prepared properly.

Look out for sick or injured animals. If the purchase money is big, it is worth the investment to have the vet out to look at your prospective animal before you buy; the seller should be perfectly happy to have you do this, and if he is not, suspect he may be hiding something.

Beware of buying a commercial dairy's cull animal! While it may be possible to get a good animal from a local dairy—a healthy, sound cow whose milk production rate is not quite up to commercial standards—it's not that likely to happen. Commercial animals are, well, commercial animals, and they have behind them a history, individual and genetic, of confinement and formulated feeds, whereas what you want is something that thrives on grass and fresh air. Don't think you can convert an animal to your uses the way you can take a city dude's pickup and make it into a farm vehicle by getting manure on the tires and throwing a bale of straw in the back. You want your

Figure 6.15. This young Jersey steer has less detrimental impact on soft pasture than would a larger beef breed.

first forays into animal husbandry to start with as many of the conditions for success as possible, and a cull animal is not usually going to be one of these.

There is no perfect animal, and no situation that does not carry with it some risk. Eventually, you're just going to have to take the leap. Consider carefully, and then make your purchase. Once you have your new ruminant on home ground, the real learning process will begin.

A Note on Cost

We would warn people against investing more in an animal than they can afford to lose; there is always the chance of loss, accident, or illness. Avoid show animals, which are not only going to be more expensive but are seldom or never representative of the *functional* virtues of the breed, in general being selected for other qualities entirely, like size, fashionable color, and form. Show animals intended for the "club" market—that is, the 4-H kids—will look good but cost about four times as much as nonshow animals that have most of the same virtues. So watch out for buying young stock in "club" sales season. But don't be too cheap; it's worth paying a little more, or even a little more than that, for healthy, gentle stock with good genetics. And if you get a chance to buy a promising animal from a local breeder or fancier who seems well disposed, you may find that your purchase price also buys you a source of advice and information that will pay dividends in the future.

How to Move Your Ruminant

If you don't have access to a stock trailer and a pickup truck, just how to move large animals may be a little puzzling. Possibly your seller will have a fee for delivering your new purchase to your farm. If not, sometimes the local abbatoir will be able to give you contact information for one or two people who provide the service of animal hauling. Smaller breeds can sometimes be moved in a pickup bed with built-up sides—we often do this—but we are careful to tie the animal securely. He or she is not going to be used to traveling at umpty-something miles per hour, and we don't want to find out what happens if he begins to jump around while we are barreling down the highway. Don't be careless, for your animal's sake, for the sake of the other people on the road, for your own sake. If you do use a pickup bed, tie the animal's head down, and we mean *down*, low, so that he can't jump partly out and hang himself over the truck side, or slip in the truck bed and hang himself there.

Daring people—we confess ourselves among them—have been known to move small ruminants in the back of a car or van. If you decide to do this, you need to make sure the animal is completely immobilized so it can't come loose midjourney and start leaping around the vehicle. Assign someone to see that it doesn't work its way out of its halter or collar, and by all means put a sheet of plastic under the animal or your car may never be the same again. And, if your new animal is coming from down the road and the distance to be traveled isn't too great, remember that animals have legs! A mile or two along a quiet back road isn't too far for most ruminants to shift their own freight.

If you do choose to use the services of a livestock hauler, take a look in his trailer before your animal is loaded. If he is a responsible person, it should be clean—well, not clean exactly, but it should not have in it manure from the last animals he hauled. Livestock trailers can be vectors for disease, and there's no point in adding to the uncertainty of a new animal purchase by exposing it to possible pathogens from the last farm the hauler visited. And however you arrange to have it moved, when you get your animal home you will want to keep it isolated for a quarantine period from any livestock of the same species you may already have, to guard against bringing in an illness or parasite and infecting the whole farm.

Handling and Gentling

Our first few days with a new animal can sometimes be a little exciting! To avoid unnecessary anxieties, make sure before you trust any animal behind electric fence that it has been trained to respect what is, after all, a purely psychological barrier. If, of course, she was being confined behind an electric fence on the farm from which you purchased her, great; if you are not certain, it will be better initially to have a conventional fence backing up your electric twine. You don't want an ignorant animal to charge your hot fence and, in the excitement of her first encounter with electricity, head out into the wild blue yonder. If your animal comes from an intensive grazing system, you know she is probably already well acclimated.

While we're on the subject of handling, there are a few other amenities to the farm superstructure that, while we went for years without them, can be

Figure 6.16. A good stanchion: head gate firmly attached to the wall, a feed crib, and a barrier to the animal's left.

of service when you need to catch and hold a largeish animal like a full-grown Nubian or a milk cow. A small paddock or corral where you can enclose her for a short time behind solid walls can be useful, for breeding or weaning, or as a sick pen for animals you need to isolate for a recovery period. And for expected vet or AI tech visits, it is nice to be able to hold an animal close to the barn or house, so you don't have to keep this person waiting while you chase the animal she has come to examine.

And when the vet is ready to get down to work, a stanchion, head gate, crush, or just a "milking gate" can be very handy. A head gate—a closable framework in which to catch and immobilize an animal's head and, by extension, its whole body—needs to be solidly built and well attached to the barn wall or fence so that it will hold even an excited animal. A crush is a narrow section of livestock chute that closes on the whole animal, and while a crush is no doubt nice, the only way we'll ever have one is if by some miracle we find it at the salvage yard, because they are expensive. Without these amenities one can get by just fine tying the cow to be examined alongside a strong gate with her head next to the hinges, and then using the gate to push her up against the corral fence or wall. You can tie the nether end of the gate with a rope across the cow's behind, to create a temporary crush; our AI training was on enormous Holstein cows that were immobilized in just this way.

Restraints: Halter, Collar, or Nothing?

There are differences of opinion about what sort of strap or collar, if any, should be left on an animal in the field, and there are good arguments for each. Whatever you decide, be sure to keep the question open, and stay observant. The natural thing to do when you buy an animal wearing a halter or collar is to leave it on her when you turn her out into the field and then forget about it, and this is just what you don't want to do. Animals grow! And any straps and ropes they are wearing don't grow with them. We know this, and yet it has happened more than once that an animal on our farm and constantly under our eye has had its halter grow tight and rub a sore on chin, nose, or forehead before we noticed anything was amiss. While we really prefer to keep a strap on our animals, for ease of catching, handling, and leading, most of ours wear collars only when they first come to the farm, before they have been gentled and accustomed to us and our ways, or else during times when we expect to be leading them often, as when a young cow is being trained to walk herself up to the dairy at milking time. Also, before a cow's first calving, or while watching for a heat before her first AI, we put a collar on her to ensure that later, when hormones are interfering with her domestic training, we don't have too much of a rodeo bringing her up to the barn.

Be careful what sort of restraining device you use. Collars are minimal, and we prefer them to halters for that reason; if you plan to bell an animal—picturesque, and also useful when you are looking for a stray—you'll use a collar. Halters, on the other hand, are more secure, and provide more traction when you are training an animal to the lead. In any case, make sure that whatever you use, fits. A loose halter can be a trap for a scratching foot (and, yes, a cow can reach her chin or forehead with a hind hoof for scratching). A cow can die from something as seemingly minor as snagging a foot in a collar or halter, if she cannot get it free again and ends up for an extended period lying on her side. You will probably want to leave the halter or collar off once an animal is sufficiently gentle, or as a compromise, you can make the collar out of something that will break under sufficient stress; plastic chain, like the stuff they make ticket lines with at the airport, while admittedly ugly, meets the requirements.

The best way to lead an animal, in the end, is to have gentle, well-trained animals that are used to being handled, and for this, the younger you start the better. Here the intensive grazier is necessarily going to have the advantage, since his animals will be accustomed to have him among them and will associate his presence with their daily move to fresh forage. But proximity is one thing, and touch is another. An animal that will stand within 3 feet of you while you are occupied with a fence reel and posts may scoot away when you start reaching for her head.

You can begin training babies to tolerate a human touch while their mother is being milked, by tying them next to her and giving them a small treat. At first most will fight the lead rope, but in a short time they become accustomed to it; once they have learned to associate the restraint time with a treat, they come willingly. The time of weaning to a bucket or bottle is also a great opportunity to accustom an animal to handling, and this is when we usually do it. And gentle mothers by their example teach their offspring to be confident of their herders.

CHAPTER SEVEN

The Dairy

. . . and the brood kine, as of yore, for thee brim
high the snowy milking-pail . . .

—Virgil, the Georgics

You can field-milk a ruminant—stroll up with a bucket and help yourself—and we know some very good dairy women who do just that, but most people will want to have a stanchion in a shed or barn for times when neither they nor the animal want to be out in the weather. A good stanchion closes behind your ruminant's head so she can't back away, but gives her freedom of movement up and down; it should be well attached to the shed wall or a post so she can't pull it over on herself (and you). A cow with a wooden stanchion hanging around her neck can get pretty excited! A feed crib or hay rack in easy reach lets you give treats or supplements and teaches her to look forward to her time in the barn.

Hand milking a ruminant includes a learning curve; we will go into milking form and method in chapter 10.

Figure 7.1. Only minutes old, this calf knows where to find his breakfast.

In the Beginning—Calving, Lambing, and Kidding

Mammals don't make milk until they have babies, so the precursor to any milking endeavor is parturition. This is something your animal will probably do just fine without you. Chances are, you'll come down to the pasture one morning to move her fence and there she'll be with a new baby beside her, or maybe still wet on the grass under her vigorous tongue. But you'll want to be on the watch; coming up on her due date you should start checking several times a day, including just before bed, and perhaps once at night, to see that all is well with her. Once she is in labor it will be desirable to keep an eye open and make sure things are progressing. If she labors for several hours with nothing to show for it, or nothing but a pair of white, waxy hooves that advance and retreat with each contraction but never seem to progress, you will need to get involved, or get the vet. Chances are her baby has its head back, or is otherwise badly positioned in the birth canal, or possibly it's just extralarge, and a helping hand, and maybe a calving chain, will be needed. Although we have ourselves assisted occasionally with a badly laid calf or lamb, we aren't giving lessons; if you intend to become a ruminant midwife, you will want to do your own research, ahead of time. And whether you plan to do your own birth-assisting or not, you will be happier when problems come up if you have already met and talked to the local large animal vet, and let him know you would like to be part of his practice when need arises—and he will probably be happier, too.

Cleansing—passing the afterbirth—is something else that seldom needs human intervention. A ruminant will usually cleanse on her own, generally within a few hours of parturition, shedding her placenta in the field where the crows and vultures can get it—if she doesn't eat it first. Instinct. Maybe it's to foil the predators that circled around the primordial herds, or maybe she derives some nutritive or chemical benefits from it. Maybe both. Suit yourself, you can take the thing and bury it in the compost bin where it can contribute some nitrogen to your garden, or you can leave it to unassisted nature. If you find it, it's considered standard-of-care to examine it for missing parts. It's going to look a lot like a large beef liver attached to a big balloon, and good luck to the beginner who wants to know if it's all there, but examine the thing by all means if it pleases you. Since our animals almost always birth in the field, it's not uncommon for us never to see the placenta; so far, that's worked just fine.

Sometimes the process of cleansing takes more than a few hours, and it is not very unusual to see a newly calved cow with her placenta hanging out, banging against her udder as she walks, for even as long as several days after calving. Unpleasant, yes. The first thing is to realize that the situation is not serious, as it would be in a human being; in a cow, a retained placenta is not necessarily a recipe for internal bleeding and infection. Opinions differ on what to do about it. The nursing of the calf is the best remedy, since it is designed by nature to stimulate the mother cow to produce oxytocin, the hormone that causes uterine contractions. You can pull on the externalized part, but not too much: you don't want to break it off short, since the weight of the expelled portion is part of what pulls the rest of the placenta out. An old farmer once told us to tie a brick in the placenta, for extra weight (we didn't try it). Don't be in a hurry. Experienced stockmen in our area give as much as a week as the limit before they begin intervening in such cases, and we ourselves have known a cow that miscarried and then took several *weeks* to cleanse completely, the necrotic placenta coming away gradually in a sort of cheesy white discharge.

When the Milk Comes In

As your new dairy animal approaches time for lambing, kidding, or calving, you want to prepare her for the big job of converting forage into lots of high-protein, high-fat milk for the baby and the bucket. Instinct prompts the infant grazier to increase the quality of her forage intake at this time, but conventional advice says in this case instinct is misleading, and the conventional advice is probably right.

Think about it; ruminants in the wild have always tended to drop their babies in early spring, when mama is just coming to the end of the lean fare of winter. Then, when the infant is born and running beside her and she needs to make plenty of high-quality milk for her baby, the spring grass season begins and she has a whole banquet of high-protein forage spread before her. In imitation of nature, while giving your ruminant the best care—security, a nice place to bed down when the time comes, plenty of clean water—you are supposed to limit the quality of her intake before parturition; otherwise, she and/or the baby may be fat enough to cause problems in calving, or when her lactation begins, the milk may come on like a freight train and give her mastitis, even milk fever. Some brown, poor quality hay, or an area of stockpiled forage left over from the winter, is about what she needs, although she won't appreciate it. And in the first week after parturition you still want to limit the richness of her food; learning to milk is hard enough without adding the complications of engorgement or mastitis.

Advice differs widely as to how much milk to take, and how often, in the first day or two of a lactation. In some cases, at least with low-production breeds, you might be able to leave the new baby to do all the milking and take all the colostrum; however, if the mother animal is of a dairy breed we would strongly advise that you do some milking yourself, to reduce the pressure in her udder. This will also allow you to obtain a store of colostrum to freeze for sick or orphaned animals (cure extraordinaire for what ails 'em) and decreases the chance you'll be dealing with mastitis right away. You should look for advice pertinent to your own breed of dairy animal as you develop your early lactation protocols. Situations vary, so we'll just share with you how we manage when a mother ruminant has just calved or kidded.

Milking a New Mother at the Sow's Ear

First, as soon as we know of a new baby—which on our small place is usually pretty soon after, if not before, the big moment—we go out to make sure the little one is getting on his feet all right and can find his breakfast. Babies seldom need any help, although we'll occasionally whip off our shirts (we seldom think to bring a towel) and give the calf a wipe-down, or guide his little nose to the teat. Then, if all is well, we leave mother and baby alone together until the next normal milking time, when the new mama is brought up to the barn with the other

Figure 7.2. Dam-raised calves learn nutrition by example.

lactating animals, either with her baby—if we can find it and if it will come—or without it, if, as frequently happens, it is nowhere to be seen.

Yes, really. We used to worry when we couldn't locate a newborn calf—they are such little, knock-kneed, wobbly guys, perfect coyote bait—but in an intensive grazing system tall grass is usually only one paddock away, and since single-strand fences, while perfectly adequate to contain an adult animal trained to respect them, are little or no barrier to the tiny baby who can slip right under, often the little one is nowhere in sight. Once his belly is full, he knows instinctively to find a quiet nook and bed down until the next nursing time. Mama approves, and while her baby takes a nap she goes back to grazing. So you go down at milking time and find a lone mama, with no clue where the calf has gone. Very disconcerting. We used to institute a farmwide search for the missing baby, anxiety growing by the minute, only to have the little beggar pop up out of the grass at his next mealtime, from somewhere quite nearby, in a spot we could have sworn we'd searched thoroughly. Now, when a new baby is nowhere to be seen, we look at the mother animal to see how we should feel about it. If she's relaxed, then we are, too.

Anyway, with or without the new calf, about twelve hours after a birth the new mama takes her place in the milking string. Her udder will appear quite full, and indeed it will probably have seemed so for a couple of days at least. With dairy breeds, in fact, she will probably be waddling, swinging her back legs wide around an enormous bag. But when milked, she won't give much. This is partly because most of the grotesque swelling you see is edema—fluid-filled mammary tissues, not colostrum-filled milk ducts—and partly because many ruminants don't, initially, see any reason why they should let down their good milk for someone who isn't their baby. But whether we get much or little, we milk the mama right out at this time—that is, we take all we can squeeze out of her, just as we do with any lactating animal.

You will see differences of opinion on this practice, with some people advising that you "milk her half out"—a confusing piece of advice, we think, since how are you to know when a hugely engorged udder is half-empty?—or to take "just a little" (how much is that?). The fear is that taking too much, too soon, will bring on calcium deficiency, or "milk fever," a condition not uncommon in high-producing dairy cows and that, untreated, can lead to prostration and death. Nevertheless, our own early lactation protocol as described here has evolved in part on the observation that most of our grassfed animals don't give much in the first day or two anyway, and in part from our confidence that grassfed animals are unlikely to be calcium deficient. We do usually administer a tube of calcium supplement (looks like a tube of plumber's caulk, and you'll find it in the farm store marked "CMPK"—if you don't find it, ask for it) either shortly before a cow calves, or at her first milking, just for security. And we keep an IV apparatus and a couple of 500 milliliter bottles of calcium gluconate solution (farm store) on hand in case any of our cows lies down and can't get up again, a classic symptom of milk fever (and of a number of other things, too). That said, we have never had a confirmed case of milk fever on our farm, although we've occasionally given IV calcium—and, as we say, when a cow comes in for the first time during a lactation, we milk her out (see chapter 10 for milking technique).

First Milk

What a cow gives right after calving isn't strictly speaking milk, but colostrum, the "first milk" which mammals make to build their babies' immunities and prepare their intestinal systems for the work of digesting milk. Colostrum is thicker than milk, and usually more yellow in color—in early spring, it may be nearly orange—and there usually isn't as much of it as there soon will be of milk. Although some people drink colostrum as an immune booster, it isn't generally used much in human food—although there *are* recipes out there—and we mostly freeze it for any future orphan or bottle calves we may buy in, or in the event a calf develops scours after he is weaned to the bottle or bucket, or for any animal that acts like it has a problem—we are sure it has saved piglets with pneumonia.

After this first time, the newly calved cow will be milked on our regular schedule (which is twice a day for most of the year, sometimes dropping to once a day in late winter or early spring). Colostrum gives way to milk over the course of the first few days after calving; by day five we consider the cow's offerings to be milk, and fit for the table. The acute edema that makes your cow's udder look like a balloon under strong inflation will pass over time (generally in under two weeks). This will make her easier to milk, since the teats and bag will not be stretched taut and hard as a rock, and from that point on it will also be easier to determine whether you have gotten all the milk out of her udder.

Milking Routines

Once your dairy animal has borne her baby/babies, you will have to decide what milking schedule is going to serve you best: once a day, twice a day, milk share, or take it all and wean the baby to a bottle. Remember that if you have a high-production dairy breed, you need to relieve her of her lactic burden on a regular basis, either by twice-a-day milking, or

Figure 7.3. Extra colostrum can be frozen for starting bottle calves.

by milking and then giving the baby access to her for half the day. By and large, milk sharing, at least for the first weeks of lactation, can be the best route for farmer, ruminant, and offspring.

Schedule

While twice-a-day milking is traditional, and undoubtedly harvests more grass energy than a single milking can, there is a strong trend among family cow people toward the single daily milking. The advantages are obvious: apart from reducing chore time by half, a single milking, unlike a ten-to-twelve-hours-apart, morning-and-evening routine, can be scheduled at any time of day that suits. The reduced production resulting from less frequent demand is, in any case, often seen as an advantage to homesteaders who find those daily gallons somewhat intimidating, and there are low-production breeds (try the dual or tripurpose breeds like Devons or Dexters, to name only two; or traditional—"heritage"—genetic lines of the dairy breeds), which do just fine on a single milking a day. Any traditional breed of cow, in fact (as opposed to the ultrahigh production strains among commercial breeds), will, after the first several weeks of lactating, adjust to a less frequent milking schedule, shifting naturally to the lower demand by producing less milk.

Where twice-a-day milking is possible, however, the higher level of production can be a source of huge advantages for the grass-powered farm. Surplus milk is the jet fuel of the small homestead: from the people, to the hogs, poultry, and guard and vermin-control animals, milk provides the highest quality food protein the farm can provide, produced with the greatest positive ecological impact. The initial intimidation many people feel at the delivery of large quantities of milk to the kitchen on a twice-daily basis gives way before the beautiful complementarity of filling the hog trough with skim milk, buttermilk, or whey; pouring some into the dog's and cat's dishes; and putting a pan of clabber in front of the poultry—all of it, food converted from yesterday's grass. A grassfed, dairy-based homestead operates at a new level of ecological efficiency, its resources narrowed to the most primary: sunlight, soil, and rain.

Milk Sharing—Full-Time or Part-Time

There are various ways to establish a milk-sharing relationship with a baby calf, kid, or lamb. It is possible, with a low-production cow, to leave her calf with her full-time, and milk once or twice a day, taking whatever the calf has left you (and whatever Mama will let down to a mere human being). It is certainly nice for mother and baby to get as much time together as is consistent with your plans for dairying, and the health and size of your babies will absolutely reflect this advantage. But sooner or later the unpredictable amount of milk the calf leaves you may seem inadequate for the farm's needs, or not worth the trouble of milking and cleaning up afterward, and then you will want to consider going to a half-time milk sharing arrangement.

It is a common practice among those who keep a family cow, and perfectly consistent with the animals' well-being, to confine a baby ruminant away from its mother for twelve hours of the twenty-four, while she makes milk for the bucket, and then let the baby run with its mother the other twelve hours, when it can nurse at will. This is no hardship for the calf/lamb/kid. A calf's pattern in any case, when she is left with her mother, is to have a good nursing periodically during the day, followed by a long nap in the high grass somewhere; a bottle-fed calf is usually fed just twice daily. Baby ruminants don't need to have constant access to their mothers' udders to be healthy and well fed, so let your preferred milking time dictate whether the calf is confined at night—preparatory to a morning milking—or during the day, for an evening milking schedule.

Calves—Feeding and Weaning

"Weaning" can mean either taking the calf from the cow, to be fed milk from a bottle or bucket, or taking the calf off milk altogether. The typical commercial dairy calf is weaned to the bottle the first day; whether the calf is a heifer, which, being valuable as a future dairy cow, is usually fed real milk, or a baby bull, which typically receives artificial calf milk replacer (dreadful stuff), the new baby never sees its mama again. This is not as we would have it, and for the homesteader it is neither helpful nor economical nor necessary. It may be a little simpler in some ways to wean immediately, but it is not in the end the best thing for the calf, the mother, or the homestead.

Feeding Calves

The easiest way to feed a calf, of course, is to leave it with its mother; she has what it needs and can give it the best possible start in life. But for the homesteader who is keeping a cow for the sake of her milk, sooner or later more will be wanted than the calf is leaving, and at this point it has to be weaned from its mother to a bucket or bottle. And if your dairy cow's excellent supply of milk has you determined to use some of it for starting an extra calf for the freezer, you may also want to pick up a little dairy bull (the cheapest bottle calf available). In either case, you will want to know how to feed him.

First, a word about calf milk replacer. Don't. It is expensive and a very poor substitute for the real thing. A good calf is too valuable to waste, and a cheap calf is too much work to be worthwhile on the poor odds that it will thrive adequately on replacer. Feed calves whole milk, skim milk, or buttermilk—real buttermilk, from butter making—but feed them *milk*. The advantages will far outweigh what might at first seem to you the extravagance of feeding to calves good milk that might have been drunk, sold off-farm, or made into cheese.

Calf bottles—just big versions of the ones some people use for human babies—can be found at the farm store. The little *X* cut in the end of the nipple will almost certainly have to be enlarged or your calf is going to get very frustrated (and so, consequently, will you). Take a razor knife down to the barn with you the first time you use a new bottle, and once you determine that the milk is coming too slowly, extend the *X* cut a little in each direction, checking to see when you have done enough to

Figure 7.4. Calf and lamb bottles.

allow the milk to flow freely. You may also have to enlarge the little air hole beside the nipple. Then when you invert the bottle to feed the calf, hold your finger over this hole until the calf starts to suck, to prevent the milk running out now through the enlarged cuts.

Calves nursing on their mothers feed with neck angled down and head arched back, to get under her udder, and that is how you want your bottle calf to nurse. Hold the bottle low—waist high or lower—so your little guy isn't extending his head up to nurse, or he may get milk trickling into his lungs, which will predispose him for pneumonia, a problem bottle and bucket calves are too prone to anyway. If your baby doesn't get the hang of the bottle right away, you'll need to straddle him, catch his head so it's poking out between your knees, pull up his chin with one hand, and use the other to force the nipple into his mouth. You may have to hold him there awhile until he tastes the good milk and gets a couple of gulps; in fact, you may have a wrestling match, with the calf flinging his head around and spraying you with milk and knocking you on your keester. When the bottle is empty you'll wonder if you got enough of the milk into the calf to sustain life. Don't worry. Some calves have to be handled like this for two or more, even several more, feedings, but they quickly learn what to do and after that you can expect to be mobbed when you come into the paddock with a calf bottle.

Eventually you will probably want to train your baby to a bucket, if only because buckets are so much easier to clean than a bottle and nipple. Or you may want to start off with a bucket, skipping the bottle altogether. Fine; it may take a little longer for the calf to catch on, but he'll get it pretty quickly. The classic method is to let the calf begin sucking on your fingers (he won't need training for this, and he'll probably still be doing it when he's half grown) and then lead his head, still sucking, down into the bucket. After a moment he will be getting milk along with your fingers, and you gradually withdraw them, leaving him a successful bucket drinker. Or he gets his nose full of milk, sneezes, pulls away, and you try again. You'll get milk all over your barn jumper or whatever you have on, he'll fight like crazy, and you'll wonder why you ever thought you would like farming, and why anyone would want to raise her own food anyway, when there's a Walmart just a couple of miles away. And after a few episodes like this—just the blink of an eye in the life of a farm, or of a farmer—he'll be off and running. Make a square frame to hold his bucket, or make the feed crib in his stall just the right size, and you can set his bucket in there to feed him and save yourself two minutes of bucket holding. Then when he's finished he'll keep sucking and butting at the bucket until he's got it knocked on the floor, and he'll step in it and chew it until it's a real chore to clean. Or you can just keep holding the bucket.

Figure 7.5. Even the youngest members of the family can help raise a bucket calf.

Scours

We don't like taking a tiny calf from its mother, knowing how superior its condition will be if it gets at least a few weeks of mama-care before it is weaned to the bottle. We do occasionally buy bucket calves to raise for beef, however, and even if we have frozen colostrum to offer for the first few days, it is easy to see that these little guys are operating at a disadvantage. One symptom of their inferior health can be scours (diarrhea). Baby ruminants on mama's milk make a thick yellow or orange poop, and you won't necessarily see much of it because often Mama cleans it up while baby nurses. But sometimes, particularly with a calf, kid, or lamb that goes early on the bottle, this thick, inoffensive-smelling poop will turn stinky, pale, and runny, soiling the animal's rump, tail, and back legs. This is scours, and you want to do something about it because it doesn't take long before dehydration makes your problem worse. Lots of people go straight to antibiotics for this one, although there is no guarantee these meds will help, and they will certainly wreak havoc among the calf's gut flora. But there are other, more holistic approaches to dealing with scours: some online research will turn up more than one suggestion, or you can try our methods.

This is what we do at the Sow's Ear. As soon as we see a case of scours (which happens rarely now, thank goodness) we offer colostrum instead of milk, if there is any colostrum in the freezer. Of course, we already have our baby in a dry pen with plenty of bedding, out of the wind, as well as out of reach of any other young animals we may have at the same time, so we are so much to the good. A few feedings of colostrum may take care of the problem completely, nothing else needed. Lacking colostrum, we'll feed a bottle of warm water with electrolytes (farm store); in fact, we'll probably go down a couple of extra times over the next feeding period and offer more water and electrolytes. After twelve hours we offer half water and electrolytes, half milk, and if these aren't slowing things down enough, we'll mix a tablespoon of pectin—same as you'd use for making jam—into the formula. (You can buy pectin in little boxes at the grocery store, at least in the summertime, or you can pick up a larger package of bulk pectin somewhere and keep it on hand for whenever you may need it.) The idea is that the pectin will thicken things up and slow the passage of milk through the calf's gut. The third feeding is water/pectin/milk again, then if things aren't looking too bad we'll ease off the water until the calf was getting just milk and pectin.

After two or three days of treatment we look for things to be firming up a bit, but it may take a while for complete recovery. Generally, the calf's affect

Feeding Orphan Calves

When we first began using Isabel's extra milk to raise baby Jersey bulls from the local dairy, the little guys were so compromised that we saw a lot of scours. We read up on the condition and found that there was no recognized, sure-fire prevention or treatment, so we experimented and came up with our own preventive protocol for baby calves on the bottle, which has served us in good stead ever since.

To begin with, we freeze colostrum early in every lactation—there's always lots extra, and we put it in plastic buckets in the freezer, where the unsuspecting take it for orange sherbet—and feed this for the first four or five days, up to a week, to any newborn on the bottle. When we transition to whole milk—skim will be adequate later, but we start new babies on full-fat—we add a raw egg to each bottle. The theory is that the albumin in the egg whites acts as a binder, slowing the passage of the milk through the animal's gut. We add an egg to each feeding for at least a couple of weeks; after that, if the calf is off to a good start, we'll drop it, but if there's any question about his tummy we'll keep on for another week or so. The experts may say, and some of them have, that this dodge can't possibly do any good, and they'll give a list of reasons why not, but we haven't had a bad case of scours in years—not since we started using this method.

during this time shows no sign that things are a bit off at the back end, and he's full of *bobance* and bounce and just as anxious for his dinner as ever. If we had a baby who was getting listless and weak, we'd follow the same protocol but offer smaller feedings more often and pile some hay over him to keep him warm. And our old farm manuals have other suggestions: we have seen yogurt, clabber, or the water in which chopped fodder beets have been boiled, all listed as sovereign remedies for calf scours.

Calves and Sucking

A word of caution. A calf's strongest instinct is to search for a teat and suck it, and like a baby human, his desire to suck is not going to be satisfied automatically just because his belly is full. On the contrary, a baby calf is going to suck on anything remotely like a teat that he can reach, which is why he'll suck your non-milk-producing fingers—and, if he can reach them, your chin, your ear, or your elbow. He will also suck, if there is another calf in the pen with him, any part of that calf's anatomy he can get in his mouth. Any part. That means ears, tail, umbilicus, scrotum, whatever. This propensity on the part of baby bovines is largely responsible for the fact that any illness in a pen of calves is going to be communicated like wildfire, and it is going to dictate some of your calf-rearing decisions.

One of our first lessons in calf care came painfully, when we had three or four dairy calves bought in to help use up excess milk from our first cow, Isabel. We put all of the calves in a small pen in the lower part of the barn, bedded them in hay, and fed them milk twice daily. After about three days they all had scours, and in a week they had pneumonia. These are both common calf illnesses, and whether the calves were ill when we purchased them—not unlikely—or picked up the germs on our farm is not the issue. The issue is that, because we had all the calves in one pen, whatever one calf had, they all got. We called in the vet that time, and although she gave them antibiotics and analgesics we lost one. But we learned a valuable lesson: keep baby calves separate while they are such avid suckers. We had seen, on commercial dairy farms, that each little calf is in its own circled stock-panel pen, but we had not realized why. Now we knew, and ever after we have made sure that very young bottle or bucket calves are not able to reach each other for sucking. When we designed four loose box stalls in the upper part of the bank barn, we put the boards between the stalls too close together, and too high, for calves to reach through. As they get older, we let an hour or more go by after morning feeding before we turn the calves out to pasture, just so they've had time to forget sucking for the time being. This isolation is only necessary, notice, for calves not raised by their

Figure 7.6. Chicory, another "weed" that is actually a valuable forage plant. Courtesy of the Franciscan Sisters T.O.R. of Penance of the Sorrowful Mother.

mamas; those calves, lucky babies, satisfy their sucking impulse on the teat.

One last word on calves and sucking: it should be taken into consideration when you are deciding how, and when, to castrate. Think about it.

Weaning the Calf from the Cow

When we wean a calf from the cow varies according to circumstances, as, for example, how big the calf is, how much milk he is taking (baby bulls can be voracious), if we have a better use for the milk ourselves, and how rough he is on his mama's teats. A big, eight-weeks-old bull calf who is golloping 3 or 4 gallons of milk a day and leaving tooth marks on Mama's tender parts is liable to be weaned to the bucket earlier than a mannerly heifer on the small side. Neither will ever be cut off from milk, mind you, before twelve weeks, and most will go to fourteen weeks or better with *at least* 5 quarts of whole or skim milk a day, but sooner or later we will separate calves from mamas, either to wean to a bucket, or to wean to good green pasture.

There are two basic methods that can reduce the strain of cow-calf separation, both on the animals and on all the humans within a mile or so. One is to place the calf in a pen where it can't reach Mama's udder, but within nose-to-nose contact, so they can comfort one another in their enforced isolation even though nursing is impossible. And it had better be a pretty good fence; a calf that wants its mother isn't likely to be balked by five strands of barbed wire, and he's almost certainly capable of hurdling 36 inches of hog wire, too, or taking it down. The other method is to move the calf completely out of sight and hearing of his mama, like, a couple of miles away. On our home farm neither of these arrangements is possible, unless we take the cow off pasture and put her in the barn a few days, where she can

Weaning to Grass

Calf weaning is the one situation where we break the cardinal rule of rotational grazing—Never Open All Your Gates—which is (almost) never to be violated. With baby weanling calves, whose stomachs are still very much in transition from digesting milk to digesting tough cellulose, and whose total daily grass intake is minute, it is extremely desirable to make sure that what forage they are grazing is of the best quality. At ten weeks our Jersey calves can still nip out under a single-strand temporary fence anyway, to wander at their sweet will over the entire pasture, and we let them, figuring that their tiny needs won't have a measurable effect on the pasture as a whole. If they nibble a clover plant here, a few blades of perennial rye there, they are meeting their dietary needs with the best the farm has to offer, without compromising our grazing plans much more than would a rabbit. Yes, you can turn a weanling out into a paddock that hasn't been grazed since April, and now it's July and the orchard grass and timothy are all dry stalks, and the red clover is a tangle of scorched stems and flower heads, and the only green plants are ragweed and Queen Anne's lace. He'll graze, after a fashion. But he won't be able to derive all he needs for good growth, even though the same paddock might be just fine for a dry cow or an eight-months-old steer. Better to spoil him a little.

lean over the calf pen gate and lick her baby. Or we can just put the calf in the barn and his mama in the pasture and put up with all the hysterical mooing for a week or so.

But of late years, when we have had the management of the monastery farm, we can move calves a full 1.5 miles away from their mothers at weaning time, which seems to reduce the trauma for the cows all right—they'll call their calves for a while and then go back to grazing. The calves, on the other hand, still moo themselves hoarse, and if they're not behind a tight fence they'll get out and end up on the county road, not once but repeatedly, and by preference in the middle of the night. Shut up your weaning calves tight, in the barn if necessary; two hours of chasing a calf through thorn bushes by flashlight will take the veneer of anyone's love affair with the rural life. Whichever method you choose for weaning, try to make the calf as comfortable as possible, and make sure he is getting his full ration of milk, or maybe a little extra, to counteract the stress he is experiencing. Give him a flake or two of your best hay, and if you have to shut him up in the barn, make the prison term as short as possible: after two or three days he should be getting used to the situation, and then you want to get him out in the sunlight, maybe with other weanlings or the sheep or pony for company, and on your best tender green grass and clover, for at least part of every day.

Calf Nutrition

Now, you're going to feed this calf milk twice a day, and the usual instructions are for a ½ gallon at each feeding, and this is okay, but begs the question why, when calves come in all sorts of sizes and breeds, there is this one-size-fits-all principle for feeding them. Actually, most calves, at least after they've done a little growing, would be taking *lots* more than a gallon of milk a day if Mama were there. The only reason this allowance doesn't go up (for bucket or bottle calves) as the calf grows is that calf-milk replacer is expensive, and real milk is valuable: in a cold, mathematical way, it isn't practical to feed milk or formula. Instead, instructions for calf raising—even from sources that claim to be about grass farming—usually tell you to make sure he has high-quality hay available; clean water, of course, at all times; *and a couple of pounds or more of high-quality dairy feed or creep feed in his crib, every day.*

Now, here's a puzzle—often the same person who will tell you that all-grass cows make the best milk and grass is the natural food for cows will give you instructions for feeding your calf grain. He will, in fact, assure you that your calf needs grain to thrive. Why? We don't know ourselves, but the answer, we speculate, has two parts: one to do with the grass or hay, and the other with the calf.

Ruminants in nature, of course, are born in spring and run beside their mothers, nursing whenever they like and getting their noses down near

Figure 7.7. Weanling with companion calf.

hers to eat, or at least explore, whatever they find her nibbling. On good spring pasture with mixed grasses and plenty of legumes, a calf with Mama's milk to carry him is going to thrive on this very thing. Even the nation's by-and-large poorly managed grasslands grow adequate forage for spring beef calves, which spend their first several months with their mamas, to thrive. But bucket calves (generally dairy animals, since beef cows raise their own babies) don't get this tutelage in good grass nutrition. And because dairy cows are bred to calve year-round, to keep the supply of milk to the market steady, dairy calves may be weaning at any time of the year, not just in spring when the pasture grasses are tender and nutritious. As it weans from milk a calf needs high-quality forage with a good mineral content, and this is something the manuals don't assume will be available.

In addition, there are the genetics of the calf itself. Of late years dairy cows have been bred to produce lots of milk, on a diet of lots of grain—in many cases nearly a *bushel* a day—plus silage, baleage, hay, chopped coarse forage with protein/urea syrup poured on, and almost anything else but plain green grass. So while dairy calves are, in fact, herbivorous mammals, which means—theoretically—that when they are young they live on grass plus milk, modern "improved" dairy calves may have a hard time getting all they need from 1 gallon of milk plus pasture access or a truss of sweet hay every day,

Working with Salvaged Parts

In nature, as in the human-constructed world, there is the ideal and there is the actual. The perfect farm, with perfect animals and perfect genetics, perfectly managed, exists—but only in our minds. Our real, flesh-and-blood, dirt-and-manure farms are made of what's available *now*, which may be and probably is leached, demineralized soil; commercial livestock genetics; and a farmer whose time, especially in the beginning, is going to be full-to-overflowing with new chores and new decisions. For the ideal farm and the actual farm ever to draw nearer together in the reality of our homestead is going to take lots of time, and labor, and patience. Especially patience: patience to see the suboptimal and put up with it; to work with it, but never to forget that *it is suboptimal*. We have to live, initially, with many things we are ideologically opposed to, while never ceasing to aim for what is ideologically preferable. Maybe we can't afford to buy organic seed for the garden this year, or install a whole-house solar array on the roof; maybe the only feed we can find for the chickens is commercial crumbles, replete with goodness-knows-what. That doesn't mean we should give up, and it doesn't mean we're failing. Until the earth is healed of all we have done to disrupt it—that is to say, this side of the Second Coming—we are going to be working with limping, lamed systems, and we must make the best of it. Meanwhile, our goal must never cease to be the healing of those systems. The minute we decide that some compromise is inevitable or permanent, or—God forbid—economical, we are doomed to follow the course that got us where we are now, in which some imaginary value we label *efficiency* or *economic necessity* is used to justify the codification of what was originally a compromise, moving it from the realm of "temporary rescue effort" to that of "permanent shortcut."

and they certainly will not grow at the same rate as a grain-fed animal.

You are going to have to make up your own mind: maybe you would rather have a bigger calf, faster; in that case, you're probably going to feed him grain. Two Hereford steers up the road are growing out nicely on 20 or 30 pounds of grower ration *each*, daily, plus forage; in November, when they go to the abbatoir, they'll outweigh our Jersey steers by a good margin. But our own preference is for an all-grass animal, even if it does dress out smaller at eighteen months of age. As a grass farmer, you'll get lots of opportunities to think about this one, and to try different approaches. There are, happily, more and more grass-based dairies out there that do not feed grain, which in fact consider it important for rumen development that calves be fed milk and forage without concentrates, and maybe you can get your first animals from one of these. One way to optimize calf health and growth, if you are bottle or bucket feeding, is to add in an extra feeding. If we absolutely have to take a young calf from its mother, we offer extra milk in this way; although the how-to books warn that too much milk will give a calf scours, we have never seen it, and of course calves that are raised by their mothers drink multiple gallons daily with no ill effects.

Calf Housing

Now, the matter of the calf's "stall." Really, if you're just starting out as a self-sufficient smallholder on a new farm, chances are the calf isn't in a stall, or anything very much like one, but he does have to be in a space that contains him, and that space needs to be dry and not drafty. If it is high summer and very warm outside this is no problem, but at any other time, take this part seriously. Your success as a calf raiser is going to go up about 400 percent if you recognize your calf's need for a nice, dry bed and a good windbreak. That doesn't have to mean a loose box stall in a draft-free horse barn, although such a thing would make an admirable calf nursery. On a lot of protohomesteads, including ours at one time, "calf stall" will mean a stock panel bent into a circle and held with a twist of wire or two or three carabiners, with a half a bale of straw or hay broken into the middle, the whole thing set in a draft-free corner of the barn—or garage—or shed. Or it could mean the potting shed with all the shovels and hoes temporarily removed, and—again—the necessary half-bale of straw on the floor; even, while the calf is young, stacked hay-bale walls in an open shed (but he'll soon learn to challenge the stacked-bale walls). Just remember the dry, draft-free part; no wet feet, and draft-free but not stuffy or close, since warm damp air in the barn can predispose him to respiratory problems. After the first few days, if the weather is warm and dry, you can turn the baby out onto a patch of green grass after its milk feeding, either inside a fence or on a tether. After a couple of weeks, he can sleep out if the weather is nice, but leave the barn door open in case it rains; with no mama to fill his belly with warm milk if he gets wet, he's better off dry.

Breeding

While rams and bucks may fit nicely on the small farm, the smallholder will not usually be in a position to keep a bull (they fall under the heading of "expensive, specialized, single-purpose tools" and, moreover, are dangerous. Enough said). This means she will have to see to the matter of breeding her cow herself, either by getting access to a bull, or by AI.

Breeding is an undertaking involving several steps. First, you must identify when the cow is in heat. This is both simple and fraught with frustration. Cows' estrus cycles average twenty-one days, give or take a day—or two, or three. A cow in heat may show very clear signs of her condition—bellowing; pacing; decreased lactation; clear, stringy vaginal discharge; and a tendency to mount everything in the pasture—or she may show few or none of these. A bull is specially attuned to cow cycles and has little difficulty determining when a cow needs to be bred, whether she's ostentatious about it or not, but we poor humans have to depend on less acute senses.

The one absolute sign (for humans) that a cow is receptive to breeding and capable of conceiving is a visible *standing heat*; that is, she stands to be mounted by another animal when she could move away. Only at this point in a heat, when one of her ovaries is shedding a fertile egg, can a cow be bred with success. You don't have to have a bull to observe standing heat, since other cows or steers in the paddock will mount, or be mounted by, a receptive cow, but you have to be there to see it, not always easy since she will only be in standing heat for twelve to twenty-four hours, and mounting events generally only happen a few times in that period.

And just seeing an attempt at mounting doesn't give you nearly enough information by itself. If the animal mounting is a bull calf or a steer, he may just be feeling frisky, with no reference to the openness to breeding of any cow in the pen. If it is a female mounting another female, and the one on the bottom scoots out, you still don't know enough, since cows not in estrus will sniff, lick, and mount a cow that is coming into estrus but is not yet open to breeding. On the other hand, the cow that is just coming into estrus is liable to mount other cows (or any other pasture mates) frequently, so if you see an individual cow mounting several other animals you should have a strong suspicion that the cow on top is coming into heat and watch her accordingly. If you are rotating your cows' paddock frequently, you see them regularly, and are in a good position to observe a heat coming on and make your arrangements—call the AI tech, or the neighbor with an obliging bull.

Sometimes you will get a cow that is too modest to give good heat signs, or maybe she's just sneaky and does most of it at night. It can be tricky to catch those few moments in her cycle when she stands still and lets another animal mount her. If your cow has been doing extra bellowing, sniffing other animals and curling back her upper lip, or has experienced a sudden drop in milk production that you haven't explained from some other cause, it's a good idea to hang around and watch her for a while. Take a cup of coffee down to the paddock, pull up a stump, and get comfortable. Fifteen minutes spent observing the cows a couple of times a day can be very useful and is good for your blood pressure—but choose a time when the animals have not just been turned onto fresh pasture, or are expecting you to give them a treat. If they are distracted by other things they are less likely to be thinking about mounting.

But, you ask, what if I have only one cow in the pen? What if there are no companion animals larger than a hen, how am I to recognize a heat? Well, all the other signs are helpful. Pacing the perimeter fence and looking longingly down the road is indicative; so is lots of extra mooing or a clear, stringy vaginal discharge. Escaping to go visit the neighbor's cows, sniffing the wind—or you—with a curled-back upper lip, or crowding up on you when your back is turned and initiating a mount, are all pretty good indications that your cow may be in heat. What then? Well, we had only one cow for years, and when we saw these signs, or at least several of them at once, we called the AI tech. Generally Isabel would have a drop of several pounds in the milk she put in the bucket, moo with extra vim, and sometimes do the little rush that made you wonder if she was going to mount you (sounds spooky, isn't dangerous). We'd call the tech, he'd come that night; then, sometimes, we had to call again three weeks later. We'd mistimed, maybe, or it was just the luck of the draw. AI at its best is only about 75 percent successful per shot, maybe because while a bull is generous with his semen, those little straws only hold a fraction of a cubic centimeter. But AI fees are usually low, bearing no comparison to the value of the calf (and the milk) you expect to get from a successful breeding, so if you have to do it more than once, it's still value for money.

Of course, if you bought your cow bred, and she has just freshened (dropped her calf), you have a good date to count estrus from, which gives you a leg up. A cow's cycle is considered to be three weeks long, give or take a couple of days, and she is expected to cycle back almost immediately after calving. Mark the calendar for eighteen days after her calf is born, and then haunt her for a while. If you see nothing this time around, mark another three weeks ahead, and haunt some more. Of course, this may have only the usefulness of giving you a target date, since if your cow calved in April and you want another April calf next year, you don't want to breed her until about July, several cycles after calving. Around here, though, we are grateful to catch a cow in standing heat almost anytime, and we'll breed regardless—except during March or April, which would give us a calf forty weeks later, in January or February, the coldest part of the winter, and not ideal for our pasture-kept cows.

Just to keep you guessing, though, a cow will sometimes exhibit mounting behavior with no reference to her state of estrus or that of any other cow in the pasture. A cow you were relatively certain had *settled* (conceived and implanted a fetus) will suddenly be jumping on every animal in reach, and you smack yourself on the forehead and go inside to scratch her due date off the calendar. Only, not so fast; in such cases sometimes the cow is just establishing her dominance in the herd, declaring herself (literally) Top Cow. This can happen if you have recently acquired a new animal, or removed one, or something else has happened to stir the pond up. We saw this one February when we bought in a couple of first-calf heifers; suddenly our oldest cow, Isabel, who had been bred six months earlier, was jumping on everything in sight. We were about ready to schedule her a trip to freezer camp, thinking she had gotten beyond the age of conception; fortunately, before we could ship her, the vet was

The Breeding Calendar

Wild ruminants like deer are hardwired to breed for a spring parturition, and for convenience we refer to spring calves, timing that, after all, mimics nature—but in reality your ruminant may come into heat, breed, and calve/kid/lamb at any time, in any season. Don't think that if you missed a June heat you've missed this year's opportunity for breeding your dairy ruminant; mark the calendar for her next expected heat, and have another go at it. With several lactating animals, we are glad to spread parturition and freshening over a relatively long season; only from February through May, months that would give us a winter-born calf, may we refrain from breeding cows in standing heat. With our often-freezing Appalachian winters, we prefer not to see baby animals arriving in midwinter.

out to do pregnancy checks on the others. After all the other cows had been through the routine, she asked, "What about that one?"

"Oh, her." We sighed wearily. "No, that one's open [not pregnant]."

"Are you sure?" the vet inquired, and we rolled up Isabel for a quick sonogram—which, lo and behold, revealed a six-month calf in there. Glory be. After all that jumping around, too. Imagine finding you'd sent *that* to the meat locker.

Heat Detection Aids

There are several products available made especially to help people who keep cows, but not bulls, detect when a cow is in heat. There are stickers and paints that can be applied to the cow's spine near the base of the tail; when she stands to be mounted by another animal, much of the paint is rubbed off, or the coating on the sticker—like that on a scratch-off lottery ticket—is worn off to show a bright color underneath. If, however, she is not in standing heat, she will shy away as the other animal rises, resulting in only a little rubbing on paint or sticker. We have used both tail paint and stickers and prefer the stickers in general; it is easier for us to determine the degree to which a scratch-off sticker has been scratched off than it is to remember just how much paint we applied to a particular tail. Neither method is foolproof; if the flies are pretty bad, a constantly flicking tail can wear off paint or scratch up a sticker pretty good. Paint is not expensive; stickers, in our neck of the woods at least, are going to cost a buck or a buck and a half apiece, and they come in packages of fifty; for the service they render, that's cheap. The farm store may not have either; we order ours through the mail.

If you end up online looking for heat detection aids, you will see that there are other devices intended to improve your chances for a successful cow breeding. Most of these use hormones to induce a timed heat; some of them are abort-and-rebreed schemes, so a cattleman can get all his cows bred to drop their calves at the same time, thus shortening his calving season and giving him a bunch of calves of the same age when it comes time

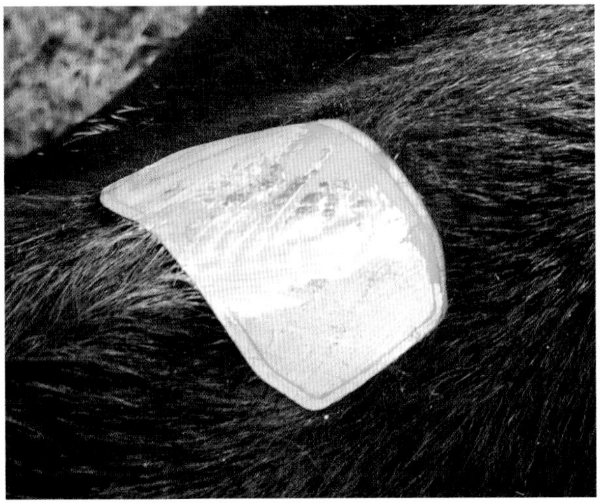

Figure 7.8. Incomplete scratch-off down the center of the sticker indicates no standing heat.

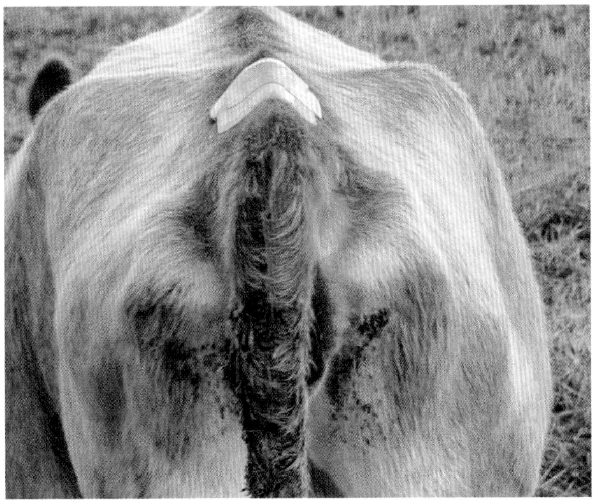

Figure 7.9. Complete front-to-back scratch-off indicates standing heat.

to market. Monkeying with an animal's fertility cycle seems unnecessary to us, as well as unnatural, and we find abort-and-rebreed programs distasteful. Our old AI tech says natural heats seem to work best anyway, so we have never tried "seeding" a heat.

Bull Selection

As far as bull selection goes, you may have lots of options, or not so many. Your neighbor's bull is what he is. If you aren't concerned about breed and all you want is a calf in your cow's belly, he can probably get'r done. Your AI tech keeps some semen available in his tank, maybe a little dairy as well as beef, or if you get hold of him ahead of time he may be willing to order you straws from the sources available to him, usually conventional genetics of commercial breeds. Or, you can arrange to have specialty semen sent to him, at some expense, from sources of your own provision. If you are keeping heritage animals, this may be the best option available to you for keeping your bloodlines pure and improving your genetics.

That said, there are considerations other than pure bloodlines when you are choosing how to breed your milk cow. For the grazier, grass genetics are a real plus, and it is worth our while to pay extra to breed to a grazing line. You can select for "high components" genetics, meaning that cows of this line should have higher-than-average protein and fat in their milk. There is a sizable crowd looking for milk with the A-2 form of the beta-casein protein, and if your cow carries the A-2 gene, you might consider it worthwhile to get semen from a homozygous A-2 bull for possible added value if the calf is a heifer.

Size of offspring is another issue. Not so long ago, farmers had a preference for bulls that threw big calves; the bigger the calf was, the reasoning went, the stronger and more viable it would be. But larger calves mean more calving difficulties, and the trend now is toward moderate-sized offspring. Some bulls are specially labeled with the term "calving-ease," for their tendency to throw smaller offspring, and if you are breeding a small cow or a first-calf heifer, one of these might be a good choice. In an older animal, and for a bigger, blockier calf to go in the freezer, you might opt to use beef semen in your dairy cow. Sexed semen is available too, more or less guaranteed to give you the gender of your choice, *if* the cow conceives. Expensive, high-tech stuff, and we don't recommend it. Don't worry: if you are uncertain about making all these arrangements yourself, explain to your AI tech what you are wanting, and she will be able to make suggestions. Don't fret too much about getting everything just right; the important thing is to breed the cow, so she will come back into milk the next year.

Drying Off

At the other end of the schedule, it's customary to give a dairy animal a rest before she has to start the parturition/lactation/gestation cycle over again, so, six to eight weeks before an animal is due to calve or kid, we stop milking her. Drying off is particularly easy when Mama is due in March or April, when an all-grass dairy animal is typically at her lowest production: we just stop milking. If we can move her to where she can't see the barn or hear the clank of the bucket in the dairy, all the better; the less she thinks about being milked, the sooner she'll stop lactating, which is what we want. If she doesn't stop—and many individual animals among the modern commercial dairy breeds are hardwired for constant lactation—we have to worry that impaction (milk produced without anywhere for it to go) will result in a low-level case of mastitis. Then when the calf arrives and the cow's milk comes in, bacteria are already present for a serious udder infection. All-grass animals, fortunately, don't usually produce such large quantities of milk late in a lactation that it's hard to shut them off.

If, on the other hand, the animal in question is producing 2 or 3 gallons a day, like maybe in mid-May on the best spring grass, we take a bit more time drying her off, dropping first to a single daily milking for a little while before we cut her off completely. Then we put her out back on one of our poorer pastures and leave her to enjoy a little vacation, away from the daily routine of the milking string. The break in schedule, as well as the low-protein forage, help reprogram her hormones and end the lactation. Then again, some excellent dairy women advocate drying off cold turkey, without the once-a-day interval, maintaining that the most important element in drying off is inducing the cow to forget about the bucket.

It has happened before, too, that an animal for which we thought we had a good due date surprised us with an early calf before we ever got around to drying her off, going directly from one lactation into the next without a rest period. We were less experienced then and weren't completely sure whether she switched to colostrum when the calf arrived or just went on making milk, but whatever she gave seemed to be just what the calf needed, and mother and calf both did just fine. We wouldn't want to make a habit of it, but it wasn't the end of the world.

And then there is the cow we are ready to dry off, certain she conceived seven months earlier, who is suddenly mounting everything in the pasture and shows every possible sign of being in heat. What now? If she's not carrying a calf, have we lost our chance of milking her for the next nine months, until she can be bred and calve again? Fortunately, no. For dairy breeds, most likely if we keep milking her, she'll go right on lactating and we can breed her in her next heat; in most cases, we'll get milk until it's time to dry her off before the new calf comes. And even if she's of a nondairy or dual-purpose breed, she won't automatically dry up after ten months; as long as she's being milked and fed she'll give something, although sooner or later it's going to be so little that we'll decide to dry her off anyway. There's even a chance that this cow we're sure is in heat will drop a calf eight weeks later, exactly according to schedule, and all the fuss and bother was just spring fever. When in doubt, have the vet out for a pregnancy check.

Health

It would be nice if all our animals were just as healthy as we know our efforts deserve, but sooner or later someone is going to show up with a limp or a cough or worse, causing us to rack our brains and examine our consciences and wonder if cows get TB (they do, but not often, and not around here) or eat poisonous plants (rarely). Better to have some idea ahead of time what your standard of care is going to be, so you know on which occasions you are going to call the vet, and what you are going to say when she gets there.

Remember, ruminants and forage comprise a natural, living system, and it follows without saying, then, that it is a system where death plays a role. As the farmers say, "If you are going to keep livestock, you're going to have dead stock," and while that may not sound encouraging, it's an important thing to remember when something "goes wrong," or dies. Having a basic routine of animal observation

Figure 7.10. Tall cow pies indicate forage that is high in carbon, low in protein.

and care helps you to feel confident, when something does "happen," that you were following good practices and giving your best efforts to the livestock's well-being.

Visit your animals often—in an intensive grazing operation, this step is built in. Then, pay attention when you are around them. Don't just shift a fence every twenty-four hours and walk away; observe your livestock. Move around among them; scratch heads; give rubdowns; watch to see who is eating, who pacing, who withdraws from the group. Make a practice of noticing the manure in the paddock just occupied, its composition and consistency; know what optimal manure looks like, and which variations are normal, which indicative of some nutritional or pathogenic problem. This is easier than it sounds. Are the cow pies thin and flat? Almost liquid? Very likely your cow is on rich spring grass, with too much protein for the present state of her rumen; giving her a couple of flakes of poor hay can be all it takes to add some carbon to her digestive cycle and slow the passage of nutrients through her system. Are her flops tall and firm, like the crown of a ball cap? Too much carbon and not enough protein; try to include more young, green forage in the next paddock. Cow pies of the proportions and consistency of an unbaked pumpkin pie are an indication that your forage is doing the most good, building strong muscles and rich milk. Pay attention to little things and you will find that you are learning what "normal" looks like without effort, as part of your daily farm chores.

Get a veterinary thermometer, and use it. Look up normal body temperature for your animal species and verify it by your own observation. Your book may list 101 degrees Fahrenheit as normal body temperature for cows, and so it is; however, individual animals may, as can humans, vary from the norm, and it is important to know what is normal for your own animals. Black or dark-colored livestock will have higher temperatures on sunny days; all animals will be warmer at three o'clock on a summer afternoon than at six in the morning. Get familiar with the temperature fluctuations normal to a healthy animal, *your* animal.

Use your nose. Most farmyard smells are pleasant, or at least wholesome; bad smells often indicate a problem. Your nose is a good indicator of whether a confined animal has sufficient dry bedding; it can also tell you when something is amiss closer to home. Foot rot, a fungal infection of the hoof, gives off a bad odor that can be very distinct; you may notice it when you sit down to milk. So does the skin irritation sometimes found in tight, sweaty folds of a full udder, where organisms akin to those that cause athlete's foot can set up housekeeping. Happily, foot rot is seldom a problem in an intensive grazing operation, for the same reason intestinal parasites don't usually trouble us: the livestock don't return to a grazing area before sun and air have purified it of bad biota. Sweaty areas subject to fungal infections should simply be kept as dry as possible, to promote natural healing.

Is someone limping? Even a little limp can disadvantage an animal so that, if stress is not reduced, it loses ground and health, particularly if it shares its paddock with competitors. Removing a compromised animal from the group for a short time is a simple, noninvasive way to promote healing. Have a quiet corner of the pasture that you can fence for a hospital paddock; good forage and less competition will cure a great many problems without need to do more.

Mastitis

Sooner or later, if you are keeping dairy animals, you'll encounter some form of mastitis, which is an infection of the udder. Nursing mothers—humans—will already know all about mastitis, and how it comes sometimes when there is a longer-than-usual interval between nursings, or as a result of pressure on one part of the breast (udder), as might happen from lying in an unusual position; or from constriction from clothing; and sometimes for no discernible reason whatever. It is the same with our dairy animals.

Your first indication that you are dealing with mastitis is usually going to be in the milk. The screen lid on your strip cup will catch thickened milk, flakes, strings, or clots that are generally indications of an infection or inflammation of one or more quarters (section of the udder associated with each teat). Sometimes, even when there is no sign in the strip cup, we will notice that the process of straining the milk after milking, usually rapid, has slowed to a crawl; then we will suspect a low-level case of mastitis.

How do we know for sure if our animal has mastitis? Well, thickening of the milk as it comes from the teat is a reliable sign, but if the milk is straining slowly, or we suspect mastitis for some other reason, there are methods of testing that can be performed in the dairy. The standard test is the California Mastitis Test (CMT), which you can order from a dairy supply source. This is a simple method that allows you to check for a probable high somatic cell count (white blood cell count) at the time of milking. An old-fashioned equivalent, we have read, is to mix equal parts milk and water in a glass jar and add dish soap solution; mastitic milk will thicken and grow stringy. Farm supply and veterinary sources carry pH-sensitive test cards; a squirt of milk from each quarter is directed over designated spots on the card, which change color to indicate possible problems. Further, an animal with a clinical case of mastitis would be likely to run a temperature.

If you suspect that your animal has mastitis, it's not time to panic; it is usually simple to treat. We eliminate grain treats, milk the mastitic quarter separately from the others, and isolate that milk. It is fine for feeding to the calf—if that shocks you, remember that the calf's suckling is the natural remedy for mastitis, just as a nursing baby is the cure for mastitis in a human mother—or as food for other farm animals. Massage the affected quarter with extra care, making sure to empty it as completely as possible. Be sure your wash water and towels are clean, and once you've used them on the infected quarter, send them to the laundry. Be patient; it can take several days, even weeks, to get rid of mastitis once you've got it, but generally we expect to see some improvement within a couple of milkings, then a gradual dropping off of symptoms over the next few days. Usually the visible signs are gone in short order, but the milk will continue to show an elevated somatic cell count (as indicated by the pH cards or CMT) for some days beyond

Figure 7.11. Wash water, milk bucket (upside down, to catch no foreign material), strip cup, and bag balm.

that. Mastitis affects individual quarters of the udder; milk from the uninfected quarters should be fine for human use.

Rarely, you'll get a full-blown case of mastitis with clear yellow serum and blobs of pus instead of milk. When this happened to us a few years ago, we made extra visits to the paddock during the day to strip the infected quarter and, of course, eliminated all grain treats. Recovery took longer than with a milder case of mastitis, and might have been faster if we'd used antibiotics; on the other hand, maybe not, and maybe we'd have ended up with new problems from trying to kill the germs instead of simply eliminating the conditions that favored infection in the first place.

Mastitis isn't necessarily a sign of failure in our farming methods, but may be an indication that some part of our routine needs attention. If our animals are on short-term paddocks, they're bedding down in fresh grass daily, not usually lying in their own manure, but in winter, if they are being barn-kept, we might suspect bacteria has invaded the udder from dirty bedding. We should be milking with clean hands, of course; having a problem could remind us to be vigilant about hand washing. Maybe our dairy towels aren't getting clean between uses: we keep ours separate from other items of laundry, washing them in very hot water to kill germs. Drying them on the line, where wind and sun can purify them, is a help, too; if you are machine-drying, you can use the warmest setting. If you are milking with a machine, take it completely apart, especially the inflations, and make sure you are getting everything completely clean. Are you being gentle when you milk? Are you sure you are milking her out completely? Often just a renewed attention to the last steps of milking, massage, and stripping will correct a tendency to mastitis. And pay attention to your hands when you milk; sloppy milking, with incomplete closure at the top of the teat, means that some of the milk is being forced back up into the udder with every squeeze, a situation that can set the stage for mastitis.

Down Cow

Sometimes a cow will lie down and, for no discernible reason, will be unable or unwilling to get herself up again. We keep a blog of farming experiences for the reference of others who, like us, are trying to rediscover the means and methods of fully local, grass-based homesteading, and in any given period of time, the posts we find most frequently referenced are those with the tag "down cow." We imagine our peers out there, desperate for information on their supine bovine, scratching their heads as they wonder what term to use in their search. "Cow lies down and won't get up" seems to come up frequently. Our first Jersey, Isabel of revered memory, was a living lesson in all kinds of cow follies and foibles, not least of which was this mysterious condition we call "down cow"; no less than five times in her twelve years did she lay herself down and refuse to get up, despite all our persuasions, for days on end.

Smarter people would have sent her to the packer; we—assuming, as novices so often do, that the situation was in some obscure way our fault—did nothing of the sort but brought her fresh-cut forage and cool water, offered massage, lifted her with a hip hoist (you need strong rafters for this) and rolled her from one side to the other (no mean trick; try rolling a full-grown cow sometime) so we could milk her after a fashion, until she finally got up again. We also, although her symptoms didn't really indicate milk fever, gave her IV calcium, but the calcium never got her miraculously back on her feet, as it is supposed to do. We did have the vet out the first couple of times—it was she who suggested the calcium—but she was clearly as stymied as we were. Eventually there was some talk of a hip joint acting up periodically, which makes as good a theory as anything. We have had many cows since Isabel, and none of these have ever—thanks be to God—pulled this stunt, but it happens, and although good care may get the animal back on her feet, if she does it twice, call the packer for a date and get a new cow, would be our advice.

Immunizations

Except for tetanus, we don't generally immunize farm animals. Whether to give routine immunizations or not is something you will have to decide for

Figure 7.12. Gentling babies starts early.

yourself, with perhaps some input (or pressure) from your cooperating veterinarian. We strongly suggest that you do your research ahead of time—talk to people who are raising animals the way you want to raise them, and whose livestock look good to you; read books on the farming principles you think sound. Find out what tests and immunizations are standard for your state or area, and ask at the county extension to find out if there are any local disease problems, like blackleg, or other pathogens from wild animal vectors. Know what you think before you call the veterinarian, or you will find yourself doing by default what she considers "standard of care." If your preferences are for a low level of medication or pharmaceutical interference of any kind, you will need conviction to stand up to possible questions or criticisms, not just from the veterinarian but from neighboring farmers, chance-met would-be stockmen, and people who overhear you talking at ag events.

Worming

It is customary on commercial farms to worm ruminants regularly, even as often as once a month, but one of the beauties of intensive grazing is the natural protection it offers against internal parasites. When animals are moved regularly and paddocks are given a long rest, parasites have little chance to reinfect the herd, or build up to pathogenic levels. Still, once in a while an animal does get a mucky tail and a rough coat, indications that there may be an overload of parasites in its gut, and there are natural or at least nonmedical ways to physic her.

Joel Salatin, arch-evangelist of the family farm and of pastured almost anything, recommends the use of a product from the Shaklee company, Basic H, which we the uninformed would call "soap" but it isn't. A small amount of this in an isolated tank of drinking water ("isolated" meaning it isn't refilling, so all the water is drunk at the same concentration) is recommended for internal parasite remediation. It's okay with organic standards, too. Look it up in his inspiring book, *Salad Bar Beef*.

Likewise, many people swear by regular doses of minced raw garlic in the food ration, or a combination of garlic and rosemary. This one is especially easy since both can be grown in quantity on any farm. For lactating ruminants, offer garlic just after milking, so the effects will have worn off before the next milking, or you'll have to give the whole lot to the pigs, because it will reek. We speculate that perhaps the onion grass that comes up in the spring and seems irresistible to cows may (judging by its pungency and its smell in the milk) serve the same purpose as a big dose of raw garlic and clear out any unnecessary critters that might have found a warm home in a bovine belly for the winter. Pumpkin or winter squash seed are also supposed to be effective against intestinal worms, a proposition which seems beautifully appropriate to us: fall, a standard time for worming (so that animals go into winter conditions without a high pathogen load), is just when pumpkins are harvested and sorted into storage grade and animal-food grade. Pumpkins intended for the livestock are fed out immediately, lasting for a couple of weeks to a month or so, which means the livestock are automatically receiving their dietary worm remedy without our giving the matter any thought. Small doses follow later in the winter as we sort and pick over the storage pumpkins and squashes. Nature, and natural farming, are wonderful; we wonder how many such automatic benefits thereof have been lost by the switch to formulated feeds and other miracles of modern ag science.

Diatomaceous earth (DE) is also a natural and cheap remedy; animal-food-grade DE is available at many feed stores and farm co-ops. For method of use, browse online to see what people are recommending for your species. If it seems weird to feed your ruminants dirt, go out to the pasture and watch how they will lick a bare spot on the soil. There are also more and more sites offering herbal or homeopathic combinations for use with animals, and some of these may be a good fit with your overall farm philosophy. When one of our dairy cows showed up with a case of ringworm—quite a widespread one—we used topical mustard applications and homeopathic tellurium, and, whether those remedies worked or she just had an efficient immune system, the places healed and the hair was growing back in a short time, without need for pharmaceutical intervention.

Finally, there are the chemical and proprietary wormers, strong, toxic antiparasiticals that are

Figure 7.13. Yarrow, an edible herb and beneficial forage plant. Courtesy of the Franciscan Sisters T.O.R. of Penance of the Sorrowful Mother.

administered by pouring a small amount down the animal's spine, back of head to base of tail, or put into the animal's feed or water ration. The pour-on variety is not, as you might intuit, only for external parasites like lice, but for internal and every other sort. Makes you scratch your head a bit—no pun intended—to imagine what sort of chemicals, when poured down the spine, kill all your bugs, inside and out. Makes us wonder what else it might kill.

We have used chemical wormers in isolated circumstances, but rarely (we remember two); we don't want those compounds in our soil, for one thing, and we don't want to breed animals whose immune defenses are inadequate without such help, for another. Nevertheless, there may sometimes be a place in the range of what might be called "rescue" techniques for the use of antibiotics, antiparasiticals, and the other wonder drugs of the modern world, if used sparingly and in cases where there seems no way to boost an animal's immune system quickly enough for self-healing, like a potbellied bottle calf going into a hard winter, and with a year to flush out the toxins before we put him in the freezer. In any case, an animal requiring more than one such chemical rescue would not be one we would want to breed.

Hoofs

If you look online there are lots of videos on getting your cows' hoofs trimmed, mostly by a professional with a tilt table to which he straps the cow so he can flip her on her side. We bought clippers and a hoof knife for trimming feet, but our sense of self-preservation never let us get that close to a pistoning back hoof, and before we made up our minds to bring in a professional the cow in question would calve and her feet would straighten themselves out without any help. Since then we've noticed consistently that any longish-looking hoofs chip back to a healthy length after parturition, so we don't worry when we see them looking a little long. Cows on pasture don't tend to get overgrown hoofs, anyway; it is walking on concrete and other hard surfaces that sends a signal to the cow's feet to grow longer and harder hoofs.

Skin Problems

Cows often develop subcutaneous abscesses, sometimes quite large ones. These are fluid-filled lumps just under the skin and can happen anywhere, usually in a spot where there has been access for germs of some kind, as perhaps with an insect bite or a jab from some thorny plant. When we first began dairying with cows, little things worried us, and we had the vet out once or twice for an abscess before we learned to take them in our stride. An abscess

Figure 7.14. Goldenrod is forage when young and, in fall, bee food. Courtesy of the Franciscan Sisters T.O.R. of Penance of the Sorrowful Mother.

will usually manage itself without any intervention, anyway; it will either recede without rupturing, or rupture just fine without help. If we feel we just have to do something, we wait until the lump develops a soft spot and then cut it open with a razor blade, a deep *X*-cut an inch or so across, and evacuate the nasty stuff—pus, blood, serum—with a gloved hand. No stitches or bandage afterward; you want it to drain. In most cases, probably better to just let it do its own thing naturally.

Lice are something we have never seen on our bovines, but we used to see them on the goats, and they can be so severe as to compromise the health of an animal. They are a more common problem among confined animals, so the rotationally grazed ruminant is naturally advantaged. Regular doses of internal garlic are supposed to be useful, and there are herbal remedies and baths, and essential oils, that are said to provide some protection. Once or twice we have had animals develop spots of bare skin that turned out to be ringworm, not lice, and these we have just treated with topical mustard, or followed the advice of a grass dairyman friend, Ken: "Let the sun cure it," he says.

Bloat

Frequent visits to the paddock, such as are inherent in an intensive grazing system, are your best line of defense against one problem that, when it strikes, can kill a ruminant quickly. Animals newly turned out onto fresh green forage after hay feeding can develop a rumen imbalance and bloat quickly; unless steps are taken to release the pent-up gases, pressure will increase until it makes breathing impossible for the beast. Death follows quickly. We have lost more than one calf to bloat (although not feed-induced), and the distended form—blood trickling from mouth, nostrils, or anus—is a distressing and depressing sight. An animal that gets into the feed bin and overeats is liable to bloat, and even just lying down wrong on an incline, or receiving a slight bump from a passing pasture mate, can tip a cow or young dairy calf over so it can't right itself; the rest will follow automatically.

Once, on our farm, a greedy calf got down on its knees to snatch hay from under the manger and got her head stuck; when we found her she was already dangerously distended. We pulled her out and went for a knife (we don't keep a trochar, which is a sharp tube for puncturing the cow's side and letting out the gas), but fortunately before we gathered confidence for the stick she was starting to deflate, and we never had to find out if sticking a cow is as easy as it is supposed to be. The rumen is that part of a cow's stomach just under the left hip bone; but watch a video before you try it yourself. A trochar is one tool for managing bloat; passing a hose down her throat and into her rumen is another, either to let out the gas, or to pour in oil to break the surface tension of a foaming bloat. Knowing how to "drench" an animal in this way is useful, and it is worthwhile to acquire and keep handy some tools for this kind of intervention.

Dehorning Calves

Whether you dehorn or not is up to you. Most farmers seem to prefer their cows without horns, seeing this as being safer for the animals' pasture mates and for the humans interacting with them, and there is no doubt that horned animals are better equipped for bullying, and a careless swipe of a horned head can do more damage than a polled one. On the other hand, if you live in an area where large predatory animals are being reintroduced, as with wolves and panthers on the northwestern plains, or even closer to home, horns are practically the only weapon a bovine has, and maybe you want to leave them. A farmer west of us, here in Ohio, lost four calves and a heifer this spring to predation, and he swears he saw a mountain lion at the carcasses.

With calves, as with goats and sheep, the farmer's options for polling are the dehorning iron or caustic paste (surgical removal of the horn bud is also possible but requires the services of a veterinarian). The iron is more painful at the time; the paste is irritating for a longer period, as you will experience the first time you get some on your skin. If you use caustic paste you will need to keep the animal under a roof for a few days if there is any chance of rain; caustic paste is water soluble, and if it runs it will burn the hair off whatever it touches and probably leave a bare spot for life. Heaven forbid it should wash into an eye.

Castration

Knife, elastrator, or burdizzo; each has its adherents. Basically, you can cut open the bottom of the animal's scrotum and pull out the testicles; you can put a tight elastic band around the top of the scrotum (making sure that both testicles have descended—if they haven't, you'll need to feel for them and push them down) and let constriction do the job; or you can use a tool to crush the blood vessels to the testicles, so that they shrink and are eventually reabsorbed. Tools for these jobs are available at the local farm store. You can do it yourself, at home, and it would cost you a packet to have the vet in every time it needs doing. We aren't going to give you a blow-by-blow description of these operations, since the Internet is such a good source of information on the subject, but make your search terms as specific as possible, if you don't want some unexpected and possibly unpleasant results.

When to castrate is another question. There are different schools of thought. It is popularly believed that younger animals are less sensitive to pain and that it is therefore more humane to castrate as young as possible. We are not able to speak to the amount of pain a baby animal feels; plenty, we are inclined to suspect, if the quantity of noise a three-day-old piglet makes when Mama steps on his tail is any indication. A significant reason for castrating relatively young if you are using a knife may be that the size of the blood vessels to a large, mature testicle can result in considerable bleeding when they are broken or severed. Banding—which, in any case, seems to be the least disturbing to the calf—can be done from the time the calf is born to a much more advanced age; there are large bands made for banding older, larger bull calves. A possible disadvantage to this method is the danger of blood poisoning if the band should break or slip off while the tissue below the band is necrotic and before it is completely dried up. Castration bands are very strong, however, and while an old, cracked one might conceivably break, it's not very likely to slip off before it has done its job. If you are a low-tech smallholder, your age limit on banding is likely to be set by your ability to immobilize the calf; eventually, you and two buddies won't be able to hold him. Doubly so for the other methods; the more pain you are going to inflict, the more securely you want your animal immobilized.

We castrate calves by banding, and we typically do it between three and six months, which puts us at odds with some humane-animal-treatment groups, which is unfortunate, but there are good reasons for our choice, especially for baby bulls bought in from a commercial dairy. These little guys are compromised from the word "go," simply by being born in all the contagion of a commercial dairy barn and then taken away from their mothers immediately. They are going to be harder to raise, more susceptible to sickness, and less thrifty in general. It has been our observation over the years

Figure 7.15. Balling gun, rumen magnets, castrator with elastic band, and drenching bottle (sturdy wine bottle).

that leaving them sexually intact while they are very young makes for stronger calves with more will to live, and we prefer to band between three and six months, when our baby bulls are already fleshed out and have an established constitution. In this case, experience and observation seem to us more reliable guides for responsible animal husbandry than a one-size-fits-all protocol established by people who don't know our farm, our animals, or our methods.

We don't enjoy castrating anything, anytime, but there may be good reasons to do it, and sooner or later you'll want to decide what works best for you. Avoid castrating when animals are already subject to stress. This could include when weaning from cow to bucket, or weaning from milk altogether, or being moved or immunized. Ideally, you are only going to give the little guy one thing to worry about at a time.

Figure 7.16. Although some sources recommend early castration, we find intact calves are more vigorous.

Things to Keep on Hand

While you may never need to use them, it can be reassuring to have certain tools and remedies in your home medicine chest for use in an emergency, or when you can't reach the vet.

One handy and inexpensive tool is the balling gun, a long plunger-operated tube for forcing pills and such behind a cow's molars and down her throat. Some cattlemen we know swear by the efficacy of an aspirin bolus for temporarily reducing pain and inflammation, thus decreasing an animal's fear or tension so that it can begin recovering naturally from injury or temporary discomfort. Veterinary aspirin is available from the farm store; for very young calves we would use ordinary drug store aspirin. Always use a balling gun to administer boluses; a cow's molars are strong and sharp and will do a bad number on your fingers.

Balling guns are also used to put magnets in a cow. Yes, magnets; we administer rumen magnets routinely, because on old farm land like ours there is a good chance for a cow, who does not chew her food thoroughly, accidentally to swallow a foreign object like a nail, screw, or wire. If one of these reaches the honeycombed reticulum, the second chamber of the ruminant's stomach, they can lodge there, or pierce the tissues and bring on infection. The magnet attracts and holds metallic objects in the bottom of the rumen, the first stomach, where they seldom do any harm. "Hardware disease," as it is often called, seriously reduces the efficiency of a cow or calf and can even kill it; it is easier, and cheaper, to keep magnets in our cows.

A drenching bottle is also helpful to have around. Ours is a liter wine bottle made of thick glass; when we need to get a liquid remedy into a ruminant, we can immobilize her in a stanchion, tip her head back, and shove the neck of the bottle to the back of her cheek to pour the remedy down. Some people drench with a piece of hose and a siphon, but the bottle works okay for us. We also keep an IV apparatus on hand, for administering calcium gluconate via the jugular vein. Although we have never had a case of milk fever per se, we have had cows go down with what might have been milk fever had they been closer to calving, and in such cases we have given intravenous calcium gluconate as a precautionary measure. Watch a video of this one, and then step out in confidence; it's not difficult. Veterinary IV rigs and calcium gluconate are available, yes, at the farm store.

Castration bands and the gizmo for stretching one open, and dehorning paste or a disbudding iron, are good to keep on hand so that when you determine it is time to emasculate or poll an animal you are ready to go. Iodine or Betadine, wound spray, and some bag balm or veterinary unguent, are also handy for what might ail her. We seldom or never stitch a wound, and our vet assures us this is "best practice," wounds healing better when they are open and able to drain, so we don't keep stitching materials around. We have tried stitching a few times, as when a young cow caught her side against the corner of the metal roof of the duck house and cut a neat 6-inch gash through dermis and epidermis. The thing gaped like an empty pocketbook and we thought we just *had* to do something; that time, we found that cow's hide is nearly impossible to penetrate with even a very sharp needle, and we had to give it up. Healed just fine the way it was, and today we can't find the scar. We do keep hypodermic syringes and needles for occasional use but seldom have need to use one.

CHAPTER EIGHT

Poultry

There is no farmwife who does not know how to rear chickens. The first requisites to provide are dung, dust and cinders.

—*Palladius,* De Re Rustica

In 1920, the typical farm flock of about fifty to one hundred birds free-ranged and got a good bit of its living that way, and not just its own living, either: the sale of eggs was a significant source of cash flow on the small farm. Poultry food was homegrown and consisted of skim milk, grains, garden vegs and butchering offal, plus whatever the birds could pick up around the farm. Popular breeds were hardy and locale specific (as witness their names: Rhode Island Red, Plymouth Rock, Buckeye), scrounging much of their diet, hatching their own eggs. A good laying hen might produce upward of 200 eggs a year and lay for six, eight, even ten or more years. Chickens raised strictly as meat animals were uncommon, the poultry on the Sunday platter falling mostly into two classes: surplus males and old stewing hens. Both were comparable in market price to sirloin steak.

A lot of things have changed since 1920, though, chickens—and chicken food—not least. Gone are the multitasking farm flocks. Today practically all peeps come from a hatchery, which sells fast-growing, ultra-tender broilers that dress out in just six to eight weeks, and industry layers that produce 330 eggs a year. These superbirds make the traditional dual-purpose breeds—larger-bodied, slower-growing layers of 200–250 eggs per year—look like underachievers, so small wonder if the hatcheries fiddle with the old farm poultry genetics, breeding nonviable Gargantuas for meat birds, and trading body mass, mothering traits, and longevity for an intense, if much more brief, laying career. Farms don't generate the same sources of poultry food they once did, anyway, so it matters little that the new chickens are bred to thrive on commercial corn-and-soy chicken feed and have a harder time utilizing natural food sources.

And as, in the interest of higher egg production, we've waived any expectation that our birds will have the instinct to reproduce, every year or two we must order a batch of replacement chicks. Of course, the price of feed, and the expense of replacing birds every year or two, drives the production costs of our homestead eggs up from not-much-because-we-grew-it-ourselves to something more substantial, and if we're getting five dollars a dozen selling surplus eggs we're not much more than breaking even over the expected productive life span of our chicken. Yes, a lot has changed since 1920: which is not to discourage the reader from raising chickens (far from it!) but to call attention to the realities of

Figure 8.1. Poultry patrol the barnyard.

modern chicken raising—changes in breed traits, the shift in feeding practices, and the expectations we have about productivity—so that when we do buy chickens we will know what we are wanting from our birds, and know how to identify success when we see it. In fact, we really have to rethink chickens from the ground up, and apply to them the principles of holistic husbandry, asking: What are the characteristics of this animal? And where does it fit into the ecology of the regenerative farm?

Chickens

Chickens are domestic birds bred from the wild Asian jungle fowl (*Gallus gallus*), a ground-dwelling, earth-scratching bird that lives in small flocks in—naturally—the Southeast Asian jungle. They are small, as farm animals go—mature adults of standard size weighing anywhere from 4 to 12 pounds according to breed—with a strong instinct to scratch for their food in dirt, plant litter, and the manure of other animals, where they find the seeds, insect larvae, bugs, and leaves that make up so much of the wild bird's diet. Their life expectancy is pretty variable; while the rare individual may still lay at ten years or beyond, the industry no longer values longevity, and even in the breeds associated more with home flocks, a six-year-old bird is out of the ordinary. Planned breeding has almost wholly disassociated the domestic chicken's egg-laying habits from the day length that triggers the reproductive cycle in their wild cousins, so that instead of laying only a few eggs in the spring, a mature chicken will lay for months during the spring and summer, and even, in more southern latitudes or with some artificial light and a warm place to roost, into or through the winter as well.

Broodiness—the instinct to incubate and hatch their own eggs—is a quality not much valued in chickens today, when the majority of people buy their chicks from the hatchery, so this trait has been discouraged in most breeding programs, but you often find the individual hen that wants to go broody, and provided there is a rooster around to do his duty, she is capable of reproducing unassisted. A hen that hatches three nests of eggs each summer can with luck provide you with two dozen or more new chickens a year; with such a short generation cycle, the species is capable of rapid improvement—or decay—through home breeding.

Niche

Just as wild birds are a vital part of the natural purification cycle, spreading wastes and controlling insect numbers, so are chickens an indispensable part of homestead ecological health. Where ruminants are kept, we would almost say that chickens—or some other species of scratching bird, like turkeys, guineas, or peacocks—cannot be dispensed with. Their manure spreading by itself provides an enormous service to the farm: pathogen and pest insect larvae are eaten, or deprived of the wet, nutrient-filled environment necessary for life, while concentrations of manure are dispersed so that instead of an overload of nutrients in one small area of the pasture and a deficit in points adjacent, even application favors widespread sward health and improvement. "Hot spots" of overfertilized forage, distasteful to the grazing ruminant, are avoided, so that forage is more evenly utilized.

Along with these benefits, scratching, because it aerates pasture thatch, promotes the biological decay of plant residues, improving the nutrient cycling and water retention of the soil. This same quality makes chickens useful partners in the compost-making business, and hens given access to the compost bin can, under controlled circumstances, provide a useful service there. Turned into an off-season garden or raised bed, they can also do a good job in a short time of cleaning up insects, weeds, and seeds. And the spilled grains and animal feeds that just seem to go with farming are all acceptable grist to the chicken's mill.

Needs

Depending on climate and circumstances, some breeds can forage for nearly all of the food they require for health, at least seasonally, but for sustained egg laying, supplementation will usually be necessary, and here the family dairy animal again

Figure 8.2. Maggots in manure pies will be scratched out by hungry birds.

shows her significance: a gallon of skim milk can provide all the high-quality protein necessary for the daily ration of fifty large-breed hens or one hundred smaller size birds. Scratching in ruminant manure provides a good deal of feed for the farm flock, as well. Your domestic poultry will also get considerable benefit from doing pest patrol over your pasture insects and will even kill and eat small rodents. Provided you keep a rooster or so, and allow a hen to brood, they can also reproduce, either more or less unassisted—with the farmer providing a private nesting box and room service feed and water—or wholly unassisted, as in the case of the hen who steals her nest, sets up housekeeping for herself in a corner of the barn or under a friendly bush, and surprises you three weeks later with a batch of fuzzy chicks.

But human beings aren't the only species that value the chicken for her egg laying and her tasty meat, and if you are going to keep chickens, whether in partial confinement or free-range, you are going to have to protect them from predators. A single gander pastured with a flock of chickens is said to be a good guard animal, and we have even seen that, in the case of very large flocks, some people are experimenting with chicken-herd dogs. But at night your birds will have to have a secure shelter, or sooner or later you'll find your head count mysteriously diminished, or a marauding dog or a bold fox will visit your barnyard and leave it strewn with chicken corpses. Some sort of shelter is necessary as well for the purpose of making your birds' eggs easier to find, and to provide them with shelter from cold or inclement weather.

Where winters are cold enough to force plants and insects into dormancy, chickens must receive most or all of their food ration from the farmer. Depending on our expectations and preferences, this may be purchased commercial feeds or grains, or home-raised vegetables, seeds, and grains, plus garden and dairy surplus. Butchering offal is a valuable source of animal proteins as well. Rationing of feeds, and protecting these from rodents, insects, and decay, are the province of the farmer, as is providing a source of frost-free drinking water.

Surplus and Services

Certainly a homestead flock will produce a seasonal abundance of eggs, and these, if the farmer does not desire or does not have the means to store them for her own future use, may probably be sold without taking necessary nutrients away from the farm. If purchased commercial feeds are being used, it is worth the farmer's while to calculate the average cost of feed necessary to produce a dozen eggs and price hers accordingly, but of course, the integrated flock's contribution to the farm economy is already great without reference to any profit on egg sales. Where organic, soy-free, non–genetically modified (GM) feeds are used, whether home raised or purchased, fresh eggs command a pretty high price—five to ten dollars a dozen—and under these circumstances it may be worthwhile to cultivate a market for your extra eggs.

If you are setting eggs under broody hens, you may hatch more chickens than you need on the farm. Where careful breeding practices are used—the isolation of superior birds as breeding stock—fertile eggs and home-hatched chicks may be valuable for sale as foundation stock for other farmers' flocks. The Livestock Conservancy and the Sustainable Poultry Network have excellent resources on the selection and care of birds for breeding.

It is not our province to discuss the raising of meat birds for profit, and it's unnecessary anyway, since this topic is already being covered by people who really know what they're talking about (as in Joel Salatin's *Pastured Poultry Profit$*, to mention one obvious resource). Every couple of years we raise fifty or so Cornish Rock Crosses (CRXs) for our own table, but while we enjoy these, we're philosophically opposed to the idea that birds that are themselves not viable—CRXs start dying of their own excessive growth by about ten weeks—can really add to our own overall health or the health of our farm. Most of the chickens we eat are either cockerels or old layers. We do sell a few of these, but they are more valuable when they stay on the farm to nourish the family; even the bones will be eaten and enjoyed, by the pigs.

The care of poultry is an area where a great deal of variation is possible, and the methods that suit one farm or one round of seasons will vary from best-practice methods under another set of circumstances. Beginning with housing, feeding, and selecting which breed or breeds to raise, the chicken keeper will need to determine what she thinks will best serve her farm, her birds, and her desired outcomes.

Figure 8.3. The area around a stationary poultry house will be stripped of all but the most persistent plant life.

Shelter and Fence

Grassfed chickens may be raised in *partial confinement*, *free-range*, *on pasture*, or a combination of all three. They may even be kept in seminomadic style, so to speak, moving from place to place on the farm as their services are needed. Husbandry decisions fluctuate over the year, depending on the season, the needs of the farm, and the present goals for the flock.

Partial confinement means "periodic confinement," according to the season or the job they are intended to do, such as brooding or garden cleanup. *Free-range* birds roam the farm at will during the day but are usually enclosed in a stationary shelter at night. *Pastured* birds are chickens in a portable pen or coupe that is moved frequently, usually once a day; unlike free-range birds, around whose permanent housing the ground, far from being pasture, will resemble Carthage in the aftermath of the Roman invasion, pastured birds are always guaranteed a daily ration of fresh grass. Pasture is a good option for both meat and laying birds, whether you use sliding pens ("chicken tractors"—that is, pens moved with traction), or wheeled houses (what Mr. Salatin calls "eggmobiles" and are on our farm known as "hen coupes"), with or without a surrounding fence to centralize animal impact and discourage predators.

Partial Confinement

A good use of temporary confinement is when you have a specific site where you want your birds to apply their tilling or gleaning powers. Mr. Harvey Ussery, author of the densely packed chicken guide *The Small-Scale Poultry Flock*, describes three excellent examples of this practice. Where ground is to be cleared for planting, a full season of concentrated chicken scratching can put the "Carthage effect" to good use. Further, the small portable pen—often called a "tractor"—can be a really useful tool when moved over off-season garden beds, allowing the chickens to utilize crop residues and clean up weeds, insects, and weed seeds before the beds are replanted. This arrangement serves two purposes at once—always a good idea on the homestead—but keep in mind that if whole grains are part of the poultry ration, the birds may plant as many as they clean up, and it will be better to avoid whole seeds for chickens housed in this way, unless they are cooked or fermented first. Third, there is the combination compost bin/chicken run, in which the poultry are confined for a period, simultaneously to forage and to aerate the decomposing organic matter. Enclosing the chickens for this job not only concentrates their scratching in the place where it will do the most good but penning also keeps the sheep and goats (cats, dogs, pet pony) *out* of scraps you intended for supplementation of the chickens' diet.

Confining birds to a run for only certain periods of the year is a good husbandry tool and may be consistent with the goals of the grazier, as in, for example, penning the birds in wintertime in cold climates when nothing but the fox is abroad. Once the soil has frozen there is less and less for the birds to forage, and when there is snow on the ground our hens don't venture far from the henhouse in any case, mostly hanging out inside or underneath. At this time we may put an electric net fence around the chicken house or hen coupe so we don't have to be so vigilant about shutting the chicken door as soon as it gets dark out. Sometimes, too, when a low egg count makes us think some of the hens may be out-laying (laying their eggs somewhere other than the nesting box), we use temporary electric netting to keep the hens close to the coupe for a couple of days, so we can get a better idea of how many eggs we should be expecting to collect.

Free-Range

Free-range is a term begging for a definition. Such is the nature of CRXs in particular that, if there is no necessity of their moving from one place to another, they won't do it; with a body mass far in excess of their bone strength and organ development, they find it easier to sit where they are than to move. Our first flock of broilers, yea these many moons ago, which we started under a brooder in our back shed, were given access to the yard when they were about four weeks old and (we hoped) hardy enough to bear the rigors of fresh air and sunlight; only one or two ever ventured out, probably by accident, and they just took a quick look around and bunked.

Truly free-range birds have more or less unlimited daytime access to the great outdoors, not being fenced *into* anywhere, although they may be fenced *out*—of the garden, for instance. We free-range our home flock, which sleeps and lays in a stationary henhouse situated in the middle of the farm and within reasonable chicken distance of about four-fifths of both home pastures. This lets them share stomping grounds with the pigs, sheep, and cattle in paddocks covering about 5 acres, so the birds do clean up duty for everyone. It also gives them unlimited access to a dense mix of grasses, forbs, and legumes, as well as to the edge of the woods (this last is a mixed blessing, since the benefits of scratching on the forest floor may be outweighed by the dangers of being picked off in a daylight fox raid). Our pastures also contain seven or eight apple and mulberry trees, so the chickens do fruit-fall cleanup for several months of the year. With the limited size of our home pastures, and the central location of the henhouse, we find a portable coupe (like we use at the monastery) unnecessary: the birds' comfort range includes virtually the whole farm. A few of the most daring birds—usually one of the Black Australorps or Speckled Sussex, labeled "good foragers" in the poultry books, but maybe more accurately termed "suicidally adventurous"—will venture all the way to the foot of the pasture and do cleanup duty there.

Pastured

Pastured chickens are more intensively managed than free-range flocks, being either confined in a pen or tractor that is moved regularly to fresh forage or housed in a portable coupe that travels over the pasture, and from which they are allowed to range during the day. The differences between these practices will be best illustrated by example.

Figure 8.4. Poultry scratching in the orchard help fight curculio beetles.

Chicken Tractors

Our pastured broilers, like the flocks of Mr. Joel Salatin and many other people who raise chickens or turkeys for meat, are confined at all times to "tractors," portable covered pens without floors. These provide adequate grazing room for the broilers' needs and are moved daily to fresh grass, thereby adding significant green food to the birds' diet, fertilizing and dethatching the pasture, and eliminating some insect pests. Chicken tractors can be easily constructed from scrap wood and poultry netting. Another option that works well for us is to make an A-frame of welded-wire stock panels and triangular plywood ends, the whole held together with zip ties; the flexible edges of these pens conform well to our wavy and uneven hillsides. One part at least of any tractor must be roofed to shed rain and provide a refuge from the hot sun, and this is the end at which to put the feed trough, so it will stay dry. We offer water, either in a Bell waterer or via poultry nipples, at the opposite end of the pen, to encourage the birds to move around and utilize all the grass provided them, and in a really hot climate they may need a shade cover at this end as well; CRXs have been known to sit in the sun and die of sunstroke rather than shift their freight into the shade.

Pastured laying flocks, on the other hand, are usually kept not in tractors but in movable chicken houses ("eggmobiles" or "coupes"). These are necessarily more elaborate than broiler tractors, which need only provide shelter for a few weeks to birds whose sole purpose in their short lives is to eat and put on weight. A coupe, while it can be small,

Figure 8.5. This hen "coupe" is equipped with roosts and nesting boxes, feed bins, a water tank, a bar of poultry nipples, and a hanging feed trough.

has to be big enough to enclose the birds at night, providing perhaps a foot of perch space per bird, and it has to contain an adequate number of nesting boxes. In addition, portable coupes need room somewhere, either inside or outside, for feed troughs, feed storage bins, waterers, and water reservoirs. There are many clever homemade hen coupes combining these functions in compact, easily portable, and attractive ways: roof gutters filling drinking water reservoirs, feed bins and nesting boxes easily accessed from outside the coupe, and feed troughs hung underneath the body of the coupe, out of the way of rain or snow. Some we have seen are even equipped with a solar-powered, light-activated door closer, which shuts the coupe automatically at dusk. A woven wire floor allows manure and shed feathers to drop through, fertilizing the pasture and avoiding manure pack, ammoniated air, and parasite-breeding environments.

In some cases, the coupe is the only confinement for the pastured laying flock, the birds being completely at liberty during the day and shut in, away from predators, at night. At the monastery we run at least one coupe behind the dairy cows in this way, the hens deriving a good deal of their diet from scratching in and spreading the many manure pats the cows leave dotted over the pasture. This works well on the larger farm (the monastery pastures measure about 40 acres, as against the home farm's 5 or 6). The hens identify the coupe as home, and as long as we move them often, and only shift location at night when the birds are all inside the house, they are not particularly inclined to settle down elsewhere (daytime moves, when the hens aren't aboard, don't work; even a move of just a few yards will baffle the birds' ability to find their way home, and they'll roost in the old site rather than climb aboard the coupe in its new location). When circumstances make it necessary for us to leave the coupe in one place for several days, or even weeks (as in spring, when the pastures are too soft for driving on), we have to make sure it isn't too close to an outbuilding, or we get hens jumping ship.

On large operations, it is common for the hens to be pastured in a fenced paddock outlined by panels of electric netting. Penning in this way has the dual advantage of concentrating the birds' impact where it is most useful to the farmer, and at the same time protecting the flock from four-legged predators. Where there are plenty of permanent electrified fence lines, the job of shifting fence panels, big enough in any case, is not made bigger by the necessity of running a power line back to the charger. On small farms, the house, barns, and outbuildings may be so close together that when the location of the henhouse varies, an unconfined hen will consider every structure a potential nest site, but if you have permanent charged fences and plenty of time, surrounding the coupe with net fencing has the advantage of keeping the hens roosting where they belong at night, not adopting the back porch or a rafter over the new hay as the place to deposit their nighttime poops. With a good electro-net fence around the coupe you can sleep at night even if you didn't remember to hike out and shut up the hens before bedtime.

Creeps

Limited pasturing might also be used in the case of a broody hen setting eggs or brooding young, or with a creep feeder for young birds being introduced to an adult flock. Broody hens can be confined to a tractor to encourage faithfulness to their eggs without entirely losing the advantage of being on grass; we don't shift the pen, so as not to disturb the mother bird, who will be spending almost all her time setting her eggs anyway. And when her chicks hatch, the portable pen is the perfect place for them to enjoy the care of their mother and the advantages of open air, sunshine, and grass, without encountering all the dangers to a small chick from predators, weather, and other chickens. When the babies are big enough to venture out of the shelter, a slatted wall or door small enough to let them in and out, but too small to admit large birds, lets chicks use the tractor as a creep, where they can take shelter from hazing, or access food without competition.

Garden Enclosure

There are exceptions to *almost* every rule, and the injunction against confining birds without access to the outdoors does not apply to temporary, seasonal housing in a greenhouse or garden tunnel. The advantages of confinement at this time are several. The space inside a transparent-skinned greenhouse or tunnel is naturally warmer than you will find inside the average henhouse under the same weather conditions, encouraging egg laying and improving feed efficiency. Usually these gardening structures will offer a greater floor area per bird than the

Figure 8.6. In winter, hens keep warm and clean garden beds simultaneously in this low tunnel.

standard henhouse; by scratching in the soil birds break up and scatter their feces (which in a standard poultry house can tend to build up, capping even a deep layer of bedding), creating conditions less conducive to the exchange of parasites, both internal and external.

We are careful to allow for adequate air circulation in plastic-wrapped structures, cutting vents at either end of our low tunnels, propping them open with notched sticks when sunshine raises the temperature inside. We forgo providing roosts to our chickens while they are in low tunnels, which doesn't seem to bother them; at any rate, the earth stays sufficiently warm under plastic covers so that they are not in danger of frozen toes. A moderately creative homesteader will contrive to provide low perches if he deems them important. Hay or straw to line nesting boxes may get scratched out and spread on your raised beds, where it can contribute to the heavy capping of the soil surface that will occur if you keep hens in one location for too long; cleanup tasks in the spring will be easier if you limit the amount of bedding provided and move the chickens every couple of weeks.

Food Landscapes

Before we consider ourselves as having covered the topic of pasturing methods, we want to touch on the broader subject of *food landscapes*, as applicable in the keeping of poultry as it is with pigs or cattle. In this age of indefinite, feel-good food adjectives, we don't want to let the very important term *pastured* be defined in a way that overlooks the many benefits of browsing and foraging that are not derived strictly from *grass*.

On the intensively managed farm, food from solar energy is all around us, and while the foundation offering at the free food bar is pasture, poultry, as well as swine and cattle, are going to derive much of their food and health from other fruits of the sun and soil. Don't overlook the benefit to both animals and ecosystem of pasture and pastoral settings that include garden, orchard, and vineyard cleanup; fruit and nut falls; berry bushes; and pasture-cropped or garden-grown grains and brassicas; not to mention animal-powered lawn care. Make sure your poultry paddock planning includes running the hens under the mulberry trees during their long fruiting season, letting them clean up the strawberry bed for a day after the harvest is over, and allowing them to scavenge the first and second fruit drops in the orchard, sanitizing this before the main crop can be bothered by insects breeding in the fallen fruit. And when a garden bed is not needed for human crops, a sowing of oats, brassicas, or lettuce sprouts can provide good food value to any hens you may choose to confine there. The sun and Earth are generous, and nature comes without fences; we homesteaders have to learn to think outside of our own mental fences as well.

Choosing the System That Works for You

Different strokes for different folks, and different housing and pasturing strategies for different farms, flocks, and seasons. What works best for you will depend on circumstances. Most likely you will find yourself using a variety of methods in your chicken-keeping projects, depending on the flock in question—whether they be broilers, layers, or breeding stock—the weather, time of year, state of the pasture, your involvement in other farm endeavors, even the attitudes of the neighbors. Sooner or later you'll most likely find yourself with a rooster and a couple of dozen hens free-ranging from the henhouse, two tractors of broilers out on the pasture, and a batch of chicks under a hover in the basement, with the occasional broody hen on a clutch of eggs to keep things interesting.

Hen Housing

There are many books and online sources available for chicken house specs—floor space, number of hens per nesting box, and so on—and we will not trouble to repeat what has already been said by those with more expertise. We will, however, offer this thought.

Nature is not uniform, and it is the reverse of monotonous. Lots of books and articles will give you hard-and-fast rules on any topic for How It Must Be Done, and it would be easy to feel you weren't doing it right unless you were following the Rules. But your animals haven't read the rule book;

Figure 8.7. Homegrown chicken feed.

they don't know that they need *X* feet of floor space, so many nesting boxes per laying hen, a certain number of hours of light per day, or free-flowing cross ventilation in their sleeping quarters. Those things are all good, no doubt; if they dropped out of the sky onto our farm tomorrow, we would probably wonder briefly what sort of meteorological event rains chicken houses and then move the birds right in. But on the grassfed farm, where animals are free to roam, scratch, hunt, and forage, the Rules don't necessarily mean a whole lot. We can provide our chickens with regulation-perfect nesting boxes and there will still be plenty of hens who want to lay their eggs in the calf manger; even when there is more than adequate roost space in the coop, some of the birds are going to want to spend their nights perched on top of the broody boxes.

Read the rule books, by all means; educate yourself about what the many farmers before you have found generally to promote chicken welfare. But then look at your own farm, see what it has to offer; watch your poultry flock and pay attention to what they are telling you. After all, what you are trying to build is a farm rooted in *this* place, *these* animals, the most local of situations and conditions—and you're not going to find instructions for that in any book. Trust your animals. Trust your grass. Trust your farm.

Like most animals, a chicken, while it can bear a good deal of cold with equanimity (ours have weathered below-zero winters in open shelters with no problems—but no eggs, either), wants to keep its feet dry, to have a roof over its head in rainy weather, and to stay out of drafts. While confinement animals need to do this with some floor space for scratching around in, the free-range bird, if its feed trough is located outside the henhouse, will spend very little time indoors and in our experience needs little more than 12 or so inches of roost space where it can settle down after dark, with a nesting box where it can make its deposits.

But human beings aren't the only creatures that like a plump chicken for dinner, or a couple of nice breakfast eggs. Henhouses need to be predator-proof, and that means *tight*. Foxes and raccoons, owls and hawks can go over your fences and walls, or under them, while minks and their kin can squeeze through an unbelievably small crevice. These last, if they get into your henhouse, will not kill just one of your birds; on a neighboring farm, one mink wiped out over thirty pullets in a couple of nights. The farmer, rightly diagnosing the small neck wounds on the dead birds as the work of this species, baited a live trap. Overnight the trap was sprung, but empty. Then on better advice he exchanged his coon trap for one sized for rats—the spaces between the wires of the larger trap had not been small enough for confining a mink—and this time he scored. Rats,

Figure 8.8. The portable coupe is safe from tunneling rodents.

by the way, while too small to take on a grown bird, are fond of chicken eggs and baby chicks, so a good henhouse should be constructed to exclude these, and examined regularly for evidence of gnawing or digging for rodent access. We don't like to be reminded of how many promising nests of eggs we've seen depredated by a night-marauding rat.

Wire netting for henhouses should have small mesh and strong wires. Not all "chicken wire" meets the requirements; a determined raccoon can tear its way through lightweight chicken wire, particularly if it is old and rusty. Or he can make a hole just large enough for one paw so he can snag a roosting hen. The hole doesn't need to be big enough for pulling the hen out; coons are strong and would just as soon spend a pleasant night picking a live hen to pieces through the wire as not. We use stout hardware cloth for any openings in our henhouses and coupes—windows, drop-through floors—while we cover the larger broiler tractors with chicken wire, the heaviest we can find, over welded stock panel sides.

Standard advice is to provide a nesting box per so many hens; it is supposed to have a small opening to keep the interior dark and be set thus far above the ground, and while our own nesting boxes are more or less consistent with this picture, we know lots of farms where the chickens lay in an open box of hay or a milk crate on its side on the henhouse floor, and these seem to work just fine—not to mention places like the space between two bales of hay in the loft, on the dirt between joists under the barn floor, or squeezed behind the stock panels leaning up against the back barn wall. Hens have their own ideas about privacy, and convincing free-range birds to lay exclusively in the nesting boxes is an ongoing, and possibly futile, exercise.

After a foray into wide-open circulation vents in winter henhouses, we're working more with wintering hens in garden tunnels, since these are the birds that lay best for us in the cold season. Ten-foot lengths of plastic conduit making low hoops over empty garden beds, and covered with 6-mil plastic sheeting, make good temporary shelters with a maximum of light and a reasonable degree of retained and solar heating. After all, we grow winter vegetables in just such hoops. Our raised beds are 40 feet long by 30 inches wide, and two may be covered by a single tunnel, making the covered space about 200 square feet. We have no interest in crowding the birds, so we only confine ten or fifteen in a single tunnel. Such compaction as the chickens cause is easily corrected with a broad fork and small cultivator. More troublesome are the sprouting seed grains from feed the birds overlook, but tilled in early these would be a gratis green manure instead of a weed issue. (We have not had a predator problem despite the thin walls of our tunnels, but this is good fortune, not good planning, and it will be more prudent to run one or two hot strands of polywire around the garden at snout level for raccoons and coyotes.)

We are also planning to put up a tunnel for the hens over the Jean Pain mound behind the big barn. The heat of slow-composting wood chips, intended to warm the barn water, might just as well warm our hens' feet and drinking water fountains, we think. Fresh chips added to the top, and the manure and scratching of the chickens, should help the mound itself maintain a higher temperature as well. Warm winter drinking water is hard to keep in front of our birds without recourse to electric fountain heaters, and we have so far held out against these, but we are always looking for new ways—passive solar or compost-generated heat—to provide warm water in the winter.

Ultimately, in our climate, there are going to be weeks when our free-range birds are effectively confined. When snow is thick on the ground they show little interest in coming outside and getting their toes cold; most mornings they huddle for a while on the bare ground under the henhouse, throwing disgusted looks at the frost-bound world, then go inside and sulk. For this part of the year, then, our henhouse is somewhat crowded, and the litter, which in the warm months receives deposits almost exclusively at night, now gets twenty-four-hour use as a privy. This is just the time when no one wants to do extra outside chores, and it is just when we need to be adding new litter to the henhouse on a regular basis. Litter in a crowded house will "cap"—get covered with a layer of manure too solid to scratch through—in a very short time, creating the conditions for parasite transmission, wet feet, and dirty eggs. To make it easier for the chicken-chore person to deal with this promptly, we bag dry autumn leaves and sawdust from a neighboring mill, and store them where they are easily accessible to the henhouse. Baled cedar chips from the farm store are prettier and smell nice, but as far as we can tell, these home-sourced litters are just as effective as the purchased kind and a lot more economical. Especially in the winter, we are also careful to provide a dusting box for our hens, since this natural means of discouraging external parasites would otherwise be unavailable to them in the cold months. We keep it supplied with a combination of diatomaceous earth (DE) and wood ashes and must fill it often since the birds really enjoy a bath. This winter we will be replacing the low-sided pans we have been using with the high-sided boxes recommended by Ussery, to keep more dust in the box.

Feeds and Feeders

How you choose to feed your hens will be a function of *what* you feed them. The usual advice in your how-to-farm book is to follow the commercial model, providing self-serve hopper feeders filled with commercially formulated pellets or crumbles. Easy, quick, and requiring no thought. Also heavily dependent on GM crops, postindustry garbage, and petroleum. Lots of it is medicated, too. The grazier who is taking pains to make a natural forage diet available to his birds will want to offer something better than this. And where commercial feeds are formulated as a whole-diet food—protein, carbs, and minerals in one balanced, monotonous ration, because it's all the birds are going to get—for the homesteader whose birds are 1) not confined but out running around the farm scratching out many of their calories and 2) receiving frequent and generous offerings in their feed trough from the excess products of the farm, there is little need that their supplemental feed should be "balanced." Homestead birds have access to a widely varied diet, so there's a good chance they can do some of the balancing themselves.

Our research into poultry diet composition before the advent of feed mills has revealed some interesting contrasts to today's assumptions. Well into the last century, domestic poultry, like the family pig, were usually fed from the waste and surplus of the household: boiled potato peelings, for example, were considered standard in the poultry ration, as was clabbered skim milk. Meat scraps provided concentrated protein and constituted only a small percentage of overall intake; green food was indispensable, either fresh or dried.

Homegrown Feeds—Greens

Lots of your chicken food is to be found right outside the henhouse door. Grass, obviously; also legumes like clover and alfalfa, which birds will harvest themselves in the summer, and which can be cut and dried as hay, then chopped and fed in winter, either as is or soaked overnight. Any of the leaves you eat yourself are delicious to your flock as well, so for the winter, if your low tunnels are big enough, save the less-choice heads of spinach, lettuce, or kale for the birds; fresh greens are said to go a long way toward boosting your winter egg production. Sprouted grains are a delicacy in the poultry house, and if we are short on greens we will sprout barley (which germinates very quickly) in trays in the basement. When the blades are three or four inches tall, we offer this as part of the birds' grain ration, thus increasing the available nutrients.

Dairy

On the grassfed homestead there is nearly always some form of dairy surplus available for a high-quality, high-protein food in the poultry house. During the spring flush, we are generally drowning in extra milk, even more than we need for the table, the bucket calves, fresh cheeses, and fermented dairy products. We keep a bucket in the basement, into which we pour surplus whey, buttermilk, and skim milk; this clabbers and makes excellent food for chickens, pigs, dogs, cats—almost every animal on the farm. In the fall and winter, hard-cheese making time, there are more gallons of whey and buttermilk. If you are maintaining kefir grains, you'll probably have excess bubbly milk, and this and other surplus fermented dairy products are as good for your hens as they are for you. Don't think that amounts have to be generous; in winter your flock will be smaller at any rate, so these offerings will go further. A quart of milk makes a significant protein supplement for twelve to twenty-five birds.

Grains, Seeds, and Beans

The first thing we think of when we think of chicken feeds is grains, and there is a good reason for this. The "scratch" grains our grandparents fed as the sole supplement to their free-range flocks in summertime were a mix of whatever was homegrown or locally available and cheap, usually containing a good measure of cracked corn with some milo or maize, wheat, and whatever got swept up off the grain room floor. It was heavy in carbohydrates; its protein content might not have come up to the industry minimum of 16 percent for laying rations. Nevertheless, there never seemed to be a shortage of big brown eggs on their farms.

Poultry like grain and other seeds and will eat them whenever they are available. These foods are also easy to store for months, if they are kept dry and pest-free. In any part of the country there are locally raised seed foods, and if you can find a farmer who will sell you a few bushels of his wheat, barley, or oat crop, these can make a good basis for your own scratch mix. (Our poultry mix is a combination of sunflower seeds, millet, wheat, oats, and barley, in a 2:1:2:4:3 ratio. If the sunflower seeds include the hulls, we double the proportion of that ingredient.) You can also grow it yourself. We grow small grains as part of our garden rotations—wheat, barley, oats, and field peas (not a grain, but intercropped with the

Figure 8.9. Curly dock, an edible weed species, has seeds that are attractive to poultry.

oats)—cutting them before the grain is perfectly ripe, drying as for hay, and putting it up loose in the barn. Fed out by the armful on the big barn floor, this is scratched through by the chickens until they have scavenged all the seeds they can find, and when they are done the straw is welcome fodder and bedding in the pigpen. Corn we feed shelled (whole kernel, off the cob), mixed with other grains or thrown down as scratch.

Specialty grains like spelt or quinoa would undoubtedly be fine chicken food, too, if you are able to grow more than you want for human use. Amaranth is high in protein, but with its tiny seeds we would feed it in a trough, not throw it down with the scratch feeds. It is a popular inclusion in garden plots grown for poultry self-harvest, but with its tasty greens we wonder how much of it is allowed to grow to maturity and set seed. We are always inclined to let our livestock self-harvest (does it show?), and we've read of people who plant a garden just for the chickens; however, in our experience even one chicken can destroy a newly planted garden patch in about fifteen and a half minutes, so we have reservations, and that particular project isn't very close to the top of our to-do list.

Peas and beans are protein-packed and easy to grow, and most legumes that may be readily grown in a home garden are more than acceptable in the poultry diet. Lentils are said to be easy to grow and preferred by chickens, turkeys, and pigeons. Larger beans like pintos may be more acceptable if they are first cracked or broken, although our hens don't seem to mind eating them whole. And while we have read that pintos, soybeans, and other kinds of beans, for that matter, should not be fed raw, we have seen opinions to the contrary as well, and our own conclusion, after seeing how our pigs and chickens devour raw beans, is that many things can be included as part of a varied diet that might not work as a primary food source. That said, in winter when the woodstove is lit anyway we throw a pound of pintos in a dry skillet and let it toast overnight before we feed it to the chickens; this may increase its available food value in any case, since beans, like many foods, are more easily assimilated when they are cooked.

Meat

Butchering large animals is usually a cold-weather task and generates lots of meat scraps and offal, a small amount of which, fed daily to our egg layers, replaces the bugs, grubs, and worms that were a principal protein source in the warm months. Since the freezer is running already (actually, since it is in the unheated garage, in the winter it is probably *not* running, but never mind), we package small amounts of miscellaneous scraps, which can be defrosted one at a time and fed out over several days. Our hens thank us with more eggs in the basket. Table scraps and cooking surpluses are always welcome in the poultry house, including extra animal fats left over from cooking—good energy food for keeping warm in the winter.

Vegetables and Roots

Don't overlook the root cellar and dry cave or attic as sources of poultry feed. A beet or mangelwurzel spiked on a nail in the henhouse wall is often recommended for chickens in the winter, for calories, minerals, and entertainment; we sometimes feed beets cooked, as well. The outside leaves of storage cabbages provide vitamins and roughage. We have read that carrots make good hen food, particularly if cooked; potatoes or potato peelings, too. And those gone-by winter squash sending up compelling smells in the dry cellar will provide good food value, being full of vitamins and fat- and protein-rich seeds. While none of these may be a complete diet in itself, they all contribute to a balanced ration.

Feeders

Most of the time, we feed dry grains by scattering them on the ground outside the henhouse, or in the winter, we may throw some on the floor inside to encourage the birds to scratch and turn over their litter. Wet foods, like milk, sprouted or fermented grains, and table scraps, are fed in long troughs made of lengths of 4-inch plastic pipe, split lengthwise, or sections of roof guttering, with wooden half-circles, caulked, at the ends, the whole thing screwed to a couple of flat board feet to prevent tipping. Save the self-feeder for when you go on vacation—not only does it ensure food is always in front of your birds, but in front of all the rats, mice, and sparrows that hang around the place looking for a free meal, as well.

The Sow's Ear Flocks

While we do not claim to have discovered the perfect chicken-raising system, having been locked for over twenty years in an ongoing tension with hens who want to lay under the barn, in the bushes, or anywhere but in the nesting boxes, and day-raid predators which snatch our unsuspecting and not-so-bright biddies while they forage, we do try to bring some common sense to our farming.

It is our principle to avoid GM crops as much as possible, since they are at best an unknown and are by definition blood brothers to Frankenstein. As a starting point, we know our free-range hens have access to an entire farm teeming, for at least seven or eight months of the year, with bugs, leaves, and grubs of every description, and so are in a position to balance their own diets pretty well. Our responsibility is to see that they get enough of all the good stuff so that they have excess energy, above and beyond the needs of health, growth, and natural levels of reproduction, to lay a reasonable number of eggs for our table. So in summertime we supplement their foraging with a half-ration of scratch, or about 2 ounces per bird per day (see page 196 for our scratch recipe), fed in two feedings: in the morning when we let them out of the coop and late afternoon, when we go down to collect eggs—the late-afternoon craw-stuffing gives the birds something to digest overnight. In addition to grain, they receive meat scraps from butchering (nonbird species), clabber or whey from the bucket of waste dairy in the basement, a little bakery waste, and sometimes kitchen scraps.

In the cold months, when we cannot rely on the hens finding anything but grit and a bit of frost-burned clover or an overlooked windfall apple, we pay more attention to the composition of their diet, since it depends almost entirely on us. Our goal is to give them somewhat more than the standard 4 ounces per day per hen recommended in our poultry books, with an emphasis on protein; and, since our henhouse is not heated, we try to provide warm food in a nutritionally available form. The birds still receive our scratch grain mix, but in winter it is usually fermented or sprouted, making it easier to digest. In addition, we offer as much high-protein supplement as we can glean from the farm, principally, as in summer, dairy waste and meat scraps, but also squash seeds and cracked, toasted beans. They get these in two feedings, as in summer, with an additional ration of dry grain thrown on the floor of the coop at sundown, when we shut them away from the prowling fox and give them a couple of hours of artificial light to keep their biological clocks ticking. But we can't say we're overwhelmed with eggs in winter—sometimes we don't get any—and every year we try new dodges in an attempt to up the take. Trying to get eggs twelve months of the year in our climate is fighting nature, in any case; eggs, like practically every other food, are naturally seasonal. In the meantime, we'd rather have fewer eggs from a locally powered system than more using methods and supplements that aren't a part of the local ecology.

We are not poultry feed experts—in fact, we're not experts on anything, unless it is our own farm, at its present state of development—and this book is not going to tell you how to mix feeds of *X* percent protein for the various stages of your birds' development, with sixteen ingredients including specialized mineral additives from some company halfway across the country. There are lots of instructions already available for doing just that; check the Internet. Maybe those methods work really well for many people, but our farm philosophy is that, if we belong here, if our animals belong here, *here* will support us; even in these, the early steps of establishing our farm community as a part of the local ecosystem—the twenty years we've been on this land is the blink of an eye, ecologically speaking—we avoid methods and materials that are dependent on outside inputs. Do we get fewer eggs than if we fed a balanced commercial ration? In all likelihood, yes—but our cost-per-dozen ratio is certainly lower as well; we also have the advantage of knowing *what* we are feeding our birds, where it came from, and that it required a minimum of nonrenewable fuel calories to produce.

How Many Birds Do I Need?

On our farm, we determine our laying flock size based on two things: how many eggs we hope to get, for the table and for hatching, and how much surplus farm-raised food we expect to have that year. It's not calculus; we figure an average laying rate of 50 percent, allowing for out-laying (a chronic issue on a farm where a cozy bush or inviting nook is never more than 20 feet away), aiming for a couple of dozen eggs per day in spring and early summer, tapering in early winter to (we hope) about a dozen. Thirty hens should do it, so we try to have three dozen, maybe, allowing for accidents. Three dozen hens, and three roosters, in a centrally located house, range our half-dozen or so open acres pretty thoroughly, but with enough elbow room that they only create a moonscape of the area right around the chicken house. Manure gets turned over, insects caught, compost scratched in, fruit falls eaten, all the jobs for which we keep scratching birds are accomplished.

Looking, then, at our gardens and dairy, we can form a loose idea of how much food we will provide for the henhouse that year. Only a loose idea, though. Some years, for example, more garden plots will be held fallow in green manures, or cropped for potatoes, than devoted to corn or small grains; if we come into the spring with fewer bred cows, we know the milk supply won't be as generous that year. Weather and other conditions, in any case, will always constitute enough of a surprise to foil any attempt to be exact. We still buy many of the seeds we feed to our poultry, even as we experiment with different chicken-food crops, but we try to make our purchased grains and seeds things we can reasonably hope to grow in our climate, on our farm, in our system, with the goal of eventually replacing the bought stuff with homegrown. And in any case, forage supplemented with milk, garden surplus, and meat scraps goes a long way toward supplying the farm flock's menu during the heavy laying months.

Figure 8.10. Chickens do spring garden cleanup.

Feed requirements are not static over the year. In October we cull all the two-year-olds that look like they're not laying and bump up the feed rations for the rest, hoping to continue getting eggs once the cold weather sets in. Maybe two dozen birds make the

Figure 8.11. Rhode Island Reds at the chicken coupe.

cut, leaving more room in the henhouse, where we cover the windows with plastic sheeting (open eaves provide ventilation) and use a light on a timer to give them a couple of extra hours each day to eat and lay eggs. This used to work pretty well; lately, what with night temperatures well below zero for weeks in winter, not so much. Last winter the hens in the low tunnels were the only ones laying in January and February, which may tell us something about solar energy and dirt being more significant than artificially extended day length, or it may not. At any rate, we'll be putting chickens in tunnels this winter.

It's pretty easy to tell when we have too many chickens! Excessive amounts of purchased supplemental grain being used, and crowded perches in the henhouse. The barnyard scratched completely bare, and hens ranging far up the field into the woods, or onto the neighbors' porch. In our woodland setting, overpopulation has a natural solution—predation—and our friendly neighborhood carnivores are happy to take excess birds off our hands and then come back for more. Better to match our poultry numbers to our farm and keep the farm ecology in balance without their help.

Breeds

We choose our breed or breeds according to what we want our birds to do, where, and on what sort of diet. Looks are secondary, if not tertiary or even downright unimportant. We think a good farm bird ought to scratch a good bit of its feed, lay an egg every day or two pretty steadily from some time in February into November or December (longer would be nice), and keep doing that for at least two or three years; she should also have a carcass worth eating when it's time to hand in her dinner pail. That used to mean Rhode Island Reds, Barred and White Plymouth Rocks, or any of a dozen other breeds.

As their names imply, these breeds generally hail from some *place*, and it's a pretty good bet that they'll do well in a climate similar to that of their place of origin. Some names are misleading: Muscovies are ducks, not chickens, but with that name we assumed they were good in cold weather—remembering Napoleon's march on Moscow—until two winters ago when temperatures dropped to record lows and we lost four. Turns out the name "Muscovy" is a bastardization of "musk duck," or something like that, and they are actually native to South America. Since then we don't make assumptions; we look it up. Oh, and just as a caution: be aware that when a breed description says "good forager," it may mean they'll fly over the fence and rip out half your garden, or roam up into the woods and get eaten by a fox.

Here's the problem you may encounter when you are selecting chicken breeds: as we noted at the beginning of this chapter, hatcheries, in response to market demand, have been emphasizing certain traits and neglecting others, and the dual-purpose bird, while it may lay more eggs now than did its 1950s grandma, by and large makes a poor, thin carcass, nothing like the 12-pound birds our grandparents taught us to butcher. And coming from a long line of progenitors raised primarily on genetically engineered corn and soy, today's birds aren't necessarily good at converting a wide variety of foods into eggs and meat, either, any more than a commercial dairy cow is necessarily going to make a good grass animal. Not to despair, though—we get on as well as we can with what's available, with an eye open at all times for better sources of purchased chicks and the intention of beginning a controlled breeding plan for our own farm just as soon as we can find time and some worthy seed stock. Meanwhile, it may help to remind yourself, when the hens just hang around the barn waiting for handout time instead of decimating the pest-insect population, or when you're looking at the defunct laying hen you've just plucked and wondering whether the scrawny carcass is worth cooking, that you're working with less than ideal materials for which you need—*temporarily*—to make allowances in your ideas of success.

Reproduction

We aren't experts, but it doesn't take a PhD to hatch eggs—after all, even a chicken can do it; sometimes, anyway. If a hen stays on the nest when you come to gather eggs and fluffs up her feathers and pecks at you when you reach under her, she's broody, and maybe if you let her have eight or so eggs to keep she'll set them for three weeks and you'll have some babies. A good hen will be faithful to her eggs, staying on them and turning them at intervals so they'll be warmed evenly and hatch well. If only a few hatch, maybe you let her keep so many that she couldn't cover them all; then when she turned them even the ones she *could* cover were turned out and took chill and died. Or maybe you have a dud rooster, or just not enough roosters for your flock. You can learn to candle eggs, so your hens don't waste time trying to hatch infertile eggs. Or you can select the two best hens and the best rooster in the flock (check the Sustainable Poultry Network website, or the Livestock Conservancy website, for standards) and put them in a pen by themselves, collecting eggs until you have enough for a broody hen or two to hatch for you. Or you can get an automatic incubator, which will hold enough eggs to make it worth the effort of brooding chicks, and make yourself responsible for monitoring the temperature and humidity of the eggs. Not only will you get a better fertility rate with a confined breeding pair, you'll be improving the flock genetics at the same time. We have done, and continue to do, all of these as we find it convenient, or as the girl children on our farm decide it should happen.

Once you have chicks (poults, ducklings, goslings), you need to decide what method you'll use to care for them while they are small. Of course if you hatched them in an incubator, this is your baby, and it'll be up to you to see that the little ones are warm and fed and dry. There are lots of good sources for information on starting hatchlings, and we won't repeat here all the details of temperature, percent protein rations, and amount of floor space; we recommend Salatin's book *Pastured Poultry Profit$* as a guide, but there are lots of others out there. If you've used a hen to hatch your eggs she's going to want to keep her chicks (or whatever—our hens are assigned to hatch duck eggs as well), and you'll have to decide whether to let her. Maybe a hen smart enough to hatch eggs *ought* to be smart enough to raise a flock of chicks, but on our farm she's generally no match for all the lurking carnivores for which a fluffy chick is just an hors d'oeuvre. Too many times have we watched a proud mother hen make her way out into the pasture with seven or eight little fluff balls following her, only to have her march back in the afternoon a chick or two shy—and so on, over the next couple of days, until her goodly hatch is sadly reduced. Consequently, we either abstract her chicks shortly after she hatches them, or we pop her and her chicks into one of our smaller-sized tractors and keep them there until the chicks are several weeks old.

Broilers

There are lots of places you can go to learn to raise broilers; again, we recommend *Pastured Poultry Profit$*, which contains lots of information as applicable to the small farm flock as to small-scale commercial production. Salatin's methods are largely consistent with homestead aims: the improvement of pastures and forage; avoidance of manure concentrations; availability of natural, farm-raised green food and insects; nonmedicated health. Our personal experience provides us with a few rules of thumb of our own devising.

Don't put the chicks' water bowl in the corner of the brooder: they'll all crowd in there and get wet. When someone forgets and you find at your last check before bedtime that you have a brooder full of wet, shivering chicks, a blow dryer, the kind they make for your hair, does wonders, drying and warming the little guys thoroughly, so when you finally do get to bed that night you can sleep without worrying. And get the little guys out on grass as soon as you can but not so early they're going be cold at night; we've seen them pile up for warmth and smother the ones on the bottom.

CRXs are fat, ungainly birds, putting on weight at a morbid rate, and incapable of even the simplest life-saving maneuvers, like getting up when they've tipped over. On our sloping acres we have to keep this clearly in mind, because if a broiler ends up on his side or his back, he'll probably be dead before we find him. We've learned to keep our broiler tractors on level ground, or what passes for level ground on our farm. Likewise, these fat birds will overheat very quickly if deprived of water on a hot day, even for short time, like two hours; there are few things more discouraging than composting a couple of dozen 8-pound broilers three days before they were scheduled to go in the freezer.

And broilers are subject to some weird conditions. Sometimes, just to keep things interesting, you'll get a bird, or a whole bunch of birds, that look normal, but when you butcher them you find breast (or other muscle) tissue and interstitial fluids of a bright, grass green. Not a little, but a whole bunch. Lots. Gave us a shock the first time we saw it. Typically this discoloration happens in the lower muscle mass of the breast meat, close to the bone, but we have seen it in the legs as well, and we have cut open birds for evisceration that were full of green juice. When this first showed up on our farm, twenty years ago, we couldn't find any resource to identify it. We naturally assumed it was our fault, somehow. Later a few references to the condition turned up online, and today a computer search gets lots of hits. Green muscle disease, as it is creatively named, is the natural result of breeding animals with breast muscles so large that when they flap their wings they restrict the blood supply to the lower-lying breast muscles, resulting in atrophy and necrosis—all that green stuff. Gross. Not to worry; commercial growers are working on a solution: a list of seventeen protocols to stop the birds flapping their wings. Homesteaders who don't want to put their chickens in straitjackets *could* just go back to raising the traditional chicken breeds.

Turkeys, Peacocks, and Guineas

As scratching birds, guineas, peafowl, and turkeys belong in this section with the chickens, and they can be just as useful in farm cleanup as their smaller relatives, or rather more so, being larger and stronger. We like them, too, because they are so big the local predators make a detour around them; at any rate the larger birds generally roost high in trees, where only a suicidal raccoon is going to invite a tussle. They are often better at raising a family than our dual-purpose chicken breeds, but once chicks are hatched the new family may need to be shut in the barn for the first week or two, until the babies' wings are strong enough to get them to an elevated roost. More than once we have had a peahen settle down on a fence rail—the highest point her new chicks could reach—and tuck all her babies away under her feathers for the night, only to find her in the morning a couple of chicks shy. On such a low perch, a coon or possum can just reach *gently*

Figure 8.12. Peacocks earn their place on the farm by sheer beauty.

Figure 8.13. Squash bugs have not troubled our garden since we acquired guineas.

underneath and snag a tender little snack without unduly disturbing Mama.

Both turkeys and peafowl have been completely compatible with our chicken flocks; one tom turkey was, if anything, a little overprotective of the smaller birds, keeping a suspicious eye even on us when we came into his barnyard. Not so our guineas. Yes, one or two guineas seem to get along fine among our assorted poultry, but when the population is larger, tensions develop. When we were keeping ten or a dozen guineas on the home farm, they would refuse to give the chickens access to the feed trough, or all of them would corner a peacock in the barn or woodshed and chatter him into fits. Some of us are prepared to swear the guineas lured peacocks out on the county road into the path of traffic.

Still, we wouldn't be without a pair or so of guineas wherever we have a garden; while chickens may be incompatible with little green sprouts, guineas, although they will peck a tomato—or two, or three—earn their keep and then some, by eating cabbage loopers and squash bugs. Anything that eats squash bugs is our friend, even if it does look like a spotted rugby ball with a zombie head.

Ducks and Geese

We keep ducks to help spread cow manure (they'll bill through cow pies looking for the good stuff) and to eat slugs and the crawdads that make holes in the farm pond dam and, especially, because we like them. No other bird we have seen grows quite as fast as a White Pekin duck, and if you hatch them yourself you might just as well give up raising broiler chickens and switch to ducklings, except that ducks are harder to pick (pluck), and their meat is all dark. Some of us prefer dark meat anyway and would happily replace CRXs with Pekins, but the balance holds out, since they like white meat best and hate picking duck feathers. Ducks used to be common in the farm flock (and on the farm dinner table), and we suspect the modern preference for chicken is just good salesmanship on the part of the poultry industry, for whom land birds are an easier commodity to raise in confinement than are ducks.

While, generally speaking, a duck's laying season is shorter than a chicken's, ours lay just as reliably in spring and summer as our hens, or even more so, and their eggs are substantially larger, with thick, waxy shells, crystalline whites, and deep orange yolks, as beautiful as they are delicious. Pekins are usually poor sitters, so we put Pekin eggs under other duck breeds, or under hens (which can't count, apparently, and will remain on the eggs the extra week necessary to hatch ducklings). Muscovies are good for brooding, although, as noted before, we found them too cold sensitive for our climate; Cayugas are also reliable brooders, and better adapted to our cold winters than the Muscovies, so they have replaced the former on our farm. Khaki Campbells are excellent layers, and we'll be adding them to our breeding flock next spring. We either take the ducklings to raise ourselves or put mother and offspring in a tractor for several weeks after hatching; the local hawks are very fond of duckling, and we have witnessed them carrying off birds weighing a couple of pounds.

Geese are unusual among farm poultry in being true grazing birds, capable of deriving the majority of their living from forage. This should make them a natural for the grassfed homestead, but we can't tell you from personal experience, because we don't have any yet: good breeding stock is not to be found just anywhere, and geese are a lot more expensive than chicks or ducklings. We don't intend to keep more than one or two breeding pairs, not wanting to put too much pressure on the home pastures, which have plenty of the clover and tender green grasses geese prefer, and which we want to keep that way. Geese grow quickly, and standard meat breeds may be over 10 pounds in as many weeks; butchering surplus birds at this point would reduce the flock to just breeding animals well before we need to start providing them a winter ration of clover hay and grains. Goose eggs are huge, and we don't know anyone who eats them, but that's not to say that it wouldn't be practical.

CHAPTER NINE

The Pig

All animals are equal, but some animals are more equal than others.
—*George Orwell,* Animal Farm

In a charcuterie-loving society like ours it may sound like a silly question with an obvious answer—nevertheless, we ask: why should the small-scale sustainable farm include pigs? After all, cows eat grass and chickens eat maggots—things we can't, or would rather not, eat—but pigs eat what *we* eat; which begs the question, why would we feed it to a pig? It seems to validate one of the vegan objections to meat consumption: that, on a planet where millions of human beings hover on the brink of starvation, meat eating is a decadent luxury. After all, if soil is plowed and fertilized and drilled, crops are harrowed and sprayed and combined, converted to commercial pig pellets and hauled to market, thence to the factory farm, in order that hogs can be bred and fed and slaughtered, so that I can have my Sunday slice of bacon, the energy use *might* be called into question. Even if there is nothing *wrong* with doing this, it is without doubt a process that uses a lot of energy. So why should the homesteader raise pigs?

Figure 9.1. Pigs on cleanup duty.

Piggy Banks

Fortunately, there is an answer: pigs have a place on the independent farmstead because they are the monarchs of waste repurposing. They can be fed from the same farm activities that produce food for other species, but they are less selective. They eat parts and conditions no one else will eat: overripe, squashed tomatoes; partly fermented fruits; stems and vines; hulls and cores; bones and offal and eggshells—all the detritus of farming for food is acceptable at the pig table. The by-products of our home food processing—fermenting, canning, freezing—are pig foods as well: whey and buttermilk, apple pomace, brewer's grains, and primary mash from wine making, all are delectable pig cuisine. Pasture, forage and fodder, hay and silage are food for pigs, as well as for ruminants. Once upon a time, a farmer didn't keep pigs just because there was a market for pork; he raised pork, whether for the table or for the market, because farming produces wastes and perishable surpluses that can best be utilized, converted, and stored, with an increase in value, by running them through the guts of a pig.

This oversupply of farm nutrients happens largely because, built into any serious effort to raise our own food, is the provision of surplus. USDA crop insurance notwithstanding, the only real insurance in food production is planting a surplus of whatever it is we need, *and* backing it up with an alternate crop, in case the first one fails completely. If potatoes, for example, are necessary to our table, we plant a lot of them; then, since maybe the potato bugs or late blight will be very bad this year, we plant a big patch of corn as well. If we are looking forward to pie this winter, and who isn't, we determine to put up lots of apples, but in case of late frost or codling moth, we don't forget to plant plenty of pumpkins as well. We don't know in advance what crops will do well in a given year, so we hedge our bets—and most years we get more than we need. What to do with the surplus? Before the pumpkins freeze on the barn floor, the cabbages begin making a discouraging smell at the bottom of the garden, the apple mash has drawn so many wasps to the compost pile that we can't get at it to turn it—we feed all of these things to the pig.

Niche

The domestic pig—*Sus scrofa domesticus*—is an oddity: although it belongs to the family of artiodactyls, which includes mostly herbivores like cows, sheep, and goats, the pig is an omnivore, a hunter and a scavenger, which in the wild will eat not only grass and roots and fruits and nuts but small animals and carrion. And in addition to eating almost anything, a pig will eat almost any *amount* of anything, matching its intake to meet supply: it can gorge when there is plenty, speeding up growth while the table is loaded, and then slow down when there's less to be had, while still maintaining good health. A pig also likes to dig, wallow in mudholes, and scavenge in litter and manure piles. These characteristics make the domestic pig an especially versatile member of the natural and farm ecosystem, and its presence on the independent farmstead is a game changer.

Pigs are large animals—a mature boar can weigh 1,000 pounds or more—that can live for ten or

Figure 9.2. Three-week-old piglets.

more years, producing one or two litters of seven to twelve or so small (1 pound) piglets per year. Since pigs reach sexual maturity before they are twelve months old, their rate of reproduction is very rapid. Given extended access to a plot of ground they will plow it up and make mudholes; managed with care, pigs can be valuable assistants in the brush-clearing, earthmoving tasks on a homestead. They have heavy impact on pasture, not only because their considerable weight is supported on relatively small, sharp hooves, but because of their strong snouts and their propensity for rooting in the soil.

Many Models of Pig Raising

Pigs are as good an example as we've come across of how different husbandry methods are appropriate to different goals, and in different climates, seasons, and situations. Take the pastoral pigs of Joel Salatin, grain-supplemented hogs used to clear land or clean up a forest understory, or his "pig-aerators," feeder pigs that spend a couple of months in the spring in the cowhouse, turning over deep-packed cow bedding for fast decomposition. The grass-hay-and-dairy-fed hogs at Sugar Mountain Farm in Vermont spend their whole lives on pig pasture, without commercial or grain supplement. In Virginia, John Cuddeback fattens his pigs in the woods, on acorns. And our own pigs move around the farm according to season, in the pasture or barn as weather and forage indicate, enjoying grass, hay, and other forages; fodder crops grown in our large family gardens; and many, many gallons of waste dairy from our grassfed organic Jersey cows.

Needs

The first responsibility of the farmer who raises pigs is to provide fencing, whether for confinement or rotational grazing, in order to control and direct the pigs' impact and to prevent these strong, omnivorous, and potentially even dangerous animals from escaping to damage your farm, or your neighbor's—or even going feral. Not only feed provision, but feed *rationing* is also an important part of the job of keeping pigs, since a pig's appetite is far in excess of what is needed for growth. Depending on a variety of factors, the homesteader can expect his pig to get only a small part, a considerable portion, or virtually all of its calories from foraging; since the diversified homestead provides food from a number of sources aside from grass, under ideal circumstances, the pig can get its whole living from the farm. *Timing* of different feed sources is especially significant: as the pig is the consumer of many different farm products, knowing when to feed which (perishables before long-storage crops) and understanding seasonality (utilizing a crop when it is most palatable and digestible) are part of efficient homestead planning.

Pigs need some shelter from excessive heat and should have a relatively dry, draft-free sleeping area for cold weather; this can be as simple as a shed roof, a corner of the barn, or just a nest in a haystack or waste round bale. The farmer is also responsible for timing breeding, and if a boar is not present, the farmer must provide for artificial insemination (AI).

Surplus and Services

Given the reproduction rate of hogs, where a breeding sow is kept, there are almost certain to be excess piglets. These may be sold young, as feeder pigs or, if of a certain quality, as breeding stock. And while it is desirable to match your number of feeder pigs to your quantity of perishable surplus, and to the number of mouths in the farm family/staff, still, in a flush year, when there is an excess of garden, orchard, or dairy waste, there may be surplus pork to sell as well.

The pig's omnivorous tastes, combined with its love of rooting and digging, make it useful in other arenas: for example, that of compost making. Pigs can be employed as four-legged plows when a plot of ground is to be broken for a new garden. Held for an extended period on a limited area, they will root it over and eat or destroy every plant smaller than a tree; in some cases, they can even be induced to help clear forested land. In the barn, pigpen bedding can be made to include almost any variety of plant materials; chewed, turned, manured, and urinated on, this, when fully composted, is invaluable garden fertilizer.

Pig Husbandry

Because our pig-feeding decisions will have a significant effect on how we satisfy our pigs' other needs, we consider their diet first, before issues of housing, fencing, or breeding. The homesteader who wants to raise pork for his family should design his methods and materials to match his circumstances; that is, our pig-rearing protocols will be dictated by what we intend to feed our pigs.

Modes of Pig Keeping: Pastured, Pastoral, and Penned

Nowhere have we encountered more variety of opinion about an animal species or its needs than in researching what is possible in the way of pig nutrition! What one farmer is certain can't be done, another farmer swears he is doing with great success. One reason for this is the enormous variety of foods from which a pig can derive benefit. Although pigs are omnivores, they can live on an exclusively vegetarian diet. This is usually grain—although under such circumstances close attention must be paid to the balance of nutrients, and especially proteins, they receive. But grain is one thing, and grass is another. Pigs are not ruminants and do not have a ruminant's multichambered stomach for the digestion of cellulose—which prompts the question: *can* pigs derive their sole nourishment from forage? The answer, it would appear, varies with the pig, the farmer, and, in all likelihood, the farm.

Pastured

Until a few years ago, our observation of pastured hog programs had led us to conclude that while pigs like being in the field, and can derive considerable benefit from being there, they are not able to satisfy all their nutritional needs on forage alone. Indeed, every "pastured pig" farm we encountered turned out on investigation to be some form of grain-feeding operation, whether the grain was farm-raised cob corn, commercial pig ration, or spent brewer's mash. The "pastured" part actually referred primarily to where the pig slept and pooped. We concluded, for the time being, that an all-pasture feed regimen didn't work for pigs.

Today, on the contrary, we know of several experienced graziers whose pigs spend virtually their whole lives (from birth or weaning, until they are butchered at between eight and twelve months; for breeding sows, a decade or longer) on pasture, with minimal or even no supplemental calories. Grass and other forages—with some roots, tree nuts and fruits, insects, and any slow or incautious small animals that come in their way—make up most or all of these pigs' intake, and they grow well, if somewhat more slowly than their supplemented brethren, with vigorous health and, possibly, a great savings for the farmer. So depending on a number of factors, pasture can supply part, most, or all of a pig's nutritional needs, at least seasonally.

Figure 9.3. These pigs, penned in woods, are supplemented with a self-feeder. Courtesy of Molly Jameson.

Pastoral

Forest-fed, or pastoral, pigs are as old as pig husbandry. As such, they were long a valuable element in many forest permacultures, having the ability to convert nuts, seeds, fruits, and leaf mast—products of the solar harvest available seasonally in enormous quantities—into firm, solid pig flesh. With portable electric fencing this practice is being reintroduced now in North America, while in Europe it is a continuous cultural tradition dating back beyond memory. In Virginia, university professor John Cuddeback fattens hogs on acorns in a few acres of oak trees, and outside New Castle, Pennsylvania, Michael Kovach uses hogs to harvest apples in his woodlot. On our own steep acres we are not yet equipped to take pigs into the woods, though; we are concerned for the integrity of the hillsides. We do run the pigs under our pasture shade trees—apple and mulberry—for calories and cleanup, or we bring fruit falls to the pigpen; fall fattening of feeder pigs in the orchard or tree lot is a good permutation of the farm ecology.

Penned

On the diverse farm, where surplus or waste nutrients are available in quantity, the advantage is not necessarily on the side of using pasture alone to feed pigs. In determining just how "pastured" our pigs should be, we keep in mind that in nature pigs are scavengers and opportunists, taking advantage of whatever nutrients are available at the moment. While chickens and other poultry will often eat from the same food sources as pigs—after all, both are omnivores—the homestead frequently produces surplus at a rate that far exceeds what any but a *very* large flock of birds could consume: gallons of buttermilk, skim milk, and whey; bucket after bucket of cores, seeds, and trimmings from canning or fermenting; bushels of vegetables and fruit not suitable for storage; and many pounds of butchering

Figure 9.4. Supplementing pigs on pasture can be a challenge when all the pasture mates want a bite, too.

offal. Here the homestead pig steps up to serve the farm in his most irreplaceable capacity, as the converter par excellence of things that would otherwise spoil in short order into sound pig flesh and baby pigs.

For the small homestead, where quantities are manageable and distances are short, a pig does not have to have the run of the pasture to be grassfed. There is nothing wrong with bringing the forage to the pig, rather than the pig to the forage. Where, and how, we choose to keep pigs will arise out of present circumstances, and what, where, and how much farm-raised resources are available at any given time. Turned for a period into the garden to self-harvest late corn and beans; put over a rough pasture with, or behind, the beef steers to knock down or plow up perennial weeds; or penned in the fall to be fed the surplus of a good harvest of pumpkins, turnips, fodder beets, and corn—and the whey, skim, and buttermilk from the fall dairying—the family porker will go from place to place, moved here and there as its services are most needed, or its needs best met.

Even when it is in the field, a pig's utilization of pasture as a food source automatically drops in correspondence to the availability of more concentrated nutrient sources. In other words, pigs forage less the more other things are available. The homesteader's job is to orchestrate the flow of energy on the farm, to plug in available nutrients where they are best utilized at any given time. Proper assignment of food resources means that pigs will receive as their principle source of calories whatever wastes or surpluses are most available at the moment, or

Forage, Fodder, and Feed

For our research into almost any aspect of animal husbandry, the online agriculture library CHLA (Core Historical Literature of Agriculture, a subsidiary of the Albert R. Mann Library at Cornell University) has been a fascinating resource. The books in this digitized library go back as far as the 1620s, with titles like *Feeding Farm Animals* (Thomas Shaw, 1908), and a quick browse is enough to show that what farm animals can, should, or do eat is as varied as the farms that raise them. Pigs lead the pack. We are especially interested in those sources that describe seasonal feeding practices, like clover pasture in the summer, orchard and field surplus and spoilage in the fall, and "succulents" (wet storage foods like roots and tubers) in the winter, and in the encomiums on fermented feeds like soaked grains and chopped roots. We highly recommend a look through these resources, which we find endlessly useful.

most perishable, be that pasture, hay, dairy, garden, orchard, or offal.

Pig Crops—Grass and Other Forages

On the regenerative farm, there are multiple vectors for directing solar energy into livestock; pasture grasses convert much, but there are many nutrient-rich forages to be intercropped in the pasture or sown in the garden as well. Some of our cover crops and green manures double as pig feed—oats and field peas, after they have smothered weeds in the spring and their roots have enriched the garden soil with nitrogen, may be cut and raked for pig fodder. Field peas or soy beans sown under the corn crop provide protein to pigs cleaning up the patch after the corn is harvested. In addition, between the beds in our home gardens we have low-growing white clover and taller red clover to provide a living mulch in the paths, discouraging weeds and at the same time adding carbon and nitrogen to the soil; this, when it grows tall, we cut or pull—leaving the lower plant intact—and feed to the young pigs. A couple of rows harvested per day goes a long way to provide a feeder pig with green food. Weeds and grass mowed or pulled from the perimeter of the garden and from the hillsides and fencerows have only to be moved a short distance to be fed to the pigs.

Plant pig food. This is simple; it need not mean opening up more of your ground to tillage. Empty places in the garden—earth that needs to be covered with plant material to prevent erosion and chemical oxidation—can be maintained and kept in service by sowing it with crops intended especially for pig food. Mangelwurzels fill one space in our garden rotation, large fodder beets that have produced as much as 3,000 pounds on 1/10 of an acre. When potatoes are harvested in July, the spaces emptied are planted to turnips, a fast-growing pig food that has reliably produced for us upward of a ton of long-storing roots most summers on less than 1/10 of an acre. By the time frost puts a stop to growth, the turnips have attained a respectable size. Roots are pulled, topped, and bagged and stored either in the root cellar or the "cave," our dry storage cellar. Fed out at the rate of 20 or 30 pounds a day, these substitute for much of a pig's standard daily ration; supplemented with hay and dairy waste, they constitute a complete farm-grown menu.

Following early vegetables, in spaces where we don't want a root crop, we sow beans, which generally germinate well even in hot July soil. Some we cut green and feed to the pigs in October, harvesting daily only what will be eaten immediately; the rest we allow to mature for the human food harvest, and after threshing, feed the haulms to the pigs. Alternatively, we may turn the pigs into small sections of the garden and allow them to self-harvest. The heritage corn we plant for the table and the freezer makes an excellent field corn, so we plant a lot, and what isn't needed for human use we either feed green, or allow to dry in the field, then cut for winter pig feed. The pigs eat the grain and much of the leaf, shredding the stalks, which are thereby added to the pig bedding. Squash and pumpkins are staples in our garden and make good pig food, so we feed damaged or underripe fruits that won't store well at harvest time, then again periodically during the winter, as we cull any fruits that are deteriorating. Crucifers like cabbage and kale are favored in the pigpen, too, as are waste potatoes, Jerusalem artichokes, sweet potatoes, and so on.

Figure 9.5. Mangelwurzels, homegrown pig food.

Pasture improvements can be planned with pigs in mind. Sugar Mountain Farm practices pasture cropping—planting a harvestable crop to grow up through a perennial pasture—which allows the smallholder to provide food crops like small grains, corn, tillage radishes, crucifers (broccoli, cauliflower, cabbage, kale), sunflowers, and Jerusalem artichokes, which pastured pigs can self-harvest. With judicious rotation these may be grazed lightly and allowed to regrow, providing for more than one harvest. Broadcast sowing these crops in the pigs' last few days on a paddock means seeds fall on disturbed ground where they are driven into the soil by the animals' feet. This year we came behind our pigs with red clover seed, at once to cover the soil laid bare by their rooting and to increase the percentage of legumes in our home pasture.

Pigs and Dairy

It would be hard to overstate the value of dairy products as pig food, or of pigs for utilization and storage of dairy nutrients. Next to the cow, the pig is the dairy woman's best friend; anyone who keeps a dairy needs a pig to turn surplus milk, whey, buttermilk, and skim milk, not to mention the occasional cheese making failure, into bacon, ham, and pork chops. Dairy products also correct the primary dietary shortfall of pigs on pasture, the amino acid lysine. Where there is a ruminant converting forage into milk on a daily basis, there should be a pig to store some of this for future use, and in the most delicious forms (find more on pigs and dairy in chapter 10).

Other Surpluses

Pigs especially like wet, sloppy foods (exactly the reverse of the dry pellets or powdery mash that are the usual commercial offerings in the pigpen);

Figure 9.6. Pomace from cider or wine making is good supplement in the pigpen.

remember Wilbur, the pig who loved to feel the buttermilk and bran slops running down behind his ears when Lurvy filled the pig trough? E. B. White clearly knew his pigs; ours always want to get their heads in the way as we pour in a bucket of slurry. When we feed shelled or threshed grain, as we still must do at times, we mix it with dairy wastes, or, lacking these, the rinse water from the milk pails and jars, water from cooking or blanching vegetables, the soaking water from a baking pan, or all of them combined. A bucket under our kitchen sink catches all these things and more, to be carried daily to the pigpen to add excitement to whatever meal they are enjoying.

Kitchen scraps, instead of going on the compost pile, are welcome as a treat in the pigpen; nearly anything, including vegetable skins, seeds and cores, bones and meat scraps (excepting pork), leftovers not good enough for human consumption, and dregs of almost anything are good pig food. On our farm virtually the only kitchen wastes that don't end up in the pigpen are coffee grounds, tea leaves, and citrus peels. Offal is also good food for pigs, and when we butcher poultry the guts, as well as the heads and feet (when not wanted for human food) are cooked afterward in the scalding water and fed to the pigs. They could be fed raw, but we don't want to encourage predation in our pigs; as it is, a chicken in the pigpen is never quite safe.

Fermentation wastes are a favorite with pigs. The fall cider making always leaves a lot of good pomace for the pigpen; windfall and wormy apples, and the cores and peels of the apples we dry, freeze, or can, are prime pig treats. Our wine making generates primary mash, and brewer's mash also has lots of valuable nutrients for pig use; friends in town who brew call us when they have a bucket of spent grains. And an overfermented batch of kimchee, pickles, or the soft layer on top of the sauerkraut crock are all haute cuisine in the pigpen.

Commercial Feeds

"Pig feed," as in a single formulated, balanced dry ration, is a modern notion developed for pigs being raised for the market. It stores easily, can be loaded into a hopper feeder for automatic feeding on demand, and is designed to avoid possible nutritional deficiencies, a perfect fit for the commercial model. If one's goal is to grow pigs big and fast, commercial feeding is the easiest answer. But when pigs are raised for home use, their diet is worked out from whatever perishable nutrients need to be consumed at the moment. Our own pig-rearing practices continue to develop out of our experiences from day to day, feeding pigs what comes available, and out of our ongoing research into what pigs used to eat, before feed mills were invented.

Don't get us wrong; factory formulated feeds have some real conveniences. When home-produced hog feed is in short supply, some purchased grain can be a blessing, even a necessity.

But today, such feeds will generally also be composed in large part of GM corn and soybeans, produced with a great deal of petroleum, and they are not cheap. For animals you intend to breed, there is research indicating that GM food may mean fertility issues as well. To work out your pig feeding strategies, look around the farm, and consider where your resources are at any given point in the calendar; on the regenerative smallholding, sufficient solar-generated surplus should be available to make purchased, petroleum-produced supplementary feeds unnecessary for much, or all, of the year.

Figure 9.7. Penned pigs are available to utilize whatever surplus nutrients the farm is producing.

Keeping It All Together—Shelter and Fence

Every animal on the farm has its own set of preferences or requirements for confinement, shelter, and comfort. Pigs are no exception.

Fencing

Portable electric fence is just as valuable for containing pigs as for ruminants and poultry, but there are some differences in method. Extra ground rods on the charger can help ensure that the snap running through the fence is a strong one. And since electric fence usually does the most good when it is hung at nose height to the animal being confined, pig fence is lower than fences for cows. The lowest strand needs to be quite low; if the first contact with the fence hits the pig behind the ears, he is more likely to push forward than to back up! Shifting paddocks before the pigs have rooted the earth heavily helps avoid grounding the fence with heaped-up soil. Depending on what kind of perimeter fence you have—if you have one—you may want a backup fence of hog wire or livestock panels around your pig paddock while you are training them to respect hot wire. And the best insurance that your pastured animals will stay inside the fence is, as always, to make sure they have what they want on their own side of the charge.

Shelter

What makes adequate pig shelter varies with the seasons. The pig's traditional mode of cooling off is the iconic pig wallow, but for those of us who object to part of the pasture being turned into a mudhole, it will be important to provide alternatives. A loop of fence thrown into the tree line on hot days gives pigs a chance to get out of the sun, an important consideration for animals that don't sweat; pigs, like people, can die of overheating. Even so, our pigs think fast and creatively, and in twenty-four hours (our usual paddock duration) our pigs are perfectly capable of chewing the pig nipple that regulates their drinking water until they've created a mud puddle.

Cold is less of a problem. Shelter for pigs on pasture is often managed with movable hutches or houses. These need to be strong, or the pigs will destroy them. A sturdy wooden three-sided shed can be built on skids and set up backing to the wind for dry winter quarters, with a bale of straw broken inside and renewed as it is eaten, scattered, or grows wet. For young feeder pigs, we use an intermediate bulk container (IBC) set over on its side, with a pig-size opening cut in what was the top (now the side) and a bale of straw broken into it; this draft-free arrangement works well, again provided the bedding is renewed when necessary. On one farm we know in southern Ohio, pastured pigs farrow year-round in nests dug out of waste round bales; in the winter, when the sow and piglets bury themselves deep in composting hay, clouds of steam rising off the bales are the only sign the animals are there. The Jeffrieses of Sugar Mountain Farm report that their pigs, although provided with sheds, often prefer bunking in the open even in subzero temperatures, so long as there is deep bedding—hay, straw, or wood chips—and no wet precipitation.

The Pigpen

If you are going to keep pigs into the winter months and plan to shelter them under cover, remember: a pigpen has to be sturdy! Nonelectric fences, walls, and gates intended to confine pigs must be built with a pig's strength in mind. It is common, in pens with a dirt floor, to use woven wire walls that extend a couple of feet below the ground, or a poured-concrete footer under the wall; with solid floors, on the other hand, digging out is not a problem. On our farm, where pigpen floors are made of weathered but sound railroad ties, we have found that

Figure 9.8. Even a big pig may be more agile than you would think. Courtesy of Nancy H. Chubb.

chain-link storm fence hung on 8-inch black locust posts set deep, with the lower edge of the fence secured under a 2-by-8 spiked to the pigpen floor, works well to contain the breeding sow, Porca, a big girl of some 350 pounds. And since deep bedding makes for comfortable, contented pigs, we make our walls high, in order to allow for substantial bedding buildup. Gates are of solid wood, with big, strong hinges and sturdy bolts for closure, hung close to the floor so there is no purchase underneath for a pig snout. And we make sure the top bar of our fences or walls can withstand the weight of a companionable porker; big as they are, it's no trick for them to rear up on hind legs and lean on the fence.

The size of your pigpen is going to depend on how you intend to use it. Standard pig care books will tell you a pig needs a certain number of square feet of room at a minimum (opinions vary); in reality, of course, since pigs of different ages, sexes, and breeds are not all the same size, a one-size-fits-all answer is too simple. Pigs need enough room so

Pig Manure

One of the great benefits of penning homestead pigs at least part of the time is the tons of nitrogen-enriched, aerated bedding they generate. On the independent farmstead, where most of the farm animals self-harvest their calories and spread all their nutrient-rich manure in the pasture, a source of natural farm-raised fertility that can be spread on the garden is almost a necessity. Unless we are going to double the land we have in cultivation, and put half of it every year into green manure crops, some members of the farm community must be producing manure and bedding for the return of nitrogen and minerals to an intensive garden in which we raise most or all of our supplemental pig food.

they can establish designated areas for sleeping, eating, and defecating; despite their reputation—well earned—for loving mud and dirt, pigs prefer to keep their domestic setting orderly. So instead of hard-and-fast rules, we consider: How many pigs are we going to keep? When are they going to be in this pigpen, and for how long—at what season of the year, at what time in their life cycle? Will they have access to an outdoor paddock part, most, or all of the time?

Our own pigpens are strong enclosures—overbuilt, even—8 to 10 feet on a side. Three are in our various barns, the fourth is open-sided but roofed and adjoins one of the indoor pens. One has access to a large hillside pasture fenced with hog wire; the others open into our main pasture, which is fenced with five strands of barbed wire (and is not pigproof). We use them in a variety of ways: sometimes as a piggy suite, with two pens run together, as when the boar is cohabiting with Porca at breeding time or, later, as a maternity ward, when Mama is nursing piglets. Sometimes we use them singly, for one or two feeder hogs, or for an entire litter of weanling piglets that needs to keep warm by snuggling up together in deep bedding. An isolated pen allows us to quarantine new animals and could double as an infirmary pen if we needed to provide peace and quiet for an injured or sick pig. How many pigs go in a given pen varies, depending on how big the pigs are, and what their relationship is to one another: a litter of piglets with a sow, or a litter of weanlings, will bunk together while babies are small; breeding animals, on the other hand, may be kept separate much of the year.

Bedding

Like most livestock, pigs are hardy animals, not much disturbed by variations in weather, but in cold, wet, or wind they need a good bed to get cozy in. Wherever your pigs spend the winter, in shed, barn, or pasture, deep bedding is their best provision for comfort. The idea of "dry" bedding needs some revision when applied to hogs, which, we have been told, will actually urinate on bedding to add nitrogen, hastening the decomposition process and the warmth it generates. But wet, soppy bedding is only fit for the compost bin and should be changed or renewed until the end result is relatively dry and somewhat fluffy. Straw or hay put in our pigpen is food as much as it is bedclothes, so we expect a good bit to be eaten and apply bedding generously. Other sources of carbon may be more available for you: dry leaves and sawdust are good bedding, if there is enough, and even shredded paper should provide insulation in cold weather and the carbon necessary to soak up nitrogen before it evaporates or washes

Figure 9.9. Straw and shucks make good nursery bedding.

Figure 9.10. Piglets eat what their mothers eat; make sure your breeding stock suits your farming style.

away. We tried wood chips one winter—free from a local tree service—but found them liable to pack down, cap with manure, and retain moisture, and later they were slow to break down even in the compost bin; since then we've stuck to hay, straw, leaves, haulms, shucks, and other such farm-generated carbon.

Troughs and Feeders

We hand-feed our pigs, which does not mean that they eat from our hands but that their feed is rationed to them in two or more meals per day. We don't use an automatic feeder ("self-feeder" or "hopper-feeder"); instead, we feed in sturdy troughs lag-screwed to the floor in our various pens. These troughs are made from 3- to 4-foot lengths of 12-inch PVC pipe, split lengthwise, the ends closed with semicircles of board 1 inch thick. A scrap of 2-by-something a couple of feet long is screwed or bolted under each end of the trough to keep it upright and give us something by which we can attach it to the floor. These feeders have served for years, needing little repair, and are an inexpensive option. Pig keepers warn against high-dollar pig appurtenances in any case, saying that, whatever it is, the animals are certain to destroy it. Troughs serve for feeding wet feeds, like fermented grains, skim milk, or whey; whole roots, like mangelwurzels or turnips, and green forage or hay, we throw on the floor of the pen at the clean end—as opposed to the end assigned by the pigs for their latrine, not a nice place to put their dinner.

How high you set your trough depends on a couple of factors. First, how big are the pigs that are going to eat out of it? And how deep are you going to let the bedding build up in that pen? A trough that is high enough in the summer, when bedding is relatively thin, may in winter be rooted full of bedding between every feeding. If, on the other hand, you set the trough high, perhaps on a raised base, in order to avoid this problem, you need to make sure that it is still accessible to the smallest pig you put in the pen. And make sure, while you are at it, that the trough is plenty long enough for all the pigs to eat at once, even allowing for competition; too short, and the weakest animals will be underfed. It may be necessary to put two troughs in a pen to make sure everyone gets his share, should competition be too determined; or separate aggressive animals in a pen of their own.

Figure 9.11. Enjoying his freedom.

Water

How you provide water to your pig is going to be a function of your paddock type(s) and water sources. Here are some of the methods we use, and some that we have seen used on other farms.

In the pasture and during frost-free weather, it is pretty easy to follow the pigs from paddock to paddock with a water tank. High-sided stock tanks may come fitted with a hog drinker, a low port in the side of the tank that provides a reservoir for access by pigs and other short-legged livestock. With a float valve to regulate water level in the main tank, these will keep water before your pig at all times. You can purchase a tank with the drinker already installed, but be warned: they are pricey. Or you could fashion one for yourself, plumbing a pig nipple or nose-operated port into the side of a full-sized galvanized tank. There are also low-sided stock tanks available that are accessible to adult pigs but that will be too high for piglets to reach; a block or short round of log to climb on may be necessary for accommodating the smaller animals.

This is probably a good time to mention one danger of stock tanks: things that fall in may not be tall enough to reach the bottom and may drown before you find them and fish them out. We lost a big tom turkey last fall in just this way. A piglet that has to balance on his hind legs and reach down into the tank to get water is a piglet ripe for being jostled into the drink; so we have to make sure not to use a tank so deep that he'll drown. Sometimes we set a cement block in the bottom of the tank, large enough to give a toehold to any animal we think could end up taking a swim; that way he can keep his head above water until we arrive to fish him out.

Actually, most of the time, instead of a stock tank for pasture water we use a step-in pig nipple supplied by long hose from one of the spring tanks. This rig is a mouth-operated stem valve that we weld to a double-spiked post that can be driven with foot pressure into the ground outside the paddock. The post sits at an angle, so that the nipple extends through the fence. You won't find a pig nipple on a step-in post at the farm store; ours was home manufactured. We've never had any trouble with the pigs pulling it up or knocking it over, and it provides water wherever we care to take the pigs, or wherever we can take a hose. One real advantage to this system, on our tilted acres, is that we don't have to find a flat place to put the tank.

Behind the barn is another of our stock water solutions, a 55-gallon barrel bermed in the ground and filled from below by a buried pipe running from our upper spring. Plumbed into the side of the tank is a pig nipple on an extension that goes through the back wall of the barn and into the pigpen; this provides constant running spring water to the pen. Hog drinkers—a small, shallow bowl attached to fence or wall and supplied, like the pig nipple, from a hose, or plumbed into the side of a tank—are also available. These do freeze up in winter, whereas the pig nipple arrangement keeps water in front of the hogs in almost any weather; only when the temperature drops well below freezing is it sometimes necessary for us to use a submersible heater to thaw the nipple. This is because although the water in the tank, since it is always moving, rarely freezes, the pig valve, on its metal extension pipe, may ice up. (Note: when we intend pig nipples for winter use, we install them at an angle, so the bottom of the stem valve apparatus doesn't hold water, which can freeze and immobilize the stem.)

Then there's the good old-fashioned water trough. One neighbor up the road uses a large, heavy steel tank of just-pig height, extremely heavy, straight-sided, and flat-bottomed to offer no purchase for a pig snout. He keeps it full with buckets from the rain barrel or the barn tap; once a week the trough is dumped and rinsed, then refilled. It's a practical option, and a good one as long as the litter in the pen doesn't get so deep it ends up being rooted into the tank. If our neighbor feeds dry or granulated feeds, he must be sure to empty and wash the tank regularly; the water grows sour quickly from the feed that sticks to the pigs' noses and washes into the tank.

Planning the Year—Piglet to Pork Chop

Paddock planning means knowing where available forage (or other food stuff) is coming from for each season and determining that we have appropriate shelter for that grazing period and that water can be made available to the area in question. On our farm, for example, the first paddock grazed in the spring is likely to be up by the barn, where the sod dries out early and the barn spring is available for drinking water; from here, in case of really inclement weather, we can let the pigs back into their barn shelter. By the time the ground firms up, however, we plan to be at the lower end of the field, because the spring that fills the tanks at that end is seasonal, and by mid-July we won't be able to rely on it.

Where pastures are shared with other species, it makes sense to keep the pigs on pastures needing remediation, saving the best grass for ruminants. We have seen how hard half a dozen pigs can hit an area of forage when the mood is on them, rooting up yards of turf while the ruminants are just getting started grazing, and we don't like to see good pasture land turned over in furrows. Sugar Mountain Farm in Vermont, on the other hand, reports benefits to the pasture from a certain amount of rooting, even up to 80 percent of the paddock; in their mostly pig operation, roughed-up pastures may be less of a problem. We are beginning to experiment with carefully timed paddocks on our better pastures, cautiously optimistic that the pigs will root out some of the spring onion grass that taints our spring milk without making a moonscape, but ready to pull them off if we aren't satisfied with the results.

Pig Pans

Supplemental feeding in the pasture can mean moving a trough from paddock to paddock, but the animals are going to turn it over and push it around when it is empty. A section of sheet metal too big to be flipped can be used instead, with the feed just poured onto it, but look out for sharp edges. Farmers are creative, and we have seen old highway signs used in this capacity, or sections of heavy rubber conveyor belt. The Judys of Green Pastures Farm in Missouri, where pigs are only supplemented while they are being trained to follow the cows, report that they pour the grain on the ground and the pigs eat it there. We're sure this works, but on our small farm we would be concerned for the level of rooting this would encourage; when we paddock pigs on the garden, we pour whole kernel ("shell") corn on places we want the pigs to hit hardest, knowing they'll root and dig to get every last kernel.

And by all means remember that those pig noses and feet are petroleum-free auxiliary plows and seed drills! While the pigs are there to root and stomp them in, throw out seeds of any forages you want to incorporate in the pasture; in the case of larger grains like corn, which the pigs will eat rather than stomping, sow seed behind the pigs, after the soil has been disturbed to make a welcoming bed for your new pasture crops. Over the weeks following, when this paddock is in R&R mode, seeds will germinate readily that could not have pushed their way through the existing thatch.

Figure 9.12. More homegrown pig food.

Seasonal Pig Food

Once you're on to the idea that pig food should be generated on the farm—and that most pig food is a by-product of some other occupation—you have the tools you need to plan for your homestead pig keeping. Make a list of the resources from the farm, and start from the ground up. Grass—how much, when, and where—and with whom will the pigs have to share it? What can you expect from the garden, and when will it be ready? What kind of surplus will you have from the dairy—and don't forget to think of whey, skim, and buttermilk, not just extra drinking milk. Don't forget, too, that if your pigs are pastured with or behind your ruminants, the pigs will derive nutritional benefit from undigested seeds—and other things—left in the other animals' manure. Do you have woods or an orchard, even one fruit tree? Windfalls will be available at harvest time. Are there grape vines on the farm? Berries, nuts? When will you be processing food for storage, and what sort of by-products will result? Look abroad—do you have access to other sources of waste nutrients, like a local microbrewery needing someone to haul away sour mash, neighbors' kitchen wastes, a restaurant or grocery store that will allow you to leave a bin for food scraps? Think creatively; our country is just full of food going to waste.

Spring

On our farm the schedule of surplus varies from year to year, but overall it is pretty consistent. In spring the ground is still too soft to have the pigs out on pasture, but this is the beginning of calving season, when ruminants that were dried off in February give birth and begin lactating again, so there will be colostrum and dairy surplus for the pigpen. Soon, too, we will be cleaning out the root cellar and dry cave of the last of the winter squash, beets, turnips, and mangelwurzels and sorting the seed potatoes, thus providing another source of pig calories. We are generally getting low on hay about now, if we aren't downright cleaned out, but early garden preparations, as the last of the tunnel-grown crops begin looking tired and weeds are getting the jump on spring planting, result in many a pig salad, complete with mineral-rich soil clinging to the roots. And if the sow has just farrowed, or we have started a batch of feeder pigs, these will be small, so the quantity of pig calories needed will be moderate.

Summer

By the end of that first pass over the pastures, the grass is getting ahead of the ruminants, so we may include the pigs in the second go-round if the soil is not too wet. Even if the ground is soppy, the pigs will still get lots of forage, since all the cut grass—mowed from the side of the farm lane, trimmed from under the electric fences, cut and raked in the orchard—can go right into the pigpen, as well as broken bales from the first cutting of hay. The cows are near the top of their milk production curve, giving more than we and the calves need, and summer is so busy we are glad to give surplus to the pigs, leaving most of the cheese making for autumn and early winter, when we have more time, and when, according to some sources, the milk quality is better for cheese making anyway. We are already jamming, canning, and pickling, generating lots of food preparation wastes; there are also great quantities of thinnings, weeds, and cull vegetables from the garden. Summer is a time when the farm reliably produces *lots* of pig food.

Autumn

In the fall there is so much to feed the pigs! Most years the garden continues to produce copiously, the orchard trees are ripening their fruits, hopefully the pastures are in good shape—and *everything* is welcome in the pig trough. By September most of the calves are weaned so there is plenty of milk for cheese making, with gallons and gallons of whey as a by-product. The protein percentage in whey is low (about 0.5 percent), but the quality is excellent, and 2 or 3 gallons can supply a medium-sized pig on pasture with all the supplement it needs. If we have raised a flock of broiler chickens, these will be butchered soon, and for a day or two the pigs will enjoy their own batch of chicken (-gut) soup. Canning and preserving leave us with big buckets of skins, seeds, cores, cobs, and shucks; all the fermentations—krauts, cider, wine—generate their own share of pig food in significant quantities as well.

Hay is available now; broken and misshapen bales that don't stack well can go to the pigs at the rate of two or three flakes or more per pig per day. (This is a good time to buy hay, if you don't put up your own.) What doesn't get eaten will make bedding; the Finneys of Spring Valley Stock Farm back a whole round bale into each of their supersize pigpens and let the animals have a saturnalia—some they eat, the rest goes to form the basis of their deep winter bedding.

In late fall we clean out the garden. Cornstalks and bean haulms are hauled to the pigpen, or the pigs brought to the garden and turned loose on them. The bulk of the field corn should be already in the barn, where, if you can keep the chickens out of it, it can be fed out as other food sources run short; if you don't have a good storage place, you can offer it immediately. Cull squash are an excellent fall pig food, not only because these vegetables have a shorter storage life than roots, and should therefore be fed out first, but because their seeds, favorite foods of pigs and chickens—cows, too—are a natural dewormer. Root crops like turnips, sugar beets, and mangelwurzels are stored for winter, but the tops go to the pigpen right away.

Winter

By December we will be offering turnips, sugar beets, and cull potatoes. Mangelwurzels wait until January, since until they've sweated in the cellar a while, they can cause intestinal upsets. After Christmas, they are a prime food for feeder pigs and lactating sows. With field corn, hay, and dairy waste these constitute our primary winter pig offerings. Demand has dropped, anyway, with only the breeding stock, and maybe a batch of February piglets, to provide for; this is the season for large animal butchering.

Pork doesn't need a long dry cure, like beef, so we butcher hogs any time the need is on us and the weather cold enough (around freezing). The weather can't be *too* cold, though—we hang the sides in the garage overnight, so the meat will chill and be firm enough to cut, but we don't want it to freeze. Where frost-free water is a problem in the barn, butchering in early winter saves the trouble of keeping water in front of the pigs over the cold months: sharpen your best knives, order a roll of butcher paper and some freezer tape, and invite your most adventurous friends to a Pig Retirement Party. When we butcher with the monastery, or with our neighbors up the road, the event becomes a three-day party, with lots of volunteers and friends turning out for the fun, the sense of community, and all the great food. When it's over, if you are not overwintering breeding stock, you can muck out the pigpen, roll up the polywire and hang it in the barn rafters (where it won't get chewed up for mouse nest material), wash out the water tank, and close up shop until spring.

Fast Pig, Slow Pig

You *can* grow a pig fast, but is he a better pig for it? If his growth rate is not that of a commercially fed hog, does it follow that he is less thriving? Pigs are, well, pigs; we find that ours are willing to think about a meal pretty much any time of the day—or night. Given the right temptation, a pig can always eat a little bit more; we can't rely on our pigs to tell us when they have had enough, because that is going to be never. The sight or sound of a human being walking by their pen or paddock is enough to bring all the pigs up to the fence, grunting and squealing in anticipation of a treat: toss them a pocketful of windfall apples, or a wheelbarrow load of weeds, and watch them dive in.

Does this mean they are not getting enough calories? The answer depends largely on the object of our pig keeping. Commercial operations have to meet strict production goals in order to make their overhead: for the mortgage to be paid, finished pigs have to be shipped out in the shortest possible turnover time. Not so us little guys, who only intend to fatten a hog or two for bacon and sausage; our concern is to build fertility, utilize nutrients responsibly, and grow the best possible pork for our tables.

We have been interested to learn from our research in Cornell's online agricultural library that the present standard of the six-months-to-market-weight (roughly 225-pound) pig has only been considered the norm for a few decades. A century ago, home pig management looked very different: a farmer with a litter of piglets might sell several as "feeders" when they weaned at a few weeks old, keeping only as many as his farm could presently feed. Of those that remained, some would probably be slaughtered or sold at around six months, or about 100 pounds weight; while only one or two would be grown to the large size desired for bacon, ham, and salt pork, the pig food available on the farm being inadequate for a larger number of hogs. Because the farm only generates a certain amount of surplus, it becomes expeditious to reduce stock numbers as the animals' size—and nutritional needs—increase.

Piggy Diet—Theory and Practice

For the self-sufficient homestead, buying in pig feed is going to be a stopgap, not a standard practice. Most of what goes in the pigpen will be perishable nutrients produced in excess in other parts of the homestead. These won't come with urea or GM soybeans or corn—fortunately—but neither will they come with nutritional-value labels. For the commercial farmer, the nutritional needs of hogs at various stages of life are calculated in terms of "balanced rations" and "dry matter," but since most farm-raised foods aren't dry and few individual food types supply every part of a pig's nutritional needs, this approach isn't going to be terribly helpful to the homesteader. How are we, feeding miscellaneous farm calories, supposed to be sure the piggy is getting what he needs? Fortunately the homestead pig, valuable converter of whatever nutrients are going unused at any given time, does not require minute calibration.

In many pre–World War II farm manuals, the nutrient categories and feed values of various farm-raised feeds are set down in general terms (for an in-print manual, Dirk van Loon's *Small-Scale Pig Raising* has some helpful charts and information). From these we've learned some interesting facts: for instance, that the use of grain as a primary pig food was far from universal, even when that grain was raised on the farm; that the acknowledged number-one high-quality protein food for domestic

Figure 9.13. Turnips produce well in the fall garden and are excellent pig food.

farm animals—pig, calf, and poultry—is skim milk (quantities of which, before the days of refrigeration, used to be generated when cream—a cash crop—was skimmed from the previous day's milking); and that it was a common practice to let hogs harvest much of their own food in the field, keeping them on grass for part of the warm season, then allowing them to clean up—"hog down"—harvested crop fields later in the summer. We found out that roots (like turnips, beets, carrots, mangelwurzels, and potatoes), crucifers (cabbage, kale, and their relatives), and cucurbits (like squash and pumpkins) were traditional farm-raised foods for pigs and cattle just as much as, or more than, corn and other grains, the root crops especially being simple to raise and easy to store; we learned in addition that, supplemented with relatively small amounts of a high-protein food like milk, and some hay for roughage, a combination of these can supply the bulk of a pig's diet.

These facts led us to reconsider some of the other things we had read about pigs. "Pigs should be fed two or three times a day, as much as they will clean up in 20 minutes," may be a good rule for pigs on mash (ground, grain-based feeds), but it will be immediately evident that the same food-value intake cannot be expected from two or three 20-minute meals of roots. And when a pig drinks whey, does that count as eating or hydrating? Or what about when he is on pasture, grazing for hours every day? For the homestead pig, duration of time at the table is obviously not going to reflect the same caloric intake as for grain-fed factory hogs.

In addition, the weight of bulky foods and dairy waste, which make up the largest percentage of what most homesteads will have to offer, is mostly water. While we can find charts that translate these foods into "dry matter" equivalents, and other charts instructing us on the exact dry matter needs, in terms of protein, carbohydrate, and minerals, of any given pig—based on its breed, age, sex, state of gestation or lactation, and condition in general—most of us aren't going to sit down every day with a calculator and Pearson's square to determine how much of whatever is most available at the moment adds up to a "balanced ration." So it is necessary to develop a rule of thumb for our own farm, our own pigs, our own crops.

First of all, we should remember that our farm-produced nutrients are superior in practically every way to the factory formulated feeds they are replacing—in their mode of production, mineral value, freshness, flavor, variety, and more. Encouraged by this knowledge, we can educate ourselves to categorize farm-raised foods as "mostly protein," "mostly energy," "mostly roughage," and so on, so we can balance our pigs' meals something like we do our own. This way, our concern will be less for the composition of individual meals than for the day's overall intake; that is to say, if one meal is heavy in protein, the other(s) that day will probably emphasize carbs, roughage, and vitamins. Finally, as we get more experience with our farm surpluses, a rhythm becomes apparent between those foods that will store quite a long time at any temperature—mostly grains and cured grasses (hay); those with a fairly long life in a cool cellar, like roots and crucifers; and the most perishable of foods, like dairy surplus and food preparation wastes. From these we can develop an economy of the farm year, an ecology of farm nutrients.

On our farm, for example, a summer day at the pig trough will often start with a bucket of buttermilk and kitchen scraps, to be followed in the middle of the day with a large weed-and-cornshuck salad, with a mess of creamy cobs from the freezer corn to fill up on in the evening. In the fall, cull pumpkins and winter squash may provide carbs and protein, with whey for added protein, and whole corn stalks—ears and all—for some concentrated calories and lots of chewing entertainment. In winter, after the squash have been used up, roots come into their own; a sow will crunch her way through 20 or 30 pounds of mangelwurzels or turnips with relish, and fill in the corners with field corn, waste dairy, and hay.

Are we certain that every meal our pigs eat is nutritionally balanced? Of course not. Are we confident that our pigs are getting what they need for health? Our experience answers an unequivocal "yes." Are our plates well provided with pork of our own provenance, grown from nutrients derived from our own sunshine, rainfall, and fertile soil? Absolutely.

Our goal here is not to present all we have learned about farm-raised pig feeds but to demonstrate some principles and techniques of the sun-nourished farm; so we will not begin listing the relative food

Balancing the Farm

One of the bugbears of the infant small farmer is the vast quantity of knowledge and experience we don't have. We imagine that there are people out there, past and present, for whom farming is an exact science of methods developed over many years, the application of which takes the mystery out of husbandry. One or more of these savants may live next door to us, from which point of vantage he can observe our efforts; his comments may be to the effect that we are Doing It All Wrong. Even if we don't live next door to this person, we may meet him at the grocery store or the feed store, or he may just lurk in our mental underbrush, ready to tell us the same things: "You can't raise cows without grain"; "You have to feed pigs a commercial, balanced ration"; and so on. Books on farming and animal husbandry, helpful as they may be, can compound this problem; this is because their goal is usually to raise animals the fastest way possible on the cheapest nutrients, and their writers don't for a moment imagine that your goals are any different. Commercial profit margins depend on maximizing the use of barn space by moving more animals through, more quickly, and by feeding either purchased or farm-raised grain inputs. If their pigs grow slowly because the lysine in their diet wasn't perfectly balanced, costs go up and profits go down; the long list of ingredients in commercial pig rations reflects this preoccupation.

We're not exactly saying none of this matters, but we are saying that the smallholder who is trying to manage an ecosystem has different goals in mind. We balance the *farm*, not just the daily ration, believing that the good of the whole is the good of the parts as well; believing also that natural systems, which have always depended on seasonal food sources and physical contact with the earth—rather than an unvarying, "balanced" diet of imported grains, minerals, and cheap protein sources—are, because they are natural, probably pretty functional also. We do try to know something about the food value of the surplus and forage our animals eat, of course, so we can use them efficiently. One thing that comes to light as we do this research is the eminent appropriateness of natural foods compared with the commercial formulas. Lest we forget: modern, dry feeds-in-a-sack weren't invented because they were a nutritional improvement on farm-raised feeds but because they are easier to handle and store, and they take the brain work out of feeding.

values of grains, roots, and forages, or the protein content of a gallon of skim milk. Our own sources are available to anyone online—one of the wonders and advantages we homesteaders of today enjoy!—and you can visit them to research the nutritional value of your own farm surpluses. We will just list here a few simple but compelling facts:

- From ¾ to 1 gallon of skim milk per day provides all the high-quality proteins needed for a pig to grow and fatten well, from weaning to butchering; while whey has only one-quarter the protein value of skim milk, that protein is of equally good quality and can be fed to appetite to provide a pig's complete daily protein requirements.
- Under the right circumstances, grass and hay can form the primary source of a pig's calories.
- Ditto, surplus storage vegetables.
- The value of "accidentals" in what is available to a pig fed from the dirt, while impossible to calculate, can contribute significantly to growing a healthy and inexpensively fattened pig.

It should be evident from this that the feed store is not the only, or even the best, source of rations for your backyard pig.

Buying Pigs

The right time to acquire pigs is when you are expecting to have a surplus of perishable nutrients to be utilized. The question is, where to buy them? Commercial pig operations, breeding pigs by the thousands year-round, aren't likely to want to sell you a piglet even if you wanted to buy it—and you don't want to. And small-scale operators willing to sell feeder pigs don't hang out signs on every street corner; after all, fewer and fewer Americans raise a backyard pig (how many do you know?), and where the market is small, so will be the supply. So the first step in getting into pig husbandry will be finding someone local to sell you a piglet, and you will have to buy when he is ready to sell: feeder piglets aren't like tennis shoes, available any time of year. At six to twelve weeks a piglet will be "feeder" size, meaning it may weigh 35 pounds or so; its cost will therefore be calculated on an imaginary weight of *50* pounds, and you can pay this, or you can wait to buy until it really weighs that much and then find that someone else picked them all up last week. Gather ye piglets while ye may; if you don't have surplus nutrients in quantity at the moment and have to feed a little grain to begin with, fine—you don't want to be without a pig when the cow freshens, or the garden comes in, or fall deluges you with blemished pumpkins and apples.

Figure 9.14. Pasture-sown cushaw squash and tomato plants.

There are lots of pig breeds out there: big pigs, small pigs, red pigs, black pigs, pigs with wattles like a turkey or bristles like a hedgehog. What is it you want? A long, lean animal that will make lots of bacon? A fat pig for plenty of lard? An energetic pig that will earn his keep as a plow and compost turner before he goes into the freezer, or a gentle mama pig that will keep you supplied with babies? To a certain extent these traits can be selected for by breed or type, but individual characteristics are going to be, well, individual, and if you aren't already just in love with a certain breed, you'll probably find you're happy with what comes available. As with all livestock, buy from a farm where the animals look clean and healthy; don't buy from an operation you wouldn't want next door. And *no pets, no rescues.*

A word about cost. You can look in the local farm paper and try to decipher the stock numbers, which will reveal how much feeder pigs were selling for last week at the livestock barn, and this puts you in a per-pound ballpark. Or call around a little, if you can find piglets in the classified section, and see what the generality are charging. One little detail to remember is that many spring pigs are sold as "club" pigs; that is, pigs for 4-H kids to raise for show. These have classy mamas and papas and will cost somewhere between twice and five times as much as—or even more than—their otherwise identical (to us) brothers and sisters intended for bacon without benefit of a pass through the show ring. Just for comparison's sake: around here, when feeders might be a buck a pound, club pigs could easily cost five.

Raw Milk and Shipping Fever

We have never given our purchased piglets prophylactic antibiotics for shipping fever. One year we bought in seven piglets, and within a week all of them were coughing and lethargic. We called in the veterinarian, who gave them megadoses of high-powered antibiotics; still, we lost three. Worse, the four that survived never grew properly; although we kept them to eight months of age, they only reached about 150 pounds, instead of typical market weight of 225 in six months, despite the fact that they ate just as much as a larger pig would (they were on commercial ration, and we were keeping records). We determined never to waste time, vet fees, and feed on such pigs again.

Next year we purchased four piglets from the same farmer, one we trusted, and again, they all had "shipping fever" inside of a week. We didn't call the vet—too much money had gone that way before—and we figured they'd die of what ailed 'em. To keep them comfortable meanwhile, we visited the pen often, taking each time about a quart of colostrum/raw milk (we had a cow freshening). At the sound of the colostrum being poured into their trough, all the pigs would spring to their feet, race over, and take about three gulps before they collapsed, panting. They were so exhausted from trying to breathe that they couldn't take time to drink or eat, but when they heard their trough being filled, for a moment habit overrode even exhaustion.

This went on for five days, and to our disbelief, after the first three days, the pigs began to get better. Within a week, all four pigs were up and hearty, squealing for their meals. They grew well thereafter, and we never had any more trouble with them. Something in that colostrum and raw milk, brought several times a day in small quantities, was able to do what antibiotics and pelleted feed could not.

Bringing Them Home

You can move a piglet in a dog carrier, and we've seen it done, more than once. One man who bought a batch of piglets from us moved them loose in the back of his Jeep Wrangler; he told us he'd moved as many as thirty at a time that way, and we saw no reason to doubt him. When we buy piglets ourselves, we usually bring them home in the back of the pick-up with the sides on, and this has worked fine, but we make sure there is a good bed of straw for them to snuggle down in, or we put the cap on the bed, because we don't want the little guys to get a chill.

Pigs—and other livestock—are susceptible to a mysterious illness called "shipping fever," which means "when you move an animal, it's liable to get sick." It should not be surprising that common commercial practices, replete with routine low-level antibiotics and with a deficit of fresh air, result in pigs with depressed immune systems. Sometimes the seller will offer to give your pig a shot of antibiotics before you load up, as a protection against shipping fever. Suit yourself. We've never done the shots, and most of our pigs have made the transfer just fine.

Give your new babies a warm bed, especially the first few nights. Even in the summer it can get chilly when the sun goes down, and if the piglets were barn kept, with plenty of other pigs around, they aren't used to keeping themselves warm. A clean bale or half-bale broken into the corner of the pen where you want them to sleep, not defecate, should do just fine. You probably don't want to start them off on some old moldy hay from under the leaky place in the barn roof; they are going to eat some of this bedding, and breathe the dust, and there is no sense in hitting them with too many challenges at once.

Moving big pigs is another matter. You'll probably need a trailer, and don't forget that while pigs are shorter than cows, they can weigh nearly as much. Loading a big pig is a matter of convincing the animal to go where you want him to go, because force isn't going to work. Three people with pig panels—half a sheet of plywood with a hole cut in one side for a handle—can box a pig pretty well: since he can't see through the plywood he doesn't realize that mere puny human beings are the reason his pen keeps moving, so he moves with it. We have also used a ring of stock panel to move a large but fairly tame pig, dropping it over his head and then shuffling it—and him—toward the trailer, but under these circumstances the pig can see, and if he gets really serious about going somewhere else, he's going to go. People say you can back a pig anywhere with a bucket over his head, but we've never tried it.

Handling and Gentling

There is almost nothing cuter than a newborn baby piglet, and at ten weeks they can still be pretty winning. They will crowd against the fence when you come to feed them, eat from your hand, and love getting their backs scratched. Gentle your pigs in this way by all means, and accustom them to having you move around among them, so that when you are transferring them from paddock to paddock you have calm, serene piggies that trust you, not skittish, nervous animals that want to charge the fence. But as they grow, remember that pigs are bigger than you are, and they are omnivores, with tearing teeth as well as grinders. Feral pigs are predators and scavengers as well as grazers, and your friend Wilbur could eat you, unlikely as that may seem. Always treat him respectfully; don't confuse him with Fifi the lapdog, whose worst infliction might be a bite on the thumb. If you need to go into the pen with a big boar, or a sow

Figure 9.15. Pigs grazing under an apple tree.

with a litter, keep your eye on the exit, and know how fast you can get there.

Minerals

We don't want to pose as experts in this (or any) area, and you can buy mineral mixes for pigs (the kind they add to commercial pig feed) at the feed mill, but we follow the dicta in our old farm manuals, in which pigs are expected to get most of their minerals from their food and from the soil. Interestingly, these books take it for granted that every pig farmer knows to give his pigs ashes and charcoal, for digestive purposes; we try to remember to keep them available in the pigpen, even though we don't know what they are for. Cooked foods and swill are generally fed salted. If this form of mineral supplementation sounds too laissez-faire to you, you can look up the details for yourself.

Routine Health Care

Actually, our pigs seldom need anything in the way of health care. Barring that case of shipping fever we mentioned, we have never had the vet out for a pig. We add diatomaceous earth (DE) to their bucket a couple of times a week, and a big dose of raw garlic every couple of months, as well as feeding pumpkins and winter squash, or their seeds, and we seldom see any worms at butchering time. We read of other people who use rosemary and garlic in the same capacity. Vet visits and pharmaceuticals add substantially to the cost of raising animals, regardless of all the other reasons not to medicate, and we only use what is necessary. We are not in a race to the fastest slaughter weight; rather, we are fine-tuning an ecosystem, and on the farm, as in nature, there are flush times and lean times. For a coughing pig, or a languid pig not interested in eating, we might call the vet, but first we would calculate how much money we already had in that animal. A vet visit and two meds will come close to the cost of two weanling piglets, and no guarantee you'll help the piglets, either. Raw milk and colostrum, on the other hand, make first-class pig tonics; we've seen them work wonders.

We do castrate boar piglets. Sometimes boar meat is just fine, and sometimes when you cook it, it smells like a bus-station urinal. We envy people with porcine genetic strains that don't carry boar taint, but we aren't yet ready to drive across six states to get breeding stock. You cut pigs, you don't band them, most people say the younger the better, but that may only be because it's easier to cut a piglet that weighs 1 pound than a young boar that weighs

100 pounds. It's not difficult to castrate a pig. Look at a video online, or get a local farmer to show you.

Breeding

Sooner or later you may get tired of looking for piglets and not finding them when you want them; there is even the chance that availability will just dry up as the demand for feeder pigs in your neighborhood dwindles. Then you are going to begin wondering whether to raise your own feeders. While it means year-round pig feeding, having a mama pig on the property also means you'll never be short of something to eat those buckets of tomato peels and seeds at canning time, and you'll have piglets to sell as well as to fatten for your table. Our pig-raising operation is on this scale. If you are considering getting a breeding sow, you will want to do your own research into pig husbandry, and we will mention here only those things we have found are peculiar to the fodder-fed, grassfed hog, especially where that experience is contrary to standard practices.

Most obviously, we make sure Mama as an individual is a good forager, and a good converter of coarse foodstuffs to flesh. Regardless of breed characteristics, you want to know that the individual genetics you are reproducing perform well under your conditions. It will be nice if you are on intimate terms with her when she farrows, too, since you'll probably want to be able to climb in the pen and assist when the time comes, especially the first parturition. Piggies aren't too flexible around the middle, and Mama can't turn her head far enough to see what's going on at the other end during the birthing process. While piglets are small enough that the discomfort of birthing—to the mother—seems minor, she is bound to be curious about the situation and is liable to do a lot of shifting about, especially the first time, with the likelihood that one or more of the little ones could be stepped on. We generally stay beside Mama while she farrows so we can pick them up from the drop-off point, wipe off the mucus and birthing sac, and move them around to the milk parlor. We usually know a couple of days ahead of time when piggies are due, because Mama starts building a big nest.

Figure 9.16. A pastured sow with her young litter of piglets. Courtesy of Jennifer Metz.

Grass—and hay—make rich milk! Also mangelwurzels and kale, recommended in old farming books as particularly excellent food for lactating animals. Piglets will nurse up to ten weeks and beyond, as long as the sow will let them. A day-old piglet will nose about in Mama's food and eat the chewed bits that drop out of her mouth, and by the time piglets are a couple of weeks old we provide them with a creep—a sidecar pen/feeder with an opening too small for the sow to get in—where they can have food of their own, inaccessible to Mama, because anything she can reach won't be around long. On our farm, this is a fenced-off corner of the pigpen where we can give the piglets a low trough with softer foods, cut up if necessary. Skim milk and buttermilk are excellent food for baby pigs.

However you manage the details, the transition from mother's milk to solid foods shouldn't be abrupt. She'll probably see to the weaning herself—half a dozen or more little ones scrambling for a teat whenever she lies down gets old after a while—or you can pull them out after a few weeks (commercial breeders wean as early as fourteen days, not that we're recommending it) and give them their own private dormitory and food service.

The Boar

If you decide to keep a sow for piglets, you're going to have to breed her. Whether you keep a boar—and remember, they eat a *lot*—artificially inseminate, or find a boar at stud will depend on your situation. Option number three is the easiest on you, *if* you can find someone local who 1) keeps a boar and sells stud services, 2) keeps a boar with genetics you would like to have in your own pigs, and 3) keeps a farm you would be willing to have your sow visit—and possibly bring home pathogens or parasites from. If you don't enjoy such a situation, you will have to select from the first two options. We are told that AI for pigs isn't difficult, but semen is not cheap and shipping it is expensive; on top of that, determining when a lone sow is in heat can be pretty tricky. It can be done, though, and if you think you have a source of semen from a boar with genetic lines that match your farm's ecology, this could be your best option. We have kept a boar, with mixed success; our intention was to buy a new boar piglet every couple of years, let him do his job for a while, then eat him, but the first boar we butchered was so rank that when we fried the first flitch of bacon—our best black-pepper-and-maple-sugar cure—not only did it saturate the house with something like eau de outhouse, but even the cast iron frying pan had to be detoxed before it could be used again. Still, until we can get to Vermont and pick up better genetics, keeping a boar has been our best option. A farm with less available surplus for pig food would want to consider another solution.

The homestead pig has been a fixture on the smallholding probably since the domestication of the species; a more ecologically beneficial means of converting and storing perishable surplus is beyond our ability to imagine, and the role pigs fill in the economy of the diversified farm makes them indispensable to its efficiency. At any given time, whether pastured or penned, self-harvesting or having its groceries delivered right to the trough, the homestead pig represents savings on the hoof, able to reproduce itself at an exponential rate; a living, breathing unit of food storage that doesn't require a freezer, shelf space in the pantry, or room in the root cellar.

And we're not sure, after all, that good home-cured bacon and ham and sausage aren't sufficient reasons to keep pigs, all by themselves.

CHAPTER TEN

Milk, Meat, and Manure—The Solar Harvest

If lynxes eat too many snowshoe rabbits—which lynxes are said to do repeatedly—then the lynxes starve down to the carrying capacity of their habitat. It is the carrying capacity of the lynx's habitat, not the carrying capacity of the lynx's stomach, that determines the prosperity of lynxes . . . one of the differences between humans and lynxes is that humans can see that the principle of balance operates between lynxes and snowshoe rabbits, as between humans and topsoil; another difference, we hope, is that humans have the sense to act on their understanding . . . a stable balance is preferable to a balance that tilts back and forth like a seesaw, dumping surplus creatures from either end.

—*Wendell Berry, 1982: "Getting Along with Nature"*

The grass-based homestead: chickens scratch contentedly at our feet, ducks splash in the farm pond, and two happy piggies have settled down in the shade of the pasture apple tree, while nearby in their small paddock Clover and Chuck Roast, the dairy cow and yearling steer, are happily grazing with maybe a couple of ewes to keep them company. The ten-year-old goes by with a bottle of skim milk for the weanling calf, while you trundle over to the fence with a wheelbarrow full of corn shucks and cobs—you've been putting up the harvest—and toss them in, to the squealing delight of all the animals. A beautiful picture of happy, symbiotic coexistence; naturally derived nutrition; and healthy, bucolic plenty.

The only question is: how does all that plenty reach the dinner table?

There in the pasture are our steaks and roasts, milk and butter and cheese and ice cream, morning egg and Thanksgiving turkey, all thriving on their forage-based diets; somehow, we have to get it into forms we recognize as *food*—or else run away and become vegetarians. But although we know in theory where animal-derived foods come from, when it comes time actually to eat them, most of us have been practicing a psychological sleight of hand between the animal and the plate. Who actually tucks his napkin under his chin thinking, "Can't wait to sink my teeth into this cow's psoas major,"

Figure 10.1. Turning grass into milk and beef.

Figure 10.2. Barnyard Napoleon (*center*), Sunday dinner.

or sucks down the last delicious drops of a milkshake dreaming of squeezing a cow's udder? To the twenty-first century diner, there's something about fur and feathers that just doesn't go with forks and knives. Somehow, if we're going to undertake to raise our own locally derived, grassfed, sun-sourced foods, we've got to bridge the gap between roast chicken and the Little Red Hen.

People have been turning animals into food for thousands of years. That's how we come to be here in the first place: generation upon generation of pastoral ancestors, deriving their living from livestock. No matter what your parents, grandparents, or great-grandparents did in the last century or so, the overwhelming majority of your progenitors were agrarian, and they managed somehow to get milk from the cow and turn it into butter and cheese, to slaughter animals and cure meats. Not only that, they did it all without electricity, and without refrigeration. How hard can it be?

Fortunately, it's a lot easier than our fears, created and inflated by the vast separation our industrial culture places between us and the plants and animals from which we derive our food, would incline us to imagine. There's nothing magical that happens at United Dairy or Tyson that makes not-fit-for-consumption animal products into food; *au contraire*, it's food already, and if you didn't believe it might be a good idea to get yours without recourse to Big Ag, you probably wouldn't be reading this book. Squeeze a mother ruminant, and you will have milk; wring a chicken's neck (after duly thanking God for her service to you and your family, past and future) and you've got the materials for chicken pie. Okay, there may be few steps before you're actually ready to pour the cream into your coffee or crimp the crust on that pie, but no part of the process is beyond the capabilities of a determined homesteader with a strong stomach.

Dairy

The most immediately available and renewable form of stored solar energy on the grass farm is *milk*, yesterday's sunlight transformed into today's food. Milk is almost endlessly convertible: into baby ruminants, of course; and butter, cheese, and lacto-ferments of many kinds; but also into eggs, poultry, and pork. A gallon of milk—whole, skim, or buttermilk—contains all the high-quality proteins necessary, daily, to grow a piglet from 50 pounds to 250, or for one hundred small- or fifty large-bodied layers. And skim milk, buttermilk, and whey, by-products of milk processing, are, when they aren't used as human or animal foods, valuable as foliar food and soil amendment.

What this means, in practical terms, is that a single moderate-production milk cow pastured on grass can go a long way toward feeding an entire small homestead.

Do the math. Take the case of a six-year-old Jersey calving in late April, a nice time, with the spring grass just beginning. When her milk first comes in, she may easily produce 5 or 6 gallons *a day*, or between 35 and 40 gallons a week. How much of that will the homesteader drink? Or turn into cheese? How much will there even be room to store? The calf will take some of the production, of

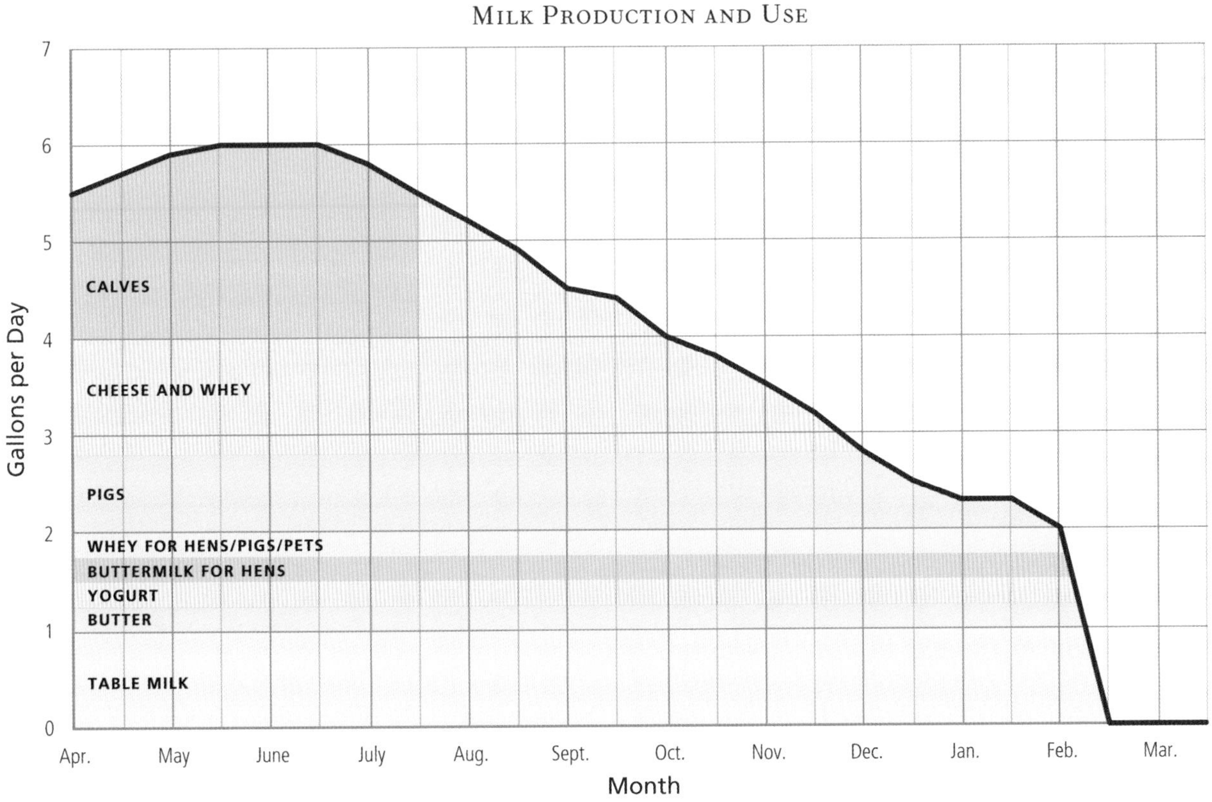

Figure 10.3. The abundance and versatility of the milk harvest would be hard to overstate.

course, but there will also be plenty for a feeder pig, and some laying hens and broilers. Over the course of the lactation, milk production will gradually decrease, but only gradually. At the same time, various demands on the total are eliminated: the calf weans around fourteen weeks, the broilers are put into the freezer at eight or nine weeks. In late fall the pig is butchered and the laying hens are culled. There is plenty of milk for cheese making—and, always, for the table. If the grass farmer considers the lactation as a whole, she can anticipate seasonal levels of production and plan for their best utilization. Of course, like everything else pertinent to the grass-based farm, good use of dairy resources takes planning, foresight, and appropriate timing.

Also, a few basic skills.

The thought of taking entire responsibility for the purity of our own diets is intimidating. An area in which we feel considerable trepidation is that of dairy products. Who hasn't had the experience of opening a bottle of milk left on the counter too long and encountering that stomach-turning odor of decay? When pasteurized milk goes bad, it goes *bad*, and your nose and eyes are there to tell you about it, so you won't consume it and make yourself sick. We have a built-in gross-detector, Nature's precursor of the USDA, telling us subtle and not so subtle things about our food. The mold in our package of grated cheese, the blue islands floating on our yogurt, seem to testify to the fragility of dairy products. If milk in clean white bottles can putrify so spontaneously, what hope have we, with just hands and buckets, of keeping our own milk sweet and pure?

Plenty, actually. Hands and buckets are simple to clean, especially when compared to the hundreds of yards of pipeline in a commercial dairy. The walk up from the barn to the kitchen is brief compared with the milk tanker's pickup route, ensuring that your home-squeezed milk will reach the house quickly. Coming in fresh once or twice daily, our milk has little time to spoil; if we do store fresh, clean raw milk over a longer period, no problem: milk is naturally inoculated during the milking process with beneficial microbes, lactobacilli native to ruminants, so kept clean, milk *gains* in digestibility and palatability over time, transformed by fermentation into numberless locale-specific superfoods: yogurt, kefir, lassi, kvark. The harvest is so abundant that this wealth of nutrition can be shared by virtually every citizen of the farm community—and of course, every twelve hours there is more milk coming. So what have we got to lose? Our grass-derived harvest is generous. And what have we to gain? A natural, renewable, and independent source of the highest-quality nutrition.

So grab a bucket. Your grass-based diet, beginning with milk, is just twelve hours away.

Parlor Manners

Clean raw milk happens in the cow. It comes from healthy animals living outdoors in the sunshine and fresh air, eating the grass and forage natural to their digestive systems, moving over pastures purified by time and sunlight between grazings. Our own responsibility for purity, however, starts with an orderly milking space and a clean bucket.

Barn, Shed, or Pasture?

Where we choose to milk is mostly a matter of what's convenient and comfortable for us and for our cow. The barn may be traditional, although in many cases dairy animals have been, and still are, milked in the pasture. But when the weather is inclement it's nice to have a roof over your head, and you don't want rain dripping off your cow's side

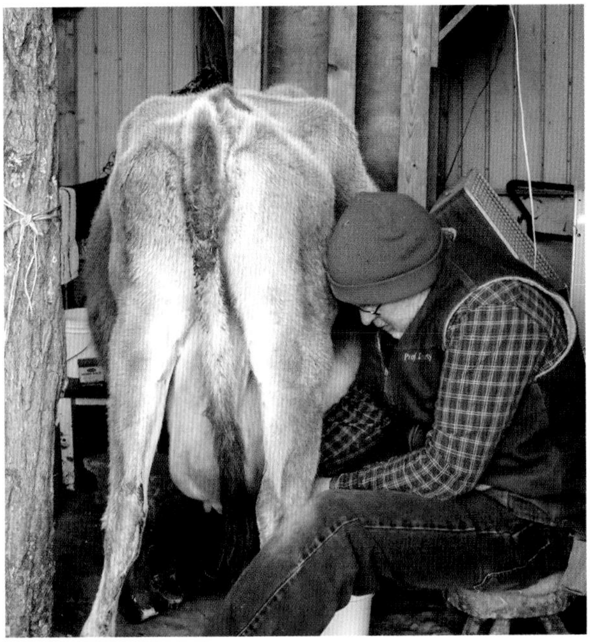

Figure 10.4. Milking parlors don't have to be fancy.
Courtesy of Christine L. Carroll.

into your bucket in any case, so eventually you'll need to have access to a shed or something where you and Bossy can get together. It needn't be fancy—our first milking parlor was a three-sided shed made of scrap lumber and salvaged sheet metal, lit with a string of solar-powered LED Christmas lights; one spring, when we were in construction mode, we milked three cows under a portable canvas awning. Briefly, we even milked in the shade of a grape arbor, picturesque but buggy; it was a relief when we got the new dairy finished and could move under cover.

We want to try to make our milking space reasonably dry and clean, with as few airborne contaminants—from mold spores to mud splashes—as possible. We're not going to say you need a concrete floor because we've never had one, but dry packed earth is better than muck and manure. Some nice, big window openings or doorways to provide light, and good airflow to help keep the dairy dry, are desirable. A bench, shelf, or hook where you can hang your bucket is a must, we think; you don't want to set a bucket on the floor where it will pick up stuff on the bottom, stuff that, later, when you tip it up to pour, can end up in your beautiful clean milk. And in any case you'll need somewhere to set strip cup, bag balm, clean water, and towels. Make sure you put your shelf in a place where nothing is going to land in it from above; i.e., don't set it under the open eaves where the sparrows roost, and be aware that if there's a hay loft, box stalls, or anything else upstairs from your milking parlor, chaff, dust, or something worse may fall through and end up in your bucket. So take appropriate precautions (we keep our milk cans covered with a clean muslin towel).

And while we know dairy women whose cows are so well trained they will stand without any restraint to be milked, ours aren't and yours won't start out that way, so you'll want a head gate, or a stanchion, or a rope tied to a post or something, so that while you're busy your ruminant won't be walking away. A head gate is easy to fashion out of a few pieces of scrap lumber, but build it strong enough to hold the species of animal in question. Nothing too hefty is required; 2-by-4s have been perfectly adequate for our needs. Put it where there's a wall or a post or some other barrier to the cow's left when she puts her head in, because you are going to milk her from the right, and the job is going to be a lot easier if she can't pivot away from you while you're doing it. (Actually, although it's traditional, you don't *have* to milk her from the right, and if it suits you better you can put the barrier on her right and milk from the left, unless her habits are already so firmly ingrained that she won't let you.) Fasten the stanchion in place strongly, or one day your cow will decide to back up while you're milking and pull it off the wall, and you don't want to find yourself in a small dairy with an excited cow wearing a stanchion around her neck. A grain box or hayrack in which to put treats is nice—a truss of hay or a small scoop of oats goes a long way toward encouraging your girl to put her head in the stanchion and stand still while you're squeezing—but leave clearance under the grain box so the barn cat can patrol there and discourage the local rodents from setting up housekeeping.

You'll want something to sit on, and its height will depend on the animal you're milking. If you make it something that lets you milk without hunching over much, that's good, and if it lets you rest your forearms on your thighs while you are milking, you're going to be more comfortable. We use a three-legged stool, or a 5-gallon bucket, or a chunk of log, depending on who's milking which animal, and how close her udder is to the floor. Padding can be nice, too; you're going to be sitting here often.

Machine or Bucket

You'll find there are several ways to get the milk out of a lactating ruminant, each with its advantages and disadvantages. Pipeline systems are for the big boys, thank goodness; let them deal with the problems inherent in a system that can't be dismantled and cleaned thoroughly. So for us little people, there are basically two options: the surge milker and the bucket.

Surge machines are powered by compressors. Rubber cups ("inflations") are attached to the animal's teats and a powerful suction draws the milk from the udder through plastic tubes into a closed tank. This reduces the problems of airborne contaminants by a big margin, yes, but increases the incidence of mastitis over good hand-milking, partly because machine milking is harder on the udder and teats than proper hand-milking, and partly because those inflations are difficult to clean thoroughly, resulting in a high bacteria load. This affects the quality of milk as well, since the tubes, bottles, and buckets associated with surge or other milking machines are always harder to clean properly than is a simple stainless steel bucket. And of course there is the expense of buying a milking machine, which can be anywhere from several hundred dollars, for a used system, to several thousand, new, in addition to the cost of ongoing maintenance. Finally, milking machines are cumbersome and noisy and leave the homesteader dependent on an electric power source, a matter of more or less concern, according to your attitude about electric power, and to the reliability of your source.

Hand-milking, while it is simple, energy efficient, and environmentally friendly, depends for its virtues on the pains taken by the dairy man or woman in the process of milking. Where care is exercised to see that bucket, hands, and udder are clean, and are kept clean, the quality of the milk will not be less than with a surge machine, and may in fact be greater, because hand-milking is gentler, and the equipment can be cleaned more thoroughly. For we who are milking only one or a few animals, the increased time expended in hand-milking will in all likelihood be offset by the lengthier cleaning process needed for a surge machine, leaving neither process with a clear advantage timewise. And there can be no measuring the value of the more intimate and multisensory information the hand-milker gets every time she sits down to milk—temperature, turgidity, and consistency of the udder; state of the teats; and so on. A good milker, one who works gently and milks thoroughly, is an asset to udder health in a way a machine can never be.

And of course, hand-milking isn't dependent on the electric power supply. On the other hand, not

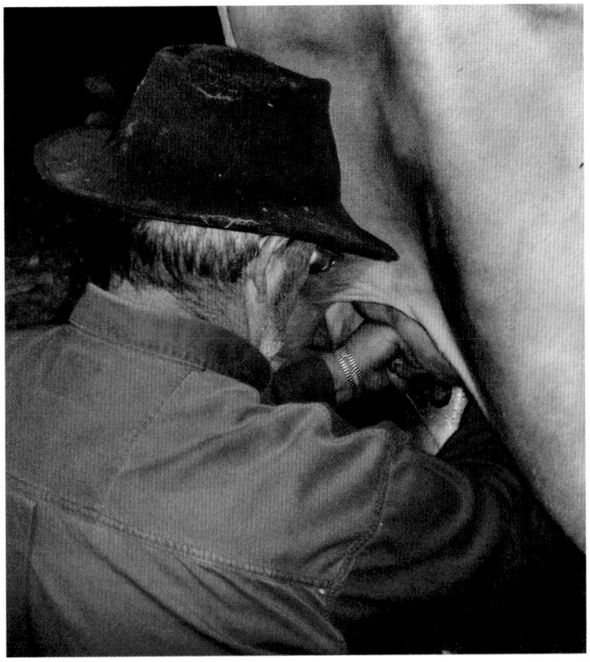

Figure 10.5. Although hand-milking is generally done from the right, ruminants can be trained to accommodate either side—or both. Courtesy of the Franciscan Sisters T.O.R. of Penance of the Sorrowful Mother.

everyone today has acquired the skill of milking a ruminant by hand, so while your best friend may be able to feed the chickens for you when you're away for a couple of days, unless you use a machine she probably won't be able to manage the milking.

Starting Clean

Whichever method or methods you settle on for your homestead dairy, the first thing you need when you start milking is a clean udder. A cow's milk-making apparatus is at the same end as her waste-elimination apparatus, and there's no industry-approved barrier between them to guarantee that the product of one will never come in contact with the operating parts of the other. In fact, they are almost certain to, at least once in a while; though the pasture-kept ruminant is far less likely to lie down in her poop than is her confinement-kept sister (who can't help it), intensive grazing paddocks, especially if they are sized for only twelve to twenty-four hours, are small, with the chances that Bossy will occasionally land in a pie being correspondingly increased. So you'll want to get her clean before you start milking, or some of that is going to end up in the bucket.

One school of hygienic dairying says to use a soft, dry brush for udder cleaning. This works fine to remove the loose hairs and flakes of dry manure that could fall into a bucket, but doesn't do much for wet stuff. Then there's the spray-and-wipe method, which uses a pump sprayer of water, or water and disinfectant, to spritz the udder and teats before a wipe-down with a clean towel. This isn't a bad way, if all you're after is a little smudge; however, for an udder that really needs cleaning it's going to be inadequate. Our own method is to wash teats with a clean cloth and a small bucket of water—warm in winter, cool in summer—the bucket held under the udder for either a quick wash-down or a thorough swilling, as needed; then drying and massaging with a clean towel. Kudos to us, we recently encountered an old farm circular from the early 1900s recommending just this approach. Some people like adding a disinfectant to the wash water, but we prefer not to tamper with the natural bacterial inoculants in the teat openings. If the cow should have mastitis in one quarter, we'll probably add Betadine to the wash water, to decrease the likelihood of its being a vector for passing infection, and wash that quarter separately.

Once we've prepped the udder, there's still another step toward making sure no contaminants end up in the bucket: the strip cup. This is a small cup with a screen cover into which we direct the first squirt or two from each teat. That way, if there is any small bit of manure or other impurity hiding

Figure 10.6. Horse nettle. Toxic and widespread, it seems to pose no danger to intensively rotated animals.

out at the teat opening, it will be pushed out and end up with the cat milk in the cup, not in the good stuff going into the bucket. The strip cup is also our first-level mastitis detector, the screen being there to catch any flocculation or stringiness in the milk and alert us to the probable presence of infection. If you don't have a strip cup, or you forget it up at the house, you can get much the same effect and information by directing the first squirt or two from each teat against the side of your black rubber muck boot.

Milking Technique

When you milk a goat in a stand, you sit next to it, and we've seen people milk sheep from behind, but when you milk a cow you've got to put your legs somewhere while you're sitting right up against her, and that somewhere is under the cow, holding the bucket. Depending on how tall she is, and how close her udder is to the floor, we hold the bucket between our knees, or calves, or both, with our legs up under her belly or even, if we are very tall or she's very short, sticking out on the other side. You may have seen pictures of people milking into a bucket that is sitting on the ground under the cow, which is fine if you want manure on the bottom, or for her to kick it over, or you want to milk with one hand while you hang onto the bucket with the other. For fast, two-handed milking, though, you want to hold that bucket securely clamped between your calves, like a viola da gamba; this way it's far less likely to get knocked over or stepped in, and held right up under the udder, it's far less likely to catch anything other than milk.

The right way to squeeze a ruminant is the way that gets the milk into the bucket! You want to be gentle, of course, but don't overdo it; think about how the calf butts and gollops—and *he* certainly gets what he wants. Getting milk out of a teat is a matter of closing off the top of the teat, usually by catching it between your thumb and one finger, then closing the other fingers successively so that the milk trapped in the teat is pushed out at the bottom. Don't imagine you're using your finger *tips* for this job—you're going to wrap your whole hand around the teat and squeeze in the action known to people who like big words as *peristalsis*; that is, pushing something through a tube by a successive tightening of the muscles encircling it, like in your throat when you swallow. It's not hard, but it does require a level of coordination that will probably not come automatically until you've had some practice. Your thumb-and-finger squeeze at the top will involuntarily loosen while you're closing the other fingers, resulting in the milk shooting back up into the udder, or you'll try closing your hand all at once, with your little finger pinching the *os* closed before you've expressed the milk from the teat. It's a skill,

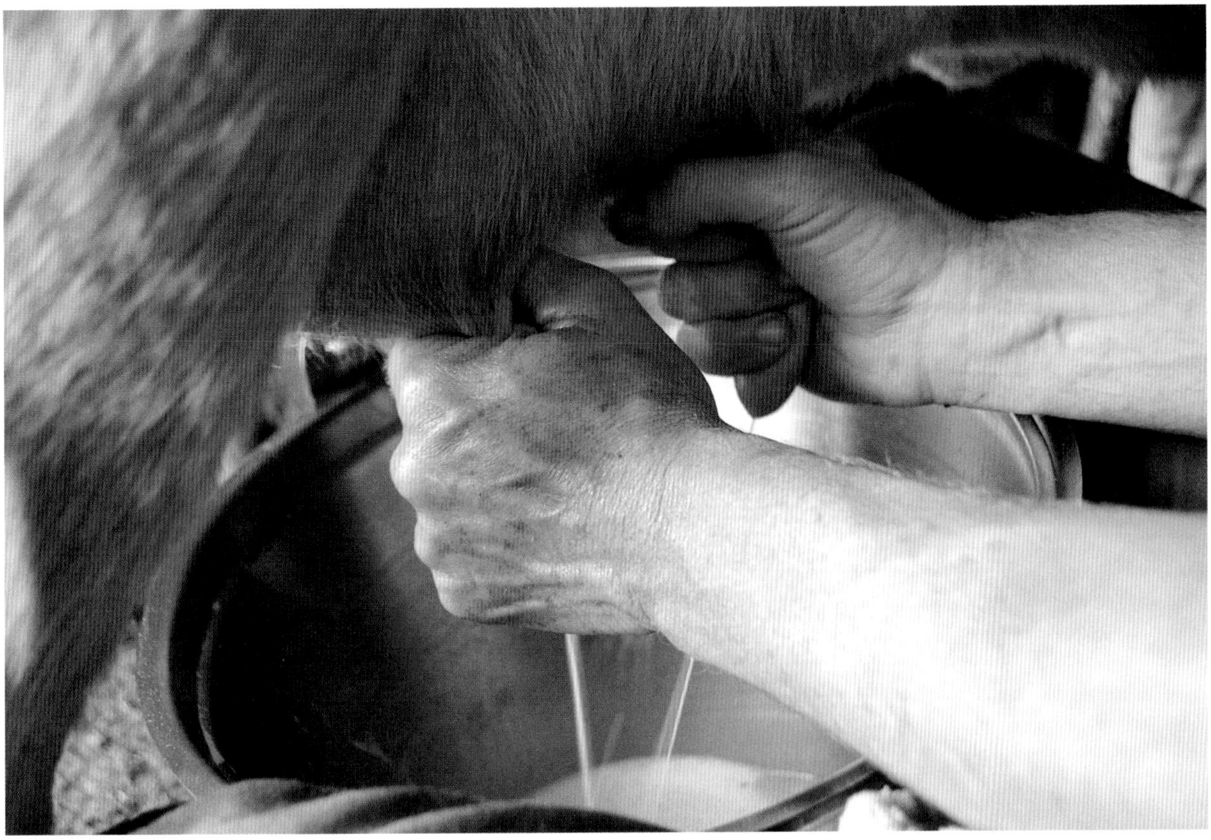

Figure 10.7. Good clean raw milk. Courtesy of the Franciscan Sisters T.O.R. of Penance of the Sorrowful Mother.

like any other skill—you begin with labor and attention, but in time your muscles learn what is wanted from them; they gain strength, establish habit, and then they do it automatically—even involuntarily, like when you're lying in bed at night falling asleep.

Now, cow's udders come in a wide variety of shapes and sizes; they can be large, fatty, and pendulous or small and businesslike, with short teats, long teats, teats like fat sausages or slender, ladylike teats. Modern commercial dairy breeding has in recent years favored the short teats that work best with inflations (but not with calves, or fingers), and the homesteader who hand-milks will learn very quickly to prefer a good, large teat that offers a generous grip and holds a substantial squirt of milk. Short teats can be a real pain, especially before time and use (yours and/or the calf's) have stretched them out a bit. They may require some experimentation at first, in order to find the best grip; however, with persistence, the milker will find whatever arrangement of finger and thumb works best to get the milk into the bucket.

Ideally, you'll be able to get your whole hand into the act, thumb, all four fingers, and palm, since this gets the biggest squirt for your squeeze and is generally most comfortable. Grasp the teat so it lies against the palm of your hand, gripping high so that when your thumb presses down it will close the teat at the bottom of the udder; this way, as your fingers, beginning with your index finger, close successively, the milk trapped in the milk canal has nowhere to go but down and out. What if the teat is too short for a full-hand grip? Smaller teats will get lost in a large hand, and the milk, when it shoots out, will spray into the fingers before it drips into the bucket. Sticky, slippery, and unsanitary. So, we find another way. If cocking your little finger out of the way like a Victorian lady drinking tea isn't adequate, try using just the thumb and the index and middle fingers, closing the fingers not against your palm but against the base of the fingers themselves, like the letter *E* in the sign language alphabet. The squirts are shorter this way, but they're faster, too, so after some practice it doesn't slow things down much. With *really* short teats you may have to start out with just thumb and forefinger until the quarter is less full and gives you a higher purchase on the teat. How, you may ask, can peristaltic action—the progressive squeezing off of the canal—be accomplished with just one finger and the thumb? With very short teats, the thumb can close the teat by catching it against the base of the forefinger, then the forefinger itself swings down on the length of the teat like a lever on a hinge to complete the squirt. The first few shots may end up in your palm before you find an angle that directs the milk into the bucket, but you'll soon get it worked out; then you pray that the teat stretches—or, failing this, that her next calf is a replacement heifer and doesn't inherit her deficiencies.

Which teats first, and which hand for which teat? Front first, back first, right or left side first? Whatever works for you and the cow. The amount of milk made in each quarter, and the rate at which each quarter empties, will vary considerably, which means that while some quarters empty quickly, others require more time or make you come back to them several times. Sometimes a pair of teats is set close together, making it easier to milk diagonally opposite teats so your hands aren't getting in one another's way. There is no right way; you just do whatever works.

Milking must of necessity be firm, but it should always be gentle, with no yanking or pummeling. Massaging the udder, even very vigorously, encourages milk let-down, as you know if you've ever watched how a calf (lamb, kid) will butt with all the petulance of a spoiled child when he doesn't think his dinner is coming fast enough; however, hard squeezing of the udder tissue, with sharp pinches or pokes, can be painful and even incline the udder toward mastitis. If you wash the udder prior to milking, you'll automatically begin the session with some relaxing massage; toward the end, when the milk is coming slowly, you'll want to massage again, which will work the milk down from higher-up *alveoli* (milk vesicles) and encourage the production of more milk. At the last, work each quarter of the udder with two hands, one to gently manipulate the udder, the other to squeeze the teat, emptying the reservoir just above the teat canal. The last, creamiest drops of milk are here, and it's a waste to leave them; in any case, leaving milk in the udder naturally signals the cow to make less milk—less demand, less supply—and can even, when substantial amounts are left in the udder, predispose your cow to mastitis.

When the cow is well and truly milked out, set the bucket aside somewhere where it won't be knocked over or have a bit of something flicked in by the cow's tail, and cover it with a clean towel—or strain it right away into the milk can, if you are using one. Now is the time to rub a bit of udder balm or other unguent into the cow's teats, softening and smoothing them (nicer for you), while at the same time protecting any scratches from infection (teats are low hanging and often get scratched on weeds and briars). And a little smear of bag balm over the *os* of the teat is supposed to discourage flagellates that might work themselves into the teat in the short time there is still a bit of milk there. Bag balm may not be absolutely requisite, but we think it makes things nicer for everyone, except maybe the calf, if she's nursing one. If she is, we just use a little less balm.

Bovine Cooperation

So how does your cow learn the milking routine? Supposing you've bought a four- or five-year-old cow from a hand-milking grass farmer, she's already familiar with having a human being snuggling up next to her hind leg and handling her intimately for fifteen minutes twice daily—but if such is not the case, you're going to have to give her lessons in interpersonal relations. Fortunately, this isn't usually a big challenge. Thousands of years of accumulated experience has made the domestic ruminant remarkably conformable to the human will, and most dairy animals, especially those accustomed to the proximity of the dairyman and that associate his presence with good things, will settle down quickly into a smooth routine in which flying feet and upset buckets have little place. There are, to be sure, certain things that may be helpful to know from the outset.

The obvious: move quietly around your animals; don't hit or yell at them (smacking a bovine derriere to get her to move doesn't count). Don't surprise her or sneak up on her, but let her know by voice or contact where you are and what you are doing. Don't make the first touch she feels be your hand grabbing her udder, unless you are ambitious to get kicked through the wall. Most of the time a bovine kick isn't such a big deal; most of the time if you stop a hoof it wasn't you she was aiming at to begin with, but a fly on her belly, or the bail of your bucket that you were unconsciously digging into her off leg, the sort of kick that is more like a brush-off, never intended to land a good blow. But surprise her with a snatch at the udder and you'll put her in self-defense mode, imagining all the ravening predators whose first move might be to sink their teeth into her tender parts, and she'll react accordingly. A better way to initiate a milking session is to start with your hand on her back, then move the stroke down her hind leg toward your real goal, so she isn't taken by surprise.

Figure 10.8. Waiting to be milked. Courtesy of Molly Jameson.

Kicking

Since it's no fun getting a hoof in the lap, or the leg, or the milk bucket, it's good to be aware of some of the other things that might make your ruminant kick, and how to avoid them. In the summer there are flies, of course, very annoying for you as well as your animal; many's the hoof that's landed in a bucket from a kick at a fly. To deter the little pests, you can dust the animal with insecticide (bad) or put a fly tag in her ear (ditto) or spray the entire dairy with fly spray (same again). We don't like toxic chemicals in general, so instead we set up a box fan next to the stanchion to keep the air moving, which seems to help some and moreover feels so good the cow minds the flies less, and so does the milker. There are organic fly deterrent sprays available; most people use these for horses, but there's no reason you couldn't keep a spray bottle in the dairy. Vinegar is supposed to be a good natural fly deterrent, but we can't say we've had notable success with it. If we've got an animal with really restless feet, we take some time and help her out a bit, stopping periodically to flip a towel at flies on her legs and belly; with some animals this extra effort really makes a difference.

Other things may irritate your ruminant. If Bossy is restless, check your fingernails; if they aren't very short, trim them, and see if she isn't more comfortable. Rings, especially big ones or many of them, can dig into a teat, or pinch it, and keep your animal jumping. Has she caught a teat on a briar and torn it, or is her calf leaving tooth marks? Scratches and sores may be bothering her, making it necessary for you to use gentler methods until they heal. Most kicking is really foot lifting or fidgeting; it's very rare for an animal to be genuinely bad tempered and aiming to hurt you. That is not to say that it does not happen, however, and if you've got one that is repeatedly aiming a hard one at you, don't wait until she hurts you to do something about it. Kicking irons are an arrangement of clamps and chains meant to discourage kicking by making it hurt her more than it does you, but we find that they keep a cow's legs too close together to allow easy access to her udder. As a last resort, sell her and buy a replacement.

Figure 10.9. Ragweed takes over on waste ground, but intensive grazing gives the advantage to forage species.

Any cow can pick up her foot and put it in the wrong place, and most of the time when this happens it has nothing to do with you; nevertheless, if her foot ends up in the bucket the least you'll have is a bucket of milk that has to be fed to the pigs, and maybe you'll have a broken bucket as well. So it's good to know there's usually something you can do about the careless foot. First of all, stay aware of those back feet; keep them in your consciousness, take them into your consideration of affairs the whole time you're milking. In the beginning this may mean a deliberate effort to watch her feet—or, more to the purpose, her hocks (the backward joint on her leg where the noninitiate might expect to find her knee, but which is really analogous to the human heel), since hocks are up where you can see them (not hidden by the bucket), and since the feet can't move without the hock moving too. After you've had some experience, you'll find that even when you're staring out the dairy door with your brain on autopilot, you'll be alerted by the slow shift of weight which means, even before the hock rises, that your animal is about to raise a foot. Forewarned, you'll be in a position to shift your hold on the bucket, or move it aside, or, if the foot raised is the one nearest you, stick out an elbow or forearm to fend the foot away. There's not usually a lot of force behind these footlings; often they're just to ease her position. She isn't really aiming for the bucket in any case; she doesn't enjoy having her foot in a pail any more than you would.

Let-Down

Other things can get in the way of successful milking, not least of which is the cow herself. Even once you've got her in the stanchion munching a treat while you're getting cozy with her posterior, there's the very important issue of *let-down*, that reflex that brings milk from the glandular tissue where it is made to the teat where you can get at it, and without which all your squeezing doesn't amount to a hill of beans. Let-down happens when a cow is relaxed and expecting to give milk, willing to give it; she gets it when her calf cuddles close, of course, and, when she is accustomed to be milked by a human, she gets it when she steps up to the stanchion with a full udder. But this latter is a learned reflex, learned through the experience of having her milk burden relieved by the hands of a human being, and it may take a few lessons before she understands that, while you're not her calf, there's still a good reason to let her milk down for you. Mostly you get this standoffish attitude from a first-calf heifer who has never been milked before, especially if you're leaving her calf on her for a while. Once the lesson is learned, it's there with every freshening; nevertheless, even in a cow of some experience you may have a few days in the beginning when your animal, despite having a visibly full udder, is holding out for the baby you're competing with. When this happens, some firm udder massage is indicated, followed by persistent efforts to get milk with your squeeze; two, or even three or four massage-and-squeeze treatments may be necessary before you get the faucet turned on. Bringing the calf into the parlor can be helpful; if necessary, you can let him nurse one side while you take milk from the other.

Sometimes the let-down reflex will be slow for other causes. A complete change of surroundings can be disorienting, and it isn't surprising if sometimes this means a slow let-down, or a drop in production. Sometimes a change in pasture mates, forage composition, even water source, especially with a newly purchased animal, will cause a cow to hold back her milk, or to reduce her production suddenly. Our own cows prefer captured water—rain or spring—over well water, a fact we notice when they are on the southeast pasture, which is watered from a deep well. Even though when they come up to the barn twice a day for milking they tank up at the rainwater barrel, their milk production drops for the period of time they are on that pasture.

Not every straining udder is a full one, however, and in the first couple of weeks after calving, lambing, or kidding, your ruminant will experience, to one degree or another, a swelling of the mammary tissues that is not engorgement, but edema. This is normal after birth, even to the extent of making it awkward for the animal to walk around her

Figure 10.10. Comfrey, a forage plant high in protein and B-12.

distended udder; in such a case the udder will be almost as large after milking as it is before—without, however, indicating that the animal has not been thoroughly milked out. How to tell the difference? If several minutes of determined massage and teat-squeezing are not initiating a let-down reflex, you are probably safe in assuming that you've gotten what there is to get. In any case, if you are milk-sharing with her calf, lamb, or kid, that baby is providing you and Mama with a safety net in case of inadequate milking and will be happy to take whatever you may have missed. In a couple of weeks, when the milking routine is familiar to her, you'll find that your cow isn't holding back on you anymore but is letting down when she comes into the dairy.

Tails

Cows switch their tails a lot and can club you on the head, sting your eyes, or slap manure in your face, in which case you can elect to tie the tail, as with a piece of string to a post or the wall, only don't forget to untie it or you can have a very excited cow pulling your knot so tight you have to cut it (we keep a sharp knife in the dairy for this emergency), and maybe leaving part of her tail behind and thereafter being afraid to step into the stanchion. We know; we've done it. If you're going to tie tails, and it's better than getting smacked all the time, be sure to use breakaway string and avoid crises.

Milk Handling

Once you've got milk in the bucket, you want to take good care of it to ensure keeping it the excellent source of high-quality nutrients it is. When milking is done, we cover the bucket with a clean cloth, if there is no lid, or we transfer to a milk can with a lid just for that purpose. Milk cans are easier to carry without spilling than are buckets, they generally hold more, and they come with lids. The ideal time to strain milk is in the dairy; any particulates—chaff, dust, hair—that have fallen in will carry a burden of surface bacteria that will have the least effect the sooner we get it strained out again. This is not some icky-clean, barns-are-dirty, scary-germ thing, but just a simple matter of good hygiene; the fewer alien bacteria there are in our milk, the longer it will stay sweet, and the more delectable the fermentations it will make. We strain in the dairy, which makes one more piece of equipment we have to schlep up and down the hill, but since we are generally milking five or six cows we know there will be some delay before the milk gets to the dairy to be strained, and we want to keep our milk as clean as possible. So as we say, we strain in the dairy into milk cans with lids, and we keep the strainer covered with a muslin towel between uses so nothing unwanted—flies come to mind—gets into the strainer.

Figure 10.11. Straining in the parlor.

Once we have the milk up at the house, we strain again (almost certainly unnecessary, but we compensate for wild abandon in some areas with overcaution in others) and refrigerate. We use widemouthed glass jars so that later we'll be able to get a ladle in there, for skimming. Another reason we use widemouthed jars is that milk is only going to be as clean as the jar you store it in, and we've never figured out how to get a jar really clean without getting a hand down in there to scrub it. Two-part canning jar lids work well for the same reason: no crevices for stuff to hide in. Old-fashioned quart milk bottles are cute, but they don't meet our easy-to-clean criterion.

At this point it's traditional to cool your milk, not because milk is a dangerous product and has to be treated with utmost caution or it will turn toxic, but in order to hasten the separation of butterfat

Cleaning Up

You read a lot about the dangers of contamination with raw milk, and commercial dairies use lots of special chemicals in their attempt to keep their equipment clean. We share here our own equipment-cleaning protocols but wish to add that for thousands of years human beings have processed and stored milk and milk products in the simplest of containers, made of the most natural materials—clay, wood, metal, basketry, cloth, animal gut—cleaned with the most natural purifiers: water (hot and cold), air, sunlight. Over the same period of time mankind discovered—we avoid saying "invented"—myriad delectable lacto-ferments directly resulting from these natural materials and methods. There is no need for fancy methods and toxic chemicals on the independent farmstead.

So. We rinse dairy buckets in cool water as soon as they are empty in order to avoid calcium build-up, also known as "milk stone." Milk buckets, strainers, and wash buckets (for udder washing) are cleaned as soon as they are empty, with a powdered soap made specifically for use with dairy utensils; they are then rinsed with hot water and spritzed with a mild bleach solution before a final hot rinse and air drying. Milk storage jars are glass, never plastic (only widemouthed jars, so we can get down into them for cleaning); these we wash by hand in ordinary dish detergent and hot water, and air dry. No matter how healthy the cow is, milk can only be as clean as the buckets and jars you are putting the milk in.

from the liquid milk; it will rise faster and thicker if the milk is cool. Actually, we can think of only three reasons to cool milk: for long-term storage (like a week) when you don't have your own dairy animal and you want the milk to stay sweet (as opposed to fermented); to get more cream to rise (as when you want to make butter); or because the milk is contaminated. Well, we throw contaminated milk to the hogs, so that one's out, and our own milk turnover is rapid (after all, more is coming in every twelve hours) so we always have sweet milk in the house, but we do share the modern prejudice for cold drinking milk (habit), and we use a lot of cream for making butter, sour cream, cream cheese, and the like. So we get our milk into the refrigerator immediately on bringing it into the house. We don't immerse the jars in an ice water bath for thirty minutes first (although we used to when our dairy animals were goats and we were trying to keep the goaty taste to a minimum); we just strain and refrigerate.

What about all the advice one sees on the necessity of flash-cooling raw milk to keep it wholesome? Well, let's think about that: for thousands of years humans have been milking ruminants and consuming the milk with great benefit to our health, when there was no possibility of refrigeration. Milk was cooled in shallow pans for cream separation, after which the skim milk was mostly used as animal food (and the best possible animal food it is) or combined with fresh whole milk for cheese making, while the cream was stored at cellar temperature until there was enough of it to make it worthwhile either to churn it or to haul it to town. Drinking milk was whole and fresh (and warm); other milk products for the table, like yogurt, acid buttermilk, and kefir, were fermented at room temperatures and consumed soon. If it were really necessary to maintain dairy products at a temperature just above freezing to make them safe for human consumption, how did our ancestors manage?

Milk for Drinking, Cooking, and Storing

What can we say of milk? It is a food above all others, so honored that the Psalmist describes happiness as a land "flowing with milk and honey," so nutrient packed that whole cultures have arisen around milk as a primary food, ultrafresh because it is constantly renewed from each day's sunlight. With milk from your own grassfed ruminant, you are as nearly food independent as it is possible to be. And despite our modern cultural ignorance, storing and processing milk for human use is a simple and beautiful art. The centuries of dairy women whose butter, yogurt, kefir, and cheese fed their communities performed their alchemy without thermometers, freeze-dried cultures, water jackets, or litmus paper, which ought to reassure us that none of these things are necessary to the making of sublimely delicious dairy foods. Whichever species of ruminant you are pasturing, you are missing the greater half of its benefits if you are not harvesting, drinking, and fermenting her milk.

Cream and Butter

Even if you don't skim your drinking milk, you'll want to know how to skim the surplus so you'll have cream for butter. Getting a substantial proportion of the butterfat from goat's or sheep's milk requires a centrifugal separator, since the fat particles are very small, but skimming cows' milk is simply a matter of lifting off the layer of cream that will rise to the top after the milk has sat undisturbed a few hours. A ladle works just fine for skimming, so we make sure to store milk in jars with wide mouths that will

Figure 10.12. Hard cheeses aging in a cheese cabinet.

admit a ladle for this purpose. Skimming with a dipper is something that anyone can do. One tries to let the cream flow into the dipper without dipping deeply enough to get milk with it. You get some anyway. It doesn't matter; a little milk won't prevent your cream from churning. In fact, we have read several times that you could churn whole milk into butter, without the bother of skimming first; it just takes a long time. We'll take their word for it; under some conditions, churning takes a long time anyway.

Skimming is not difficult, and practice makes proficient. The top of the cream tends to clot into a thick skin, which will stick to the bottom of your dipper and make it hard to gather; this can be avoided for the most part by wetting your dipper at the sink before you start, and cutting down through the top layer with the dipper edge instead of trying to immerse the entire bowl of the dipper. You can also use the bottom of the dipper to sort of edge the clotted top cream to one side so that you can get underneath it. In glass jars you can see the cream line, so you will know how far down the jar to go; after a while you won't need this clue, because you will see the cream flowing in over the rim of your dipper, and when it runs in streaky blue you are getting milk. There is no need to take every last bit of cream, anyway; it will make decent drinking milk if you leave some of the cream on, and if you are going to feed the skim milk to the pigs and chickens, they will convert it into bacon and eggs, and who would consider that a waste? Even if you take the whole layer, there will be a substantial proportion of butterfat still in your milk, something you can demonstrate for yourself if you leave a jar of skimmed milk in the refrigerator a couple of days: there will be a new layer, less thick, of cream on top when you come back to look at it. If you really want to get all the cream out of your milk you can buy a centrifugal separator, and then you'll be wasting all the cream you have to wash off the umpteen parts of it.

Making raw-cream butter can be more complicated than the lady at the dairy council table at the county fair, who uses centrifugally separated, pasteurized, homogenized, whatnot-ed cream, makes it look. On the other hand, it is not difficult; there are just more permutations when we are using living cream. Butter is the natural product of bashing cream around until the fat molecules meet up and stick together, and you can do it in a mason jar, if you don't want much, and if you have lots of people with nothing better to do than stand around shaking the jar; you could use your stand mixer, which will splash a good deal; or you can invest in that pernicious thing, a single-use tool, and find out how invaluable a good churn is to the busy dairy woman. We have always used an electric churn with a propeller dasher and find they work well, some models better than others.

We usually collect cream from a couple of days' milk before churning, so that we can make several batches at once, thus saving ourselves cleaning the utensils after every pound of butter. One-half gallon of cream makes about a pound of butter and leaves something more than a quart of buttermilk. This is not the cultured product you buy in the store, which usually isn't buttermilk at all, but just milk cultured with mesophilic lactobacilli and thickened with carrageenan or something like that; real buttermilk is the by-product of churning cream into butter, and made with sweet cream it is sweet and lowfat and delicious. Made with more cultured cream, it is acidic and may also be delicious, but this will depend upon how clean your culture was, and how often you took the cream jar out of the refrigerator in the interim to pour some in your coffee. The butter, we hasten to assure you, will be good regardless, but the taste of the buttermilk will vary the way yogurt does when you have kept a culture going for a long time. It will still be good for cooking with or, worst case, for feeding to the hogs. And in any case, the quality of a batch of buttermilk is not something to get uptight about, as the keeper of a family cow will have lots of buttermilk.

So we wait until we have collected a gallon (or 2, or 4) of cream before we churn. Then on the day we intend to churn, we set the cream out on the counter for an hour or so to warm and culture. An hour is sufficient for summer cream and summer butter; in winter it may take longer if your house is cool. You may read that the "right" temperature for churning is 50, 60, or 65 degrees, and somewhere in there should be about right. The jar should still be cool to your hand when you touch it. The churning process is childishly simple. We pour cream into the (it goes without saying, but we'll say it) clean churn jar until it is half full, or maybe just slightly more, put on the lid, and turn it on.

What we were waiting for while the cream sat out on the counter was for it to warm and acidify enough that the globules of butter fat present in the cream will gather together in clumps when they are bashed about. If, indeed, this has happened, we will churn our butter for something between five and, say, twenty-five minutes, while the cream passes through several stages; first liquid cream, then whipped cream (that's why our jar can't be more than half full), then overwhipped cream (rather runny). After this is a stage in which the milk fat globules are gathering into small grains of butter; these should, as we continue churning, gather into big lumps, which will be bright yellow in the summer and paler in the winter. (Sometimes, however, the butter stalls between the last and next to last stages, and we have to wait awhile and try again when the cream has further acidified.) The gathered lumps of butter will clog the dasher, and since our churn does not have an automatic shutoff, it just slows down, sticks, and starts burning up the motor. Bad idea. So far we haven't ruined the churn, but we try to stay in the near vicinity when we churn, to catch the butter as soon as it comes.

Oh, and the churn jar is going to vibrate pretty hard while the churn is running, so if you don't want it to dance right off the counter and onto the floor, put something under it—a towel, maybe—to absorb the jiggle and hold the jar in place.

To drain butter you can wedge a slotted spoon or a ravioli turner or something of the sort into the opening of the jar and pour out the buttermilk, holding the butter back with the spoon. Or you can

Figure 10.13. A wooden bowl keeps butter cool while it is being pressed.

reach down into the jar with your bare hand and just gather the butter into a lump, so none of it will wash out with the buttermilk when you pour it off. We pour out as much buttermilk as we can, saving it for cooking or for the animals. Then we add cold water to halfway up the side of the jar, reinsert the dasher, and churn the water and butter for a few seconds, or use our bare hand to knead the butter in the cold water. It will gather and firm up; after a bit we drain the milky water off, refill the jar with cold water to the halfway mark, and churn or knead again. This serves the purposes of 1) washing the (very degradable) buttermilk out of the (only slightly degradable at cool temperatures) finished product, the butter, and 2) cooling the made butter so that it may be packed firmly and the water pressed out of it. In all, we wash in at least three waters, or maybe four; this is almost always enough to ensure that the water runs mostly clear, indicating that the pockets of buttermilk in the butter have been replaced by

Salting Butter

How much salt you add to your butter is completely a matter of taste, so long as you are not planning to store it for an extended period without refrigeration. For table use and cooking, it's up to you. Some people prefer unsalted butter; it has a subtle sweetness that can be delicious on a thin slice of coarse wheat or rye bread. For everyday use, we salt our butter lightly, less than ½ teaspoon per pound (and some of this will be worked out, dissolved in the last drops of water forced out of the butter from the washing).

Heavy salting of butter is one way to help it keep longer without refrigeration. Larger doses of salt are worked into butter that will then be packed tightly in a stoneware crock (nonconductive and insulating) and stored in a cool cellar or better, a spring house, if one is so blessed as to have such a thing. The butter must be packed tightly; exclusion of air is as important as heavy salting or earth-stabilized temperatures in preventing the butterfat from oxidizing and becoming rancid. The larger dose of salt won't affect its use: when a portion of the stored butter is wanted for the table, it is taken from the crock (the remaining butter is then pressed down and covered tightly) and washed again in cold water. Just put it back in your butter bowl, add cold water to cover the butter, and begin working it over with the back of your wooden spoon; the excess salt will dissolve and be poured off with the wash water.

pockets of cold water. If the butter we have made is slick, soft, and shiny—the result of overwarming our cream—this cold-water washing will sometimes firm the butter up, remedying the situation. Sometimes not. Of which, more anon.

Once the butter is washed, we drain off the water and put the lump of butter in a wooden bowl (wood is a poor conductor of heat, and not so slick that the butter will slide around instead of sticking to the sides of the bowl) and then use the back of a wooden spoon to press out whatever water remains in pockets in the butter. We're pressing, not smoothing this butter, and there's not much we can say about how except you'll find your own way. We like to use a bowl with high sides so we can press against them. Extracting all the water isn't something to get uptight about, since it won't spoil our butter if there is still a little water in it. We just work the butter until we can no longer pour off any water, then add salt, work the butter to mix the salt in, and there it is: butter. We shape it into ½-pound rolls with a wooden spoon and refrigerate them if they are not to be used immediately; if we want to freeze them, we wrap them in waxed paper and place them in a plastic bag first.

Troubleshooting

Sometimes, despite all our efforts, we end up with a product that can't be washed, can't be shaped, maybe can't even be gathered into a lump. Sometimes our butter won't "come," which means that no matter how long we churn it, we don't get a lump of butter we can fish out of the churn and process. We may end up with a churn full of overwhipped cream with small, soft grains of butterfat, or ditto with small, firm grains of real butter that for some reason won't adhere to one another and make a manageable lump. Sometimes when our butter comes, it is a slick, shiny, fluffy mess impossible to wash or shape. Rarely, we've ended up with a churn full of something we're glad to give to the pigs and forget. Rarely, not often; mostly, we can convert the variations we encounter into real, firm butter, with just a little extra manipulation.

Take the instance of the churn full of tiny firm grains of butter over a shallow layer of thin, sweet buttermilk. We have let the churn run for some time after this stage was reached, and still the butter won't gather. The contents of the churn, when we shut it off, resemble something like runny Cream of Wheat. There are two possibilities here that we know of: one is to give the churn another hour on the counter, while the culture continues to grow, and then try again. Maybe by that time the cream will be acidic enough to gather itself into a ball when we churn it, maybe not. But in case this doesn't do the trick, if our grains are firm enough it will sometimes work to drain the grainy stuff through a piece of cheesecloth for several hours, chill it, and then put it through the mixer using the paddle attachment to express the buttermilk. This means we will have skipped the stage of washing the buttermilk out, so we keep this butter in the refrigerator and use it soon, but otherwise it is butter, and no harm done.

Regarding summer butter, the commonest problem we have met with is one we call fluffy butter. This happens when we let the cream get too warm during ripening, and the churn, instead of gathering the firm granules of butter as they form, whips the soft lump of butter into a fluffy mass. This is difficult to drain, since it tends to adhere to the slotted spoon we drain with, clogging the holes; it would be very difficult to wash or press. The remedy we use in this case is to shove the whole kit 'n' caboodle into the refrigerator for a while, until the butter is firm enough that we can move on to the next step. Often by thus chilling the butter we can right the whole situation; we will be able to drain our butter, wash as we would normally, work the water droplets out, salt, and shape. It won't be a bit harmed; butter it still is, and no one meeting it on the street would ever know it had a checkered past.

At other times, especially in winter, the problem will be that the butter doesn't come at all; the cream will pass through the first stages, of whipped cream, and runny, overwhipped cream, but at the stage where it should form small grains that will rapidly gather into large lumps, the process just stops. Beat the cream as we may, the butterfat simply will not gather. We have no degrees in lacto-chemistry, but we speculate that in these cases the cream has not yet cultured enough to achieve the pH necessary for the tiny globules of butterfat to adhere to one another; this speculation is supported by experience, which tells us that we can usually remedy the situation either by giving the cream another hour or so on the counter and then trying again or, better still, adding a few ounces of mesophilic starter (buttermilk), waiting the additional hour, and churning again. Most of the time, the process will then go on as intended. On the other hand, there are times when—again we speculate—unfavorable bacteria have gotten the lead on us, and before our mesophilic culture can acidify the cream adequately, the cream has spoiled. In this case it will smell cheesy or ugly, look foamy, and be fit only to be fed to the pigs. So what. Tomorrow there will be as much cream again, and the pigs will be a little bit fatter.

We just love pigs.

Lacto-Ferments

Clean raw milk that sits on the counter for a day or so becomes fermented milk, and what you call it seems to depend more on where you live and what food culture you hail from than on anything else. Commercial yogurt uses one or a few strains of lactobacilli—usually *Lactobacillus acidophilus* or *bulgaricus*—to acidify and clabber milk under rigidly controlled, hermetically sealed conditions, but your milk is not limited to a narrow range of milk bacteria; it comes preinoculated with your cow's personal array of mesophilic (loving midrange temperatures) and thermophilic (preferring to grow in warmer temperatures) converters, ready to grow under whatever conditions your kitchen has at any given moment. Lower temperatures (70 to 90 degrees) incline your ferment more in the direction of what in the United States would usually be called buttermilk, and which has little to do with the making of butter but only means "milk fermented at 70 degrees or so." Placed in a warm spot (say, 90 to 110 degrees), like at the back of a wood cookstove, if you have such a beast, or on top of the refrigerator (the motor running is supposed to keep it warm there), or wrapped in a towel next to a jar of warm water, or in any other spot where it will stay consistently warm and undisturbed for eight hours or longer, milk's heat-loving bacteria will be happy, and they'll do you the favor of converting your milk into a sweet, thick acid ferment Americans would likely call yogurt. And somewhere in eastern Europe, thousands of years ago, there developed a combination of midtemperature bacteria and yeasts that clump themselves into chewy grains in your milk jar, grow at room temperatures, and make foamy, sharp, fermented kefir, and if someone you know is growing kefir you can beg a few grains from her and start your own colony. Creating the conditions that favor your desired end product will incline your milk in that direction, as will "backslopping"—adding a dollop of a previous batch of ferment you particularly liked.

There's no need to be afraid of our milk. Modern American taste buds have been trained by commercially produced food to expect uniformity, and,

Starting Milk Cultures

Sometimes the yogurt or buttermilk culture we have been using may get contaminated in some way: maybe someone reached into the jar with a spoon that was not clean, maybe the lid picked up something off the counter, maybe we just left it too long without feeding it and it declined, like a sourdough culture may do. Mold may form on top, or the culture may acquire a sweet or cheesy smell. Using it to make a new batch of ferment, we may find that gas bubbles form, indicating that a yeast has gotten in with our lactobacilli. We won't want to use this culture as the starter for a hard cheese in any case; its denizens are too unpredictable. So when this happens we need to get a new culture started.

There are several ways you can supply yourself with a new cheese culture. One would be to buy a factory produced, freeze-dried culture from a cheese making house. This will include a narrow range of lactobacilli whose behavior and preferences will be very predictable, and it's what most cheese recipes will call for, and it will probably work just fine for you, for a while, at least. But these lactobacilli will not necessarily be the strains that like your cow, your forage, your cow's milk, or your dairy conditions, so if you want to keep working with this culture, you will probably need to buy new starter from time to time, to renew it.

A second option would be to steal a dollop of fresh starter from a cheese making friend, as you would steal kefir grains if something had happened to yours. This may be a good option, particularly if the friend lives nearby and has the same breed of cow you do, and similar dairy practices.

If, however, you don't think you know someone who can give you a good raw milk starter culture, don't despair; in reality, she's no farther off than your pasture. Your dairy ruminant is the perfect person to start a new clean culture for you, and the process is simple.

Wash a widemouthed mason jar—a pint jar will do—in hot, soapy water, and rinse well; then sanitize it by pouring it full of boiling water from the kettle and letting it sit a minute. Wash and sanitize the lid as well; two-part lids are better than those white plastic ones they make now, which have ridges on the inside that are difficult to get really clean. Drain the jar upside down for a couple of minutes and then put the lid on, and take it down to the dairy with you when you go down to milk.

Now bring the cow (goat, sheep) up for milking and clean her udder well, as you normally would do before milking, making sure to brush off any dust or chaff in the hair around her udder, drying carefully, and stripping two squirts out of each quarter. Take the lid off your jar (set it carefully on a clean towel or somewhere else where it won't pick up any contaminants) and milk directly into your clean jar until it is half full of milk. Then put the lid back on, set the jar aside, and finish milking the cow. When you take that jar of clean, fresh milk up to the kitchen and put it in a warm place for twenty-four hours, it will grow you a nice garden of thermophilic lactobacilli, lots of strains, all specially fitted to your cow, dairy, and forage. If you put it in a midtemperature place, like on the counter, you will emphasize its lovely mesophilic cultures. These will be very nice for cheese making; what's left over you can stick in the refrigerator if you're not going to refresh it every day or so.

To keep our cultures active, we always take the first (hence cleanest) couple of tablespoons off for a new jar of culture, growing it, of course, in a clean, sanitized jar with a clean, sanitized lid, and with our newest raw milk. Refreshed in this way, we find that our cultures actually improve over time, the best-fitted lactobacilli coming to dominate the culture—until we forget to refresh it for a while, or we run out of yogurt and eat our starter up.

conversely, to suspect that any variation or deviation from a strictly defined, industrially determined norm must mean that food is bad, spoiled, and unhealthy. Perish the thought. Factory-produced milk products, made of heat-killed milk, must be carefully guarded from colonization by the legions of nasty bugs just waiting to jump in and contaminate them, so the process of making commercially fermented milk is rigidly controlled, and the product is predictably uniform. Fresh, clean raw milk, varying in composition with the change of seasons, inhabited by its happy family of beneficial lactobacilli, offers a much wider range of flavors and consistencies, subtle variations on a delicious theme. Homemade yogurt made with fresh raw milk won't be identical to the commercial stuff in your grocery store plastic quart, but it does not follow that it is less successful as yogurt; in fact, with its greater array of probiotics, reason would indicate the contrary—and it's delectable. It's important to stay open to new experiences, not letting our personal food production labors get bogged down at the very point of success by some lingering vestige of commercial-food-industry brainwashing.

Figure 10.14. Greek yogurt.

Cheese

The next permutation in the process of milk fermentation, after products intended for immediate consumption, is cheese. Such a misunderstood food! In the arena of cheese making, modern Americans have been deluded, goodness knows how, into thinking that a food easy to make and natural in process is really the complicated and mysterious end product of an arcane and esoteric alchemy; being so deluded, we have been deprived of the pleasure and benefit of enjoying the limitless range of excellent cheeses that might come from our own dairies and cellars. It is as though someone had convinced us that the only safe and tasty pasta sauce was the kind that comes in a jar. Cookbooks for the hobby cheese maker, few as we have found them until recently, have, in our experience, with their instructions involving titrations and litmus paper and freeze-dried milk cultures and mold

Figure 10.15. More tools of the trade.

spores and finicking variations of timing, temperature, and humidity, only reinforce this misunderstanding. For people who are already conditioned to fear any milk that doesn't get born in the refrigerator, no wonder the idea of home cheese making ranks in intimidation factor just above unmedicated childbirth and alfresco winter bathing inside the Arctic Circle.

Let's banish this fearsome impression, now and forever. Cheese making may be complicated and difficult for the novice whose goal is to reproduce a standard factory-type cheese using commercially produced, pasteurized milk full of low-level antibiotic residue and recombinant bovine growth hormone (rBGH)—and not only complicated but expensive. But for the homestead dairy woman into whose kitchen gallons of raw milk are coming every day, grass derived, their proteins intact and with their natural lactobacilli sitting up and wagging their tails, turning milk into cheese is not only not difficult—it's almost inevitable.

That's right, inevitable. Cheese makes itself. What is cheese, after all? Curds of milk from which most of the water has been removed, flavored by bacterial acidification and the growth of molds—just what happened when, as a child, you forgot your thermos bottle at school over a long weekend and brought it home full of clear yellow liquid with a ragged whitish lump in it; only, since that was pasteurized milk, the bacteria which had performed this transformation included bacteria of decay, so the smell and taste would have been putrid instead of delicious. On the other hand, this is also just what would happen if you left a bottle of your good, clean raw milk in a warm spot on the counter for a week or so: the white liquid would be converted to a lump of tart, soft curd in a bath of whey. Beneficial bacteria in the milk, left to their own devices, have converted some of its sugars to acids, making a curd that shrinks over time to release moisture but that contains most of the fats and proteins from the original milk. Deliberate cheese making is just a refinement of this process, in which control is exercised over the type and degree of acidification of the milk, the size of the curd and its firmness, the rate of moisture loss, and, for aged cheeses, the conditions under which it is stored.

Fresh Cheeses

The fresh cheeses we enjoy (*queso blanco*, *panir*, cream cheese, cottage cheese, the range is very broad) are made by acidifying milk—by fermentation, as for yogurt, buttermilk, or kefir, or, in some cases, with an agent like vinegar or lemon juice, as with *panir*—warming it and draining off some of the whey. The tart, thick yogurt we call "Greek" is the result of draining yogurt through cheesecloth; sweet, lemon-curdled *panir* gets its firmness when the curds are cooked to temperatures just below boiling. Cottage cheese, with its distinct texture, has a long fermentation time—up to a couple of days—to develop flavor, combined with ultralow-temperature cooking to set the curds.

The next level of cheese making on our farm is fresh, cooked-curd cheeses like mozzarella, which uses rennet, an enzyme derived from the lining of the first stomach of a young ruminant, to clabber the milk more quickly, and heat to stretch the curd. We usually make mozzarella with skim milk instead of whole, for two reasons: first, we like the mozz we make with skim milk; secondly, when we have used whole milk for mozzarella, we seem to lose all the cream in the stretching process. This is not to suggest that someone else might not make superior mozz with whole milk; it's just our preference.

Figure 10.16. Herbed soft cheese.

General proportions for mozzarella making are as follows: 2 ounces of thermophilic starter (yogurt) and ¼ teaspoon rennet per gallon of milk. Ideally, 1 gallon of milk will give you 1 pound of cheese. That's the ideal, but we hasten to add that we seldom achieve the ideal (and don't actually know anyone who does), considering the savings of time in our method more valuable to us than would be the slight increase of yield we might experience were we to adopt more time-consuming, finnicking methods. Let us say that 1 gallon of milk should yield somewhere between ¾ pound and 1 pound of cheese.

In a vessel of adequate size, we warm the milk on the stove to 90 degrees and add 2 ounces of yogurt for each gallon of milk we are working with. We are aware of the quick and easy citric acid methods, but we prefer to culture our cheese, partly because when you culture milk it increases in food value and flavor, and partly because we don't grow citric acid powder. The yogurt added, we mix it in, pretty vigorously, since we know what we ought really to have done is to have mixed the yogurt with a little milk in a bowl before adding it to the milk in the pot: this would ensure even and complete assimilation of the starter culture, but we know we can cut this corner without compromising the end product, and we're glad to trim a couple of minutes off the job.

Once the starter is mixed in, we dilute ¼ teaspoon liquid rennet for each gallon of milk, in about ¼ cup cool water. We don't measure the water; we just use a small cup and guess. Doesn't make any difference. The rennet, on the other hand, we measure, because it matters, but not so much you have to get uptight about it. Dilute the rennet in the cool water and stir it into the milk, too. We stir about ten strokes, up and down, not around and around. All the recipes we've seen (until recently—see David Asher's excellent book, *The Art of Natural Cheesemaking*) tell you to stir for a couple of minutes, which puzzles us because if we stirred that long our curds would begin to form while we were stirring and we'd end up with a pot of grainy mush. Ten strokes will do it. If you stir a little too long, your curd, when it sets, will have swirling cracks in it, but it doesn't seem to hurt anything. Then we set the lid on the pot, put the pot to the back of the stove (or anywhere else where it will stay warm and won't be disturbed), and go away for half an hour or so.

What we are waiting for when we go do something else for half or three-quarters of an hour is for the milk to clabber into a big, shiny cake or slab, floating in a very thin layer of whey. The whey will be clear and light yellow, and the clabber, or curd, white. To see if it has set a firm enough curd, you stick your clean finger into the curd at a 45-degree angle, and lift it straight up. If the curd is of the desirable firmness it will make a clean break over your finger, not leaving blobby bits of soft mush. Well, maybe just a few little ones. This is not an absolute, either-it's-there-or-it's-not

Figure 10.17. Stirring by hand is gentler on curds.

condition. If you test your curd and it seems to be mushy, just give it another fifteen minutes or so and try again. You might also check the temperature; the rennet enzyme doesn't work efficiently if the temperature is much below 90 degrees. If the temperature has dropped significantly below 90 degrees, you can set the whole shootin' match in a sink of warm water to help it along a bit. Depending on a number of factors, like season and weather and how acidic our yogurt was, the curd could take as long as an hour and a half or so to set firmly. Maybe longer. Sometime around thirty minutes to two hours after we add our rennet, we should get a more or less clean break—which initiates the next step, cutting the curd.

Now that great big, white, slightly revolting curd we have made has to be cut into smaller curds, which will shrink and allow the whey to escape, preserving most of the protein from our milk in the curd, and releasing most of the water. Cheese curds are cut different sizes depending on the type of cheese you are making; for mozzarella, recipes may call for anywhere from ½-inch to 2-inch square curds. We cut our curds about an inch square; it's more important that the curd size be somewhat consistent than that they be a certain size. There are fancy and expensive cheese knives you can buy for this job, if you have as much money as Croesus and an enormous kitchen in which to store one-use gadgets; as we have neither of these things, we use the longest of our sharp kitchen knives and make cuts about an inch apart, first going across the cheese pan, then perpendicular to our original cuts, then at a slant downward, trying to visualize angles and spaces that will produce at least some inch-square curds, but knowing, as anyone would, that what we'll end up with will be a wide variety of shapes and sizes. We don't let it bother us; we'll remedy the situation in about ten minutes, which is how long we're going to let the curds sit undisturbed to shrink.

Ten minutes pass. Now it's time to cook the curd. The books say to use a hot water jacket and slowly raise the temperature of the curds and whey from 90 to 100 degrees, taking five minutes for each 2-degree rise, stirring, they add, constantly, and preferably with your bare hand. Let's do the math: that's 10 degrees, at 2 per five minutes, twenty-five minutes they want us to spend with our bare arms in a vat of curds and whey, stirring constantly. Well. Supposing we don't have a morning to devote to stirring our curds, it's not a problem; there's a lot of forgiveness when you are making mozzarella.

Forget the water jacket; forget timing the temperature rise with too slavish care. We have found that we get completely satisfactory results by warming the curds and whey directly on the stove, smallest burner, lowest setting; if we don't have the time to stir *constantly*, we disperse the heat a little using what we have always known as a "flame tamer" between the pot and the burner. A flame tamer is an invaluable gadget like a sort of metal baffle, meant to be placed between the bottom of the pot and the burner flame, and used to keep things from scorching if they have to be on the stove unattended for a long time, and you can find one at the kitchen store or, failing that, somewhere like Lehman's Hardware.

We actually do stir with our bare arm, normally, or if we are going to need our hands intermittently for other things, we may use a nicely worn wooden spoon—the rounded edges are gentler on the curd than a metal spoon would be—and stir very slowly and gently, pulling the curds up from the bottom to distribute the heat and, incidentally, giving us a chance to further cut the giant curds that are inevitably formed when you cut on the diagonal. We stir either constantly, if there's plenty of time, or about every three minutes if we're juggling several jobs that morning, and for the first two or three minutes we have a knife in one hand for cutting the oversize curds. We could break them smaller with our fingers and no harm done, but curd cut cleanly gives a higher yield. On the other hand, the whey is going to the pigs anyway, who love it and fatten on it beautifully, so there's no real waste however we do it. When the curds and whey reach 100 degrees, we shut off the heat, shove a lid on the pot, put it in a warm place, and go do something else for forty-five minutes or so.

Meanwhile our curds will be settling to the bottom of the pot. If they don't, but instead float to the top, there is yeast or unwanted bacteria in our curd. This is not punishment for our sins but probably means we somehow got foreign material in our milk or culture. When it happens we throw it all to the pigs, and maybe our starter culture too, and try again in the morning. Probably, however, your curds will have settled to the bottom of the pot without incident and have knitted into something resembling a thick, white, rubbery impact mat like the black ones you sometimes see on playgrounds. The individual curds will be distinct but all stuck together, which is fine, whatever you may read to the contrary. When these stop feeling like lumps of gelatin when pinched, and begin feeling more like soft tofu, we start testing the curds for stretch.

Figure 10.18. Cultured raw-milk mozzarella.

This is done by scooping out a small amount of curd—maybe as big as a walnut, or less—and heating it very hot to see if it melts together and gets stretchy and stringy, or if it just makes rubbery lumps. This heating can be done in a number of ways. The traditional old wood cookstove method was to touch a lump of curd to the hot stove top for a second or two, then pull it away. If it stuck and formed strings, the cheese was acidic enough. Very easy and convenient, if we happen to have a hot wood cookstove handy. If, however, we have not, we may pour a little very-hot-to-boiling water over our curd, wait five or ten seconds, and mash it with a fork; if the curd gets soft and shiny and stretchy, we have successfully reached the stretching stage of our mozzarella. If instead we have rubbery or gelatinous lumps, we will check to make sure our curds haven't cooled much below 100 degrees. If they have, we either shove the pot in a sink of warm water and recheck after fifteen minutes, or add hot water directly to the pot to warm up our whey. We'll give the curds another fifteen minutes, or half an hour, or even longer, checking periodically to see how they are coming on. Sooner or later, if we keep the pot warm, our curds will acidify. (Actually, even at cooler temperatures we find the curd will acidify eventually, and if we're pressed for time we may use this to our advantage, starting a mozzarella after dinner, leaving the whole thing overnight, at least in winter, and finishing the process in the morning.) On the other hand, if we heat our curds and end up with thin white runny stuff, we know we have overacidified our cheese, and we give that batch to pigs, chickens, dogs, or cats, and start over the next day with new milk. But if our curd is about right, the white lumps of curd, rubbery now instead of gelatinous, will mash together and take on a consistency somewhere between that of taffy and that of Silly Putty. It will stretch nicely and should be shiny and smooth.

Now we are ready for the last step in mozzarella making: stretching the cheese. Simple. Only, if we say that, you'll probably find it impossible, and then get discouraged. Okay, not too simple, because of course there are fine shades of readiness for curd, and you won't always hit it just spot-on. Sometimes—these are the perfect times—you will heat your curds and the result will be something smooth and shiny and absolutely yielding in your hands; sometimes you will instead find yourself working with something rather tough and rubbery, or else with a tendency to be grainy, as though someone slipped a little tapioca into the cheese pot when you weren't looking. There are many variations on this, but the essential thing is all that matters: if the curds can be heated and gathered into a cohesive lump that you can knead or stretch, you are a successful mozzarella maker.

You can heat curds by pouring hot—but not boiling—water over them, or by immersing them in hot whey (around 150 degrees seems to be about right, but there's lots of forgiveness here), or even by sticking them in the microwave, but why do that to your good raw-milk cheese? At this point what we do is to dump most of the whey into another pot (for making ricotta), leaving just a little bit more whey in the pot than it will take to immerse the curd. The curds we take out of the pot and place in a colander or steamer rack, which is in turn set in a bowl to catch the whey draining out of it; the pot of whey we put back on the stove and heat, without the curd. Then we try to remember to come back before the whey boils over—a kitchen timer comes in handy. We want the whey hot, but nowhere near boiling.

Once it is hot, we immerse the curds and shut off the heat—or, if we have a lot of curd to stretch, lower the heat as far as it will go—and wait a couple of minutes. Then we fish out our lump of curd—with two spoons if it's too hot to handle—and see how soft it has gotten. Usually the outside is beginning to be soft and stretchy, and it may be that we can now fold the mat of curds over and mash it together, to begin to distribute the heat through the whole mass. But if it is still so firm it won't fold without cracking, we turn it over—or break it into several smaller lumps, if our quantity of curd is very large—and put it back in the whey to warm some more. It will probably go back into the hot whey several times, heating, kneading, folding, heating some more, kneading again. The goal is to soften the curds completely and develop stretch by pulling or kneading until the whole mass looks kind of like taffy, if you've ever stretched taffy, or kind of like Silly Putty, and hasn't everybody stretched Silly Putty at some time in his life?

The Silly Putty state reached, we are morally done. Mozzarella has been achieved. If we want it to eat fresh, we pinch it into little balls, smooth these by sort of stretching the skin back and tucking it under (if you've ever made clover leaf rolls, you already know what we mean), and set them to cool in a bowl of chilled water, salted if we want our cheese salty. Marinated in olive oil and garlic these make an excellent lunch with some bread and dried sausage. If we plan to use it for pizza, we smooth the whole thing into a single ball, chill it as above, and then set it out for a few hours on a rack to dry before we try to grate it. If we anticipated the stretching stage by a little, our cheese may be a little soft, the texture a little loose and stringy, and a little whey may gather in our bowl of grated cheese. On the other hand, if we stretched our cheese too much, it may be rather firmer than we were hoping. These eventualities, unless one is hopelessly uptight, are no problem at all; we just make our pizza, shove on the grated cheese, and when it cooks, it will be just as good as it should be.

Aged Cheeses

Using rennet to set fresh milk into firm curds quickly allows us to make cheese without a prolonged fermentation and, consequently, without excessive acidification. Cheeses made with rennet can host a wider range of bacteria and molds, whose transformative action gives us the delicious subtleties we enjoy when cheese is aged, as with hard cheeses like cheddar, Gouda, or Parmesan. While many erstwhile cheese makers have some experience turning yogurt into yogurt cheese, or concocting a quick citric-acid mozzarella, the making of hard cheese has in recent years been shrouded in mystery, isolated from the homestead kitchen by a long list of expensive, single-use, and unnecessary tools and ingredients.

Figure 10.19. A fresh hard cheese set out to dry.

The actual process is simple. A short fermentation is followed by the addition of rennet; when the resultant curd is of a sufficient firmness it is cut, "cooked" (warmed), and "pitched" (a period of rest and acidification), pressed into a single mass, and salted. It is then held at a relatively cool temperature, like the north side of the cellar, for a period of weeks or months, to allow the fermentation process to continue developing the flavor and refining the texture. The various classes of cheese are determined by such things as the strains of lactobacilli, molds, and fungi that are encouraged and the methods used in heating or gathering the curd. Lots of variations can be introduced: mold and bacteria cultures, smoke, vegetable flavorings like caraway seeds or jalapeno peppers, and so on. Nothing about this process is inimical to production in a farm kitchen; most of the types of cheese we know and love originated in one.

An easy cheese to start out with, and one that doesn't require any special tools, is Gouda. Gouda is termed a "washed curd" cheese because the warming process includes the removal of some whey and the addition of hot water, lowering the acidity while heating the curd; it is a pressed cheese that may be aged for anywhere from two months to a year, maybe more, and we have found it simple and delicious.

We generally don't make a hard cheese unless we have 5 or so gallons to work with, to make a large enough cheese that the finished product won't be all rind and no middle. About 5 gallons of milk—we may use up to half skim milk for Gouda, but the more whole milk, the creamier the cheese—are warmed to 90 degrees or thereabouts and cultured for ten minutes with roughly 1¼ cups of our best mesophilic starter (buttermilk). Then we add 2½ teaspoons of rennet diluted in a couple of ounces of cool water. A quick stir up and down—maybe twenty strokes—to disperse the rennet, and half an hour or so to allow the curds to set before we use the longest kitchen knife to cut approximately ½-inch curds (they won't be), followed by ten minutes stirring gently with our bare hand, lifting the curds from the bottom of the pot, and cutting the overlarge ones into more or less uniform size. Meanwhile we've set the kettle on the stove to heat; the next step will need something around a gallon of hot water (the recipe we used originally said the water should be 170 degrees, but anything approaching boiling or two or three minutes off the boil will do just fine). Is it apparent yet that this is not an exact science?

The curds will begin to shrink and firm up as we are stirring. After a five-minute (or so) rest to let them settle, we pour off whey to about three-quarters the original level in the pot and begin adding hot water, a little at a time, stirring as we go. When the temperature in the pot reaches 92 degrees (or close to it), we let the whole thing rest again, this time for ten minutes (or so); then we pour off more whey, down to the level of the curds, and stir in more hot water to bring the curds and whey up to 100 degrees (or so). We'll keep stirring for fifteen minutes (give or take a few), and then shove a lid on the pot and let it sit somewhere for in the neighborhood of half an hour, after which we'll drain off as much whey as we can before we load the curd into an 8-inch cheese ring lined with cheesecloth, or a peanut-butter bucket with holes drilled in it also cloth-lined, or something similar, and press it for twelve-plus hours or overnight, starting at around 15 pounds of pressure and flipping and rewrapping the cheese three or four times, increasing the pressure each time until it's up around 30 pounds. Actually, the recipe calls for 50 pounds pressure, but our cheese ring has an open bottom, and we find the curds want to squeeze out the bottom if we get much over 30 pounds. It doesn't seem to affect the cheese; the pressure isn't really to expel whey so much as to knit the curds into a single mass, and it seems to do that all right at 30 pounds.

In the morning we take the cheese from the press, unwrap it, and float it in about a gallon of saturated brine solution for a day and a night, flipping it now and then so the cheese will take on salt evenly. ("Saturated" means the brine is carrying all the salt it can hold. You make a saturated solution by boiling water with sufficient salt [any kind but iodized] such that some is still left in the bottom of the pan. Just a little is enough; more will precipitate out as the solution cools, and as long as there is always some undissolved salt in the pot, you know the solution is saturated.) When the cheese is salted, we take it out of the brine and set it on a wooden rack or a sushi mat or something to air-dry for a day or two, until the surface is dry to the touch, flipping it every so often to promote even drying.

We used to age cheese in a spare refrigerator, which took a long time, and the humidity was so low that we had to wax the cheeses, or wrap them, to keep them from drying and cracking. Now we age in a cabinet made of cement board and embedded in the earthen walls of our cellar, which works fine most of the year, staying around 50 to 55 degrees until the low-slanting southerly sun of late summer warms up the earth and we have to put the cheeses back in the cooler for a couple of months. Every week we turn them, beginning with the youngest cheeses and working backward, wiping down any mold that forms on the outside with a clean bare hand; the 4- to 5-pound cylinders mottle gray and white, with sometimes spots of red, yellow, or orange. During the summer and fall we put up two or three cheeses a week and begin eating them when the first rounds are something shy of two months old. They are delicious, varying slightly in flavor and texture from cheese to cheese, according to how much whole milk was used, as well as a thousand other quirks—temperature, timing of this stage or that, what forage was principally grazed on any given day. The rind with its mosaic molds may be delicious, but if it's too "earthy" for our taste, we trim it. We have been making hard cheeses like this for about eight years, and we haven't encountered any flaws in the process; our worst problem is that we all like cheese so much that the mere couple of hundred pounds we are presently making is barely adequate for our family.

Fermenting milk, cooking curds, aging cheeses—no part of any of these processes is beyond easy reach of the homesteader in an ordinary kitchen. After all, how did Great-Great-Grandma do it? She didn't have UPS to bring her freeze-dried milk cultures from some city supply house, or a hygrometer to measure the exact humidity in her cheese-aging space, or even a thermometer to tell her the precise temperature of the milk in her pot; neither does the modern dairyman, whose goal is long-term storage of clean raw milk, need these things.

Hitches, or, The Jungle in Your Kitchen

While cheese making is a natural process, there can be some hitches along the way to turning your kitchen into a successful *kase haus*, notable among which are the lack of anything approaching a guide or mentor for the journey—after all, how many people do you know who make raw milk cheddar in their kitchens?—and the presence of those airborne molds and bacteria already living in any kitchen, with which your raw milk lactobacilli are being placed in competition. With no one to show you what a clean break or a squeaking curd is like, you're

The Art of Natural Cheese Making

Only recently have we come across a cheese making book we can recommend with real enthusiasm. David Asher's *The Art of Natural Cheesemaking* is the manual we would have liked to have when we began making cheese, a guide that really trusts milk, and forage, and ruminants, and the fermentation processes that are responsible for converting, preserving, and giving incomparable flavor to our milk-derived foods. Chapter by chapter he explains, in clear, lucid, and entertaining prose, all the elements and steps of a natural cheese making process adaptable, it seems to us, to any homestead kitchen and any dairy. Reading it is like reading a manifesto of food liberation, and we highly recommend it for cheese makers of any degree of experience or inexperience.

left to your own judgment; then when your cut curds seem to shred instead of making nice cubes, or float instead of sink, you throw all the blame on your lack of experience and wonder if you should throw in the towel as well. "Look at all this milk I'm spoiling," you think. "What a terrible waste." Two factors should come to your rescue here. One is that there will be more milk coming into the kitchen in twelve hours or so, and you're going to have to do something with it. You can't just keep shoving those jars into the refrigerator; you'll run out of space, and you'll run out of jars. The other factor in your favor is out in the pigpen smacking its lips in anticipation of your next cheese making fiasco, and it's just waiting to turn whatever you give it into bacon and ham. This is one of the beauties of holistic homesteading: there is no waste, there's only a rearrangement of the nutrients—which is a great boon for those of us on the lower slopes of the learning curve.

As for those floating curds and all the other finesses of cheese making that don't seem to be coming right, we advance this theory for your consideration, and maybe consolation.

Bacteria are present anywhere there are both water and nutrients, but we tend only to notice them when something gets out of whack. When you first undertake to ferment raw milk in your kitchen, you are introducing both a new food substance and new bacterial cultures into a space already populated with your normal kitchen biota,

Figure 10.20. With careful grazing, grass is moving into the woods.

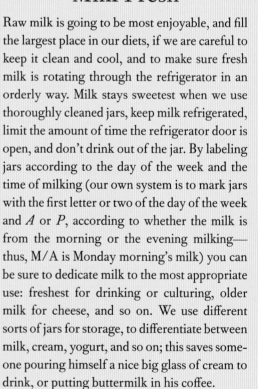

Keeping Milk Fresh

Raw milk is going to be most enjoyable, and fill the largest place in our diets, if we are careful to keep it clean and cool, and to make sure fresh milk is rotating through the refrigerator in an orderly way. Milk stays sweetest when we use thoroughly cleaned jars, keep milk refrigerated, limit the amount of time the refrigerator door is open, and don't drink out of the jar. By labeling jars according to the day of the week and the time of milking (our own system is to mark jars with the first letter or two of the day of the week and *A* or *P*, according to whether the milk is from the morning or the evening milking—thus, M/A is Monday morning's milk) you can be sure to dedicate milk to the most appropriate use: freshest for drinking or culturing, older milk for cheese, and so on. We use different sorts of jars for storage, to differentiate between milk, cream, yogurt, and so on; this saves someone pouring himself a nice big glass of cream to drink, or putting buttermilk in his coffee.

those little guys who like your foods and like your food handling methods. Enter raw milk, a perfect growing medium for myriad bacteria: those that are wafted into the milk by a stray breeze, or that are already benignly colonizing your cooking surfaces and utensils, have a field day. The original milk lactobacilli, the new guys on the block, are often outnumbered and lose control of the game: hence floating curds, funny smells, and peculiar mold populations. Over time, however, the microbes of lacto-fermentation earn their own place in the schoolyard, as the presence of raw milk and raw milk processing introduce and foster them, and once they are no longer so outnumbered, they can do what they are best at: culturing milk and outcompeting foreign bacteria and molds. In farmhouse cheese making, we promote good culture not by killing off all the other bugs—which may have their own jobs, important, if invisible to us—but by creating good conditions for those we need to help us in our work.

All of which is not unlike what is happening outside, in our holistically managed pastures.

What's for Dinner?

The exercise of taking an underproductive piece of land and turning it into a human-symbiotic ecosystem is beautiful and worthwhile, but if it doesn't feed us, sooner or later we're going to give up. We are, after all, a significant part of that ecosystem we are fostering, and we need our nutrients too. Learning to think Food—developing a part of our brain that plans meals from the ground up, seed to harvest to dinner plate—is a key part of learning to homestead, and as much as this applies in the garden, where the most beautiful eggplant is merely decorative until we harvest, cook, and eat it, it applies with geometric intensity to milk management and the dairy. In milk-based cultures, milk is what keeps the wolf from the door, each day's food provided in that moment as the ruminant, consuming forage from which we could not derive nutrition, converts it to the most highly digestible and palatable of forms. On our grassfed homesteads, we want to develop a domestic food culture that takes advantage of these sources.

So how do we turn milk into breakfast, lunch, and dinner?

Moving past the obvious—a big glass of creamy milk, full of proteins, fats, and sugars—yogurt comes to mind, fresh yogurt stirred into drinkable form and sweetened with maple syrup, or added to a smoothie of homegrown apples and fresh harvested spinach leaves. Then, of course, there is thick Greek yogurt topped with homemade granola, blueberries and walnuts, or diced dried apples and toasted sunflower seeds. Steel cut oats soaked overnight in raw milk cook quickly in the morning and are delicious with butter and cream, honey, maple syrup, or fruit preserves. And what could be better than an omelet filled with grated cheese and a slice of whole meal toast spread thick with yellow butter?

At lunchtime there's no difficulty putting a meal on the table when there's plenty of milk, sweet or fermented, liquid or solid, in the house. Greek yogurt topped with diced cucumber and tomatoes and dressed with a garlicky vinaigrette is a quick, light lunch if our activity level is moderate. Or when we are expecting to do heavy work outside, we may accompany the loaf and butter on the table with a wheel of aged Gouda or Colby and a stick of home-cured sausage, while rounds of *panir* fried in butter and spread with yogurt and curry are surpassingly delicious. In cold weather a pot of cream-of-almost-anything soup is always welcome: leftover cooked meat or vegetables, or quick-sautéed onions, garlic and grated potato, simmered briefly in milk and thickened with a little flour or the mashed vegs themselves, make a delicious, warm, and hearty meal; leftover cooked grains, like rice or quinoa, can be eaten hot or cold, topped with grated cheese, sour cream and salsa, or curry yogurt and raisins—the possibilities are many.

And planning dinner when the house is full of milk foods is just a question of deciding where to start. Fresh spinach with feta and thinly sliced onions make a delicious salad, or shall we combine those ingredients with some ricotta and eggs and bake them in a piecrust, or stuff them into a pasta shell? Beans and rice are a sustaining meal all by themselves, but with grated cheese, sour cream, and a garnish of salsa and chopped green onion they rise to gourmet level. That grassfed beefsteak just coming off the grill gets even better when you top it with a dab of garlic butter and a crumble of feta, while the accompanying potato, in whatever delicious form, is even better with butter, sour cream, and some diced Gouda. In a home where fresh mozzarella is always available, what to put on a pizza becomes a delightful conundrum; it hardly matters, when the cheese is so delicious and plentiful. On meatless menus milk is even more primary to meal planning, often in the form of thick milk- or cream-based soups like borscht or chowder, topped with sour cream and grated cheese, accompanied by a crusty loaf and butter and a salad of fresh greens and grated Parmesan; perhaps as a cream and garlic sauce on pasta, topped with mushrooms, Parmesan, and mozzarella; or in a simple quesadilla. When dairy products are homegrown, they no longer have to be treated as luxury foods; they gain in flavor, food value, and delight even as they become a common part of every basic meal.

Figure 10.21. Aged Gouda.

Milk All over the Farm

As much as milk is a central source of calories in the house, it is equally important in the barnyard. The year can be divided into seasons according to the destinations of the milk.

Early lactation, typically springtime, is the flush time for milk. With the production of perhaps 30 or more gallons a week—easily within the range of our six-year-old Jerseys—there are a lot of nutrients to be managed. Supposing the house needs a gallon of milk a day, which is moderate, and the calf is allotted 6 quarts a day, which is generous, there are still 12-plus gallons a week to play with during this season. These will easily provide a feeder pig and fifty-plus chickens with their daily protein supplement; in addition, some of the house milk will be skimmed to provide cream for butter making, with a good chance that the busy farmer will send the buttermilk down to the pigpen, giving a bit to the guard and predator-control animals, too (see figure 10.3).

At the end of three months the calf is weaned. Now, although the cow's milk production has naturally declined somewhat over time, the sum result will be that more milk is freed up for the table, or for a second calf, if one is bought in. If some is used for cheese making, it will generate gallons of valuable whey to further supplement the pig and poultry. In addition, some may be used as side dressing or foliar feed, to give a midsummer boost to the vegetable garden. Broiler chickens are put in the freezer in late summer, which again decreases the demand for milk in the barn, leaving more for the cheese kettle and other projects, like soap making.

In late autumn when the pig is butchered, still more milk becomes available to the house, despite the natural decrease in overall lactation, moving cheese making up on the list of seasonal chores. Then, since making cheese means gallons of whey are produced, it may be time to pick up another piglet—or two. Later, if both are not needed for the farm table, the extra may either be sold as a started pig whenever the quantity of available farm-raised nutrients decreases, or butchered at around 100 pounds.

Finally, by late winter milk production will be at low ebb, maybe as little as a gallon a day. By this time, hopefully, the careful homesteaders have put back plenty of hard cheese and butter, to take some of the sting out of eight weeks without milk; they, and the pig, if there is one, will have to limp through the next few weeks as best they can without their primary source of solar nutrients. The pleasure of a break from the milking routine soon pales as we grow more and more impatient for fresh dairy products, and before our new calf arrives and the cow freshens, we resolve to buy a second cow and never again be without fresh milk.

As important as the milk harvest is in the barnyard, it is of significant value in the garden as well. Was the first person to discover the fungicidal properties of a weak dilution of milk a dairy maid, emptying the rinse water from her bucket over her favorite rose bush? Milk in a solution of no more than 10 percent is an effective fungicide, remedy for black spot on roses, leaf spot on blueberries, mildew on grapevines—reportedly as effective as the leading nonorganic antifungal sprays. Whey can be used in the same capacity. Maybe these will prove the solution to our local propensity to early blight in tomatoes and powdery mildew in cucumber vines. Whey is said to be a powerful foliar food as well, and an excellent fertigation for garden plants. Good luck deciding who needs it worse, the pigs, the hens, or the vegetables; how could the home dairy ever have too much whey?

Whole Milk Conversion

Jersey milk may be as much as 25 percent cream by volume, a bit more in periods of flush forage, less on stockpile or hay.

- 4 gallons of milk = 1 gallon of cream, 3 gallons of skim milk
- 1 gallon of cream = 2 pounds of butter, 3 quarts of buttermilk
- 3 gallons of whole or skim milk = 3 pounds of cheese, 2-plus gallons of whey
- 2½ gallons of whey = daily high-quality protein requirements for one pig
- 1 gallon of skim milk = daily protein requirements for one pig or fifty to one hundred chickens; adequate but not generous allowance for a bucket calf

From this incomplete list, it is obvious that the calf, pig, and chickens need not have their own allotment of milk, if sufficient buttermilk and/or whey is generated by butter making or cheese making.

Meat and Butchering

Fresh milk, cheese, and butter are so readily available on the grass-based farm that, once you've begun, it takes little forethought or planning to build a meal around one or more dairy foods. The provision of meat for the table, on the other hand, necessarily requires some forethought and, for most of us, a conscious decision.

It is an interesting thing that, while we live in a society where vegetarianism is most certainly in the minority, and most people eat meat or meat products at least once a day, there is such a prevailing squeamishness about any part, prior to removing it from its grocery-store, styrofoam-and-plastic packaging, of the process of actually putting meat on the table. It is a symptom of how remote we have become from the earth, and from the plants and the animals that feed us, that even while a diner downs his ancho-cherry glazed buffalo wings he can shudder fastidiously at the idea of killing, plucking, and eviscerating a barnyard bird, and even feel he occupies a sort of moral high ground by being sensitive about it. We who farm for food cannot afford to be so inconsistent, and the hands that feed and care for our animals must also be those that cull—and kill—them.

Take it or leave it. The grass-based farm is automatically and by definition also an animal-based farm. If we are to foster the best farm ecology, then for many—even most—of those animals, a happy, healthy, and natural spring of life will end, quickly and respectfully, at the chopping block, or with a carefully aimed .22-caliber bullet in the head.

Where people become the caretakers of animals, loving care and respectful slaughter are natural bedfellows. Our consideration and planning of forage use and provision, breeding, and winter

Figure 10.22. You don't have to have perfect pastures to harvest perfect milk and beef. Courtesy of the Franciscan Sisters T.O.R. of Penance of the Sorrowful Mother.

management—the labor that makes this piece of land a sustainable farm instead of an abandoned, weed-infested lot or a forest glade—are the same faculties that will dictate when and how an animal life is to end, to achieve the best outcome not only for the farm and the herd as a whole but for the individual animal. A lean and sickly old age closed at last in the jaws of a predator may be a more natural punctuation to animal life, but is it therefore to be preferred to a short life of enhanced health and well-being? And when death has removed any concern the animal might have been supposed to feel about it, is it wrong that its body should become a further source of well-being for the farm that gave it its privileged, pampered, brief life? Every life, plant, animal, or human, subtracts something from the community in which it is lived—nothing can change this; what we *can* affect is the manner in which we subtract that something, and whether—and how much—we will give back.

As for meat, and getting it on the table: visualize dinner, thank God for this animal and all it has done and will do for the farm, and proceed.

Butchering Birds

People differ about the best way to kill poultry. Beth's great-aunt Drew wrung necks; she was a strong Arkansas farmwoman, practiced in the swing-and-jerk necessary for snapping the vertebrae of an excited 12-pound Barred Plymouth Rock rooster. The cleaver or hatchet and wood block, on the other hand, are quick if your instrument is sharp, and allow the bird to bleed freely; proponents of hanging the bird upside down and making a slit in its jugular assure us that such a method of dispatch guarantees quick unconsciousness and a long bleed out. We have not learned to prefer any single method for its superior results but instead opt for whichever is going to be easiest on the practitioner. Aunt Drew's way was traditional and effective, but hard on the wrist, so if the cleaver is good and sharp we go with the chopping block method, or if we are only killing a couple of birds for roasting or for chicken pie, we may find a really sharp knife—*really* sharp—and proceed as indicated.

You can dry-pick a single bird, or a pair, warm, just after they are dead, when the feathers will come away easily, but when you have a lot of birds to process you want to scald them first, which loosens the feathers and makes the job go much faster. We used to do twenty birds or so before lunch, if we had several people to help pick them. Then we built an automatic tub picker (see resources) and reduced the necessary workforce to one adult and two youngish children. Getting the temperature right in the hot water bath makes a big difference, and we usually try to maintain something close to 150 degrees, but the water temperature will fluctuate a good deal with a bunch of 95-degree birds being dunked in it, and this doesn't cause undue difficulty. Or you can skin poultry, which we think is a good option for an old stewing hen or a venerable rooster intended for canning; when we can birds we prefer them skinless because otherwise the fat that will render out of the skin may, if it gets under the edge of the lid, prevent the jar from sealing. Skinning an old bird can be a real chore, though, especially the wings, and we have to scrap the terminal joint (try skinning a wing tip sometime) when we process birds this way.

There are lots of instructions out there on how to get the guts out of a chicken; once again we recommend Salatin's *Pastured Poultry Profit$*, for inspiration as well as information, and there are some good

Figure 10.23. Grass seeds indicate plants reaching maturity, after which protein content diminishes.

videos online—but there's nothing like getting in there and getting your hands messy. The trick you really want to master is that of getting the guts out without spilling feces on the carcass, and this will be easier once experience enables you to visualize the bird's insides. If the unthinkable happens, though, and a tiny bit of poo does get on the bird, you can build a bonfire and incinerate the body to ensure no one accidentally eats it; or you can compost it, motive as aforementioned; or you can wash it thoroughly, cook it completely, and proceed as nature would indicate. We've always followed the latter course without incident. A little experience at gutting birds soon teaches you how to avoid the problem.

You can freeze a bird whole, or cut up into pieces. We're all familiar with the standard chicken parts, and with a little experience and a sharp knife one soon gets the hang of piecing birds. Poultry shears come in handy if you want to spatchcock them—cut the spine out whole—and we like to do this and then cook the spines separately and can the broth. For the freezer, from this point it's just a matter of bagging the chicken (use strong bags or the wing tips will poke holes in them), labeling, and freezing. If we are doing more than a couple of birds, we hold them, plucked and gutted, in a cooler full of cold water until we are ready to bag; this brings the carcass temperature down at once, better for the bird and easier on the freezer. Never stack warm meat in the freezer to chill—spread it out in a single layer; otherwise, it can take so long that the meat in the middle has time to go bad before it freezes. If we are only killing a couple of birds, as when making a chicken pie, we'll probably shove them in the back of the refrigerator for two or three days to soften up before we plan dinner around them. We have heard experienced people complain that old roosters are too tough for any kind of preparation to make them fit for the table, but we have always found that a short roasting—maybe thirty minutes, just enough to develop flavor—followed by an hour or so at 10 pounds pressure in the pressure cooker, can make even the most cantankerous old cock tender and delicious. Taking the meat from the bones and dicing it avoids long, stringy muscle fibers when the meat shreds in further cooking.

Canned chicken is a luxury and a convenience. Once it's in a jar, it doesn't use any energy for preservation, and it will keep for several years; where there is canned chicken in the pantry, dinner is never more than half an hour away. Broilers are too big for a quart jar, and anyway they are so tender that on our farm they are welcome to freezer room, but the defunct laying hens generally called "stewing hens" are smaller—much smaller, in these days of "dual purpose" breeds that lay three hundred eggs a year but look about the size of quail when you take their feathers off—and the eight principle parts (drums, thighs, breasts, wings) will fit nicely into a wide-mouthed quart. Our book says eighty minutes at 10 pounds' pressure, and that's always done it for us. Save all the backs, necks, and wing tips (and heads and feet, if you are up for that) for a long, long simmer, and then can the broth as you do the hen; oh, so delicious, and oh, so convenient to have around.

We butcher turkeys and ducks as we do chickens, but the latter are a real chore to pluck, due, we think, to the difficulty in saturating the feathers of a waterproof bird.

Large Animal Butchering

Birds are one thing, being so small that the job isn't too intimidating; being so small, indeed, that, the novice can comfort himself, should the whole show go south, we can just bury the evidence in a compost bin and forget. Much more sobering are the really big animals, like hogs and steers: harder to kill and to lift, much more work to break down, and a much greater loss, both in valuable grassfed meat and in self-esteem, if somehow it doesn't work out. Also, our fears whisper to us, due to its great size, a hog or steer could poison lots of people if it turns out Tyson Foods is right and only the industry can really safely prepare meat for the table.

But the truth is, breaking down a large animal into its primal parts isn't any more complicated than cutting up a chicken; it just takes more muscle.

The Custom Slaughterhouse

If you've never killed something and eaten it, you may feel out of your league when it comes to slaughtering a four-legged something that weighs 100 or more pounds, not to mention getting it skinned, gutted, and into forms recognizable as "meat." Well, you can always take the animal to the butcher; that is, if you have a local custom butcher, and if he will schedule your animal for a date this side of next year. Custom slaughter is getting hard to find, at least judging from our own experience and what we hear from others in various parts of the country. As meat production and processing grows more centralized (we read that more than 80 percent of the beef in the United States is controlled by just four packing firms), the smaller operations are dying out, with, at present, few coming in to replace them. If you are lucky enough to have a

Figure 10.24. Next year's beef.

butcher who will process your animals, and he is less than an hour away, dandy, only get your date well ahead of time or you'll discover that he has work lined up for months ahead and can't take your superannuated wether goat until four months after county fair season.

And just a heads-up: the cost of custom butchering—we expect custom slaughter to add about seventy-five cents to a dollar a pound to the cost of a meat animal, more if there is any curing or smoking involved—and the difficulty of getting a date for butchering are not the only disadvantages of sending your livestock out to slaughter. The disturbance of being loaded and hauled to a strange place is more than adequate to reduce the quality of meat when your animal is butchered, not to mention the not-inconsiderable chance of injury while being moved, and the cost of hauling. And not only may you experience a decrease in the quality of the meat, but you'll probably lose something in quantity as well, at least as far as fat (conventionally, much is trimmed in the cutting), soup bones, and ephemerals. Animal fat is a valuable product, for people food, as food for livestock, and for operations like soap making, and we are always chagrined when circumstances prevent us butchering our own animals, because we are sure the fat won't be harvested as efficiently. You can ask the butcher to save things for you, but if it's not a normal part of his routine don't be surprised if he can't or won't do it. And if you plan to keep the hide for curing, make sure the butcher knows this. You'll want to pick it up on the slaughter date (*not* the same as the drop-off date) so that you can salt it right away.

Home Slaughter

Anyone can learn to home-butcher large animals, the most economical choice by far, and only more difficult than slaughtering a chicken by reason of its greater size and the desirability of hanging the carcass up somewhere while you are working. Oh, sure, if you just *have* to end up with neat packages marked PORTERHOUSE and CHATEAUBRIAND, you probably need to take a class somewhere first, or maybe a couple of classes, but if all you require at this early point in your homestead education is *meat*, slaughtering a goat or sheep, even a moderate-size steer, is no more difficult than killing, gutting, and skinning a chicken—only make sure your knives are sharp or you'll think we lied to you. And maybe look at some diagrams or watch a couple of videos first, for a boost to your confidence.

Figure 10.25. Dry curing beef. Courtesy of Rose Dougherty.

We started large-animal butchering with deer, happy just to get the meat off the carcass and into chunks for freezing or canning. We were not the least concerned that we didn't know how to produce traditional cuts, since we didn't want them anyway. We find our local, brush-fed venison more palatable if highly seasoned, so we use it in dishes requiring ground or diced meat, either of which can come from any part of the carcass, and both of which are easy to freeze or can. With deer, as with goat and mutton, a lot of the gamey flavor is in the fat, so we cut off as much of this as we can before we dice and can or freeze it, which mitigates the ranginess as well as improving the rate of seal on our canning jars. Starting with deer was a great promoter of confidence that we could manage to butcher a large animal successfully.

From deer to goats and sheep is but a small leap, mostly having to do with the fact that now we may have money in the animal—and, of course, there is the matter of killing something in cold blood, something we raised and (probably) petted. Well. Our vegan friends assure us that we aren't supposed to get over this reserve. Nevertheless, if you hadn't already found answers for your own vegan friends you'd have given up on this project long ago, so thank God—or whatever force within or without the universe to which you ascribe credit for its great beauty and order—for this creature of His which by its very being has benefited and will benefit the farm, and proceed. A .22-caliber bullet will meet the case with most large animals, exactly where to apply it being a function of which species you are presently dispatching (and hogs sometimes require more than a single shot). Or, a sharp knife in the jugular (you're going to do this anyway) is more economical and bleeds out longer. Give it some time to drain; the quality of the meat is better when the carcass is bled out thoroughly. We have friends who catch the blood for blood pudding or catfish bait, but we don't care for blood pudding and don't do much catfishing, so I'm afraid we let the blood fertilize the lawn. It would be better to catch it and apply it to the compost bins, but we never think of this in time.

You can break down a medium-to-large animal with simple tools—sharp knives, a hacksaw, and a hand-cranked grinder—or if you have access to a good electric grinder and a band saw dedicated to cutting up meat, these can speed things up quite a bit. Either works fine, the difference being that with only hand tools it may suit you to be less ambitious. Again, what is the goal? If you don't have training as a butcher and you are satisfied to grace your table with large roasts of this delicious

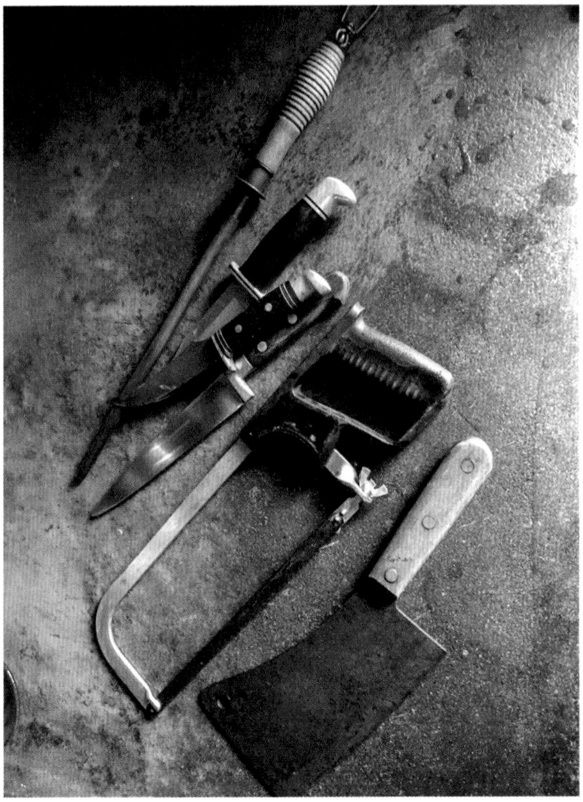

Figure 10.26. Simple tools are adequate for butchering large animals. Courtesy of Rose Dougherty.

The Skinning Cradle

There is a way around some of the hoisting and hanging of large-animal butchering, provided you have several strong friends on hand to help with the initial lifting. We find that a skinning cradle is simple to build and very convenient to use. Basically, a skinning cradle is an arrangement of bars or timbers that will hold a medium-to-large animal securely on its back, at working height, leaving room for one or several people to move around the carcass. Picture three parallel horizontal bars, the middle one lower than the other two, and those two slightly less far apart than the body of the animal you are skinning could slip through. It goes without saying that a skinning cradle must be very strong; also, that it should have a very secure base, not likely to tip. You can gut on a cradle, too; the only disadvantage we can think of is that the innards don't automatically fall out, as they do when the body is hanging, so if you make a mistake and puncture something you shouldn't (gastrointestinal tract, urinary tract, gall bladder), there it is, all over your meat. Skinning in a cradle lets more people help at once, and we find it easier on the back.

Figure 10.27.

grass- and forage-nourished animal, and many pots of stew, then sail in with your knife and hacksaw and as many hands as you can muster and break it down. Have plenty of butcher paper and freezer tape available, ample table space, and clean buckets for bulk meat and offal. The garage may be a good place to set up this operation; we generally butcher in the garage. Bones you don't want for soup you can bury, or give to the dogs, or haul to the back forty for the coyotes; guts (what you don't keep for yourself) can be composted or portioned out to the poultry or pigs. As always, avoid same-species feeding.

Sheep and goats, even large ones, can pretty much be handled by one strong principal and a helper, and the breaking down may be done on a stout table or a half sheet of ¾-inch plywood over a couple of sawhorses. The same can be said of a steer of one of the smaller dairy breeds, if it is under a year (as when you determine that you don't want to deal with providing winter water for your assistant in pasture improvement). But for larger steers and hogs, you'll need a means of hoisting the carcass off the ground while you're dealing with it, and there are lots of reasonable ways of handling this. If you plan to haul it to the garage rafters with your block and tackle, please make sure first that your building is strong enough for this purpose; a 500-pound steer, not to mention the force of the down-haul, is a lot of strain to put on a beam, and it'll spoil your day to compound the new task of butchering with the necessity of repairing the garage roof. Consider ahead of time what else you have for lifting heavy objects; we have seen people use the bucket lift on their tractor, a skid steer, or a stout branch on a sturdy tree, with perfect success. You hang a large carcass head down, to facilitate bleed out and a clean gutting. For breaking down a larger animal, you will be happier if you have read some instructions or watched someone else do it or, even better, have an experienced friend who is willing to give assistance and advice.

Figure 10.28. Sausage drying in a cool cellar.

Beef is much improved if you can give it a long hanging in a cool place first, to tenderize the meat and develop the flavor. The USDA insists on very controlled conditions and intense monitoring of carcass temperature and humidity, and it is up to you to determine what you feel good about. We butcher beeves in late fall or early winter in an unheated shed, and we usually age ten days or more, if temperatures stay moderate to low (below 50 degrees in the daytime, not much below freezing at night). If we get a warm spell we have to cut and wrap early, or drop the sides into a hog-scalding trough (a tarp would work if we didn't have a trough) and ice them down. In a cold snap, we light a heater in the garage so the sides won't freeze. The resulting beef is delicious, tender, and wholesome, even without the imprimatur of the USDA.

Rendering Lard

Few products of the harvest are more valuable than fat. It is emollient and unguent, deep-fried doughnuts and scented soaps. Waste not an ounce of your hog's valuable fat! Leave, of course, a generous rind of fat on all your chops and roasts, but whatever is extraneous, husband with anxious care.

A big part of the work of breaking down a hog is the separation of fat from anywhere it is not wanted, and everywhere there is surplus. Keep a bucket—several buckets—for stray pieces of fat, and set the children and the inexperienced to cut up all the bits pretty small, as small as may be, or run all the extra pieces through the grinder. The smaller you cut them, the more thoroughly they will render; and the more lard you can harvest, the better.

Put all the pieces in a big kettle, or freeze them in manageable packets and do just a little at a time. Put the kettle—a cast iron Dutch oven is perfect—in the oven, or on the back of the stove, and let it heat slowly, stirring from time to time. It shouldn't smoke, and it shouldn't boil; you just need the fat to melt, and what little connective tissue there is to fry up and float to the top. Take this out and let the liquid fat cool. If you do this perfectly, you will be left with a kettle of lard, pale tan in color; if you have heated it rather too much, it will be brown, but still perfectly serviceable. Warm it up until it runs and pour it into quart mason jars and cover them up. Save the fried bits, the "cracklings"; you can mix them in things, if you like it, or crumble them up and use them as a supplement to add fat to your chicken mash.

When there is plenty of lard and butter, the house shouldn't need any other fats, the whole of the year.

Curing and Smoking

Butchering hogs is more complicated because of sausage making and the curing and smoking of bacons and hams, but in the long run it is no more difficult than butchering a steer, with the advantage that pork, unlike beef, is not aged. Pork bellies for bacon go in a salt cure for a week or more before you smoke them; after that they can be sliced and frozen, or frozen whole and sliced when you are ready to cook. We haven't yet tried to dry-cure and hang a belly, although we have dry-cured hams and hung them in the garage rafters, where they kept perfectly well through two hot summers and two freezing winters and when we took them down were just delicious. To make sausage casings you squeeze the contents from the small intestines, cut them into manageable lengths, and scrape the pinky-gray lining out against a plank with the flat of a table knife. Run water through them—just slip one end over the faucet—and keep them underwater or salted down until you have your sausage ground and seasoned and are ready to fill links. Or save the small intestines of your next steer, scrape as for hog casings, and put them away layered in salt (and it doesn't hurt to freeze them) for larger sausages like salami.

Pig Butchering 101: Scraping Casings

Find one or two friends who like good, clean pork and working with manure and guts, raise a few pigs with them, making arrangements to share costs and chores, and when the pigs reach roughly

Maple Sugar Bacon Cure

Our favorite way to cure bacon is with a black-pepper-and-maple-sugar rub. We mix 2 cups of brown sugar, ½ cup of kosher salt, and ½ cup of ground black pepper with 1 cup of our home-produced maple syrup; to this are added 2 teaspoons of curing salt, otherwise known as pink salt or Prague powder. This is the stuff with nitrates and nitrites in it, and you can leave it out if you like and your end product will be just fine, but it won't really taste like bacon. We go ahead and use it, being a little vague on the dangers posed by nitrates and nitrites in the diet, and really fond of bacon. Anyway, with or without the Prague powder, we stir the above ingredients into a paste and spread them all over two pork bellies, or maybe four if they are small (20 to 30 pounds total). We rub this in well and put the bellies in a large rectangular tub with a lid (a big roasting pan would work fine, if you can cover it reasonably well) and put them at the back of the refrigerator for a week or two, turning them in their cure whenever we think of it. If we forget about them, of just get busy, they'll do all right so long as we spread the cure well over them in the beginning; in fact, if we only remember to turn them once or twice and don't get back to them for three weeks they'll be just fine, in all likelihood, and if there's a little mold in one or two spots we just wash or cut it off and go on to the next step, which is to smoke them. We generally rinse them down a little and let them dry on a rack while we build a smoldering fire in our makeshift smoker. This is an old metal trash can with a hole in the bottom and some wire loops in the lid, the former to let smoke in and the latter for hanging meat from, set on top and to the back of an old wood cookstove in our summer kitchen and propped up on bricks to decrease the heat conducted into the smoke chamber. (The chimney pipe is unhooked and the trash can set over the opening to catch the smoke.) We use maple, mostly, and smoke for about six hours, counting all the times the fire goes out and has to be relighted. Take it down when you think it's done, and try not to eat half of it in the first day.

the proportions of an upholstered sofa and the weather takes a turn for the cold, gather all available forces and butcher them. You can find a manual to describe how to locate a point equidistant from the pig's ears and its snout for the dispatch, details which we will not describe here because you can find them lots of other places; here we mean merely to describe the operation known in polite circles as scraping casings and among the rest of us as cleaning hog intestines.

This job is smelly, tedious, and long, and reserved for the sissies who spend hog butchering in the house making coffee and cinnamon rolls for the real men of both sexes who work outside with the blood and guts. These latter are the people who find the point equidistant and so on, and their jobs include blooding, scraping, gutting, hanging, splitting, piecing, grinding, wrapping, salting, and smoking. The inside crew are kept to a tight schedule delivering doughnuts, coffee, coffee cake, cocoa, and other more substantial fare promptly at ten, twelve, three, and six o'clock to the real workers. And as I say, scraping casings—less important than the jobs above-named, perhaps, but indispensable to the making of sausage links, and an art in danger of being lost to posterity. So it is described here.

First, before the intestines come inside, the real butchers have to separate the hog's intestines from the other parts of the pluck—parts taken from inside the animal—and clean them out. They can hook one end to a hose and turn it on, first making sure everyone within a radius of 15 feet is out of the way, or they can drop one end into a bucket and start squeezing down from the other end. If they elect to use the tap, it is allowed to run until what issues from the nether end of the intestines is clear water. Then the intestines are caught up in big loops and cut at two roughly opposite points, severing each and every loop at both points. (If this is not clear, imagine running over a loose coil of garden hose with a circular saw, leaving a bunch of short lengths of hose of more or less equal length [2 or 3 feet], and you will have the right idea.) The resulting mess is submerged in a pan of cold water and sent to the house, where the people are warm and so they deserve it.

What to do with such a thing in the house? First of all, don't let it be brought in the kitchen. The smell is not fecal, but it is visceral, and it tends to linger. Maybe use the basement sink, or the one in the laundry room; somewhere you won't be preparing food anytime soon.

The method of cleaning is this: first, get a big scrapple pan or loaf pan, or maybe three loaf pans, and put one in the sink. Now take a wooden cutting board or a length of smooth maple or beechwood plank long enough to span the sink diagonally, the right end sitting in the baking pan and the left end resting on the edge of the sink. If you are left-handed, reverse the sides. Take a length of intestine, to which we will refer hereafter as "casing," and hold it at the top of the cutting board slope so that 3 or 4 inches of casing hang down the board. Now, with a (preferably old and unloved) table knife with a flat cutting edge held at an angle of about 45 degrees to the board surface, scrape firmly down the thick, squishy, pinky-gray tube.

Keep in mind, at this point, that this tubing was thoroughly cleaned by the people outside. Stuff will come out the end. Lots of stuff, thick, gooey stuff, the lining of the hog's intestines. The guck you are looking at is pig, not pig excrement; tissue, not poo. Scrape the casing again, more firmly. Keep drawing the knife blade angled down the board until the casing is clear, or faintly pink. You will probably be surprised to see just how thin—much more than paper-thin—pig casings are, and you will certainly be surprised to find out how very tough they are. When you think about what they do, you will realize that this must be so. I mean, suppose you were designing a system where a hose of some really toxic guck was going to run through the interior of a living being, you'd make the hose strong, wouldn't you? And so it is.

You are going to scrape the whole length of the casing, a few inches at a time, and you are going to scrape it hard. Push the soft tissue lining out at the bottom end (this is the reason for the baking pan). For reasons that will be obvious if you think about it, you will probably want to reverse the ends when you get halfway up the tube. When nothing more will come out, hold one end of the tube over the faucet and run water through it until it runs clear.

Keep cleaned casings in another baking pan of clean water. If no one helps you with this job, you are going to have a sore arm at the end of the day; if you've never done this before, it may take you all day, too. But the effort is worth it; natural casings are the only kind worth having, and if you buy them they are costly as sin. Scrape your own casings.

Or be a real man and join the people working outdoors.

Manure

While it may seem odd that we are including manure in a chapter dealing, so far, with the food harvest, in reality it's natural and appropriate. The grassfed homestead is a community of microecosystems in which sunlight is captured, stored, and converted to other valuable forms: plant leaves and stems, of course; animal flesh, milk, and eggs; but also, and in great quantities, manure. A cow may produce upward of 10 tons of manure a year, and an equal volume of urine—no inconsiderable harvest—and anyone who has seen how well deep bedding breaks down with pigs to manure and turn it, knows how valuable pigs are as part of the composting system. Free-range chickens spread much of their manure themselves, but chickens in general seem to save a lot of their bathroom duty for the night hours, which means the litter from the poultry house will be supercharged with nitrogen. We apply chicken litter to the garden in the fall or winter, giving it time to break down before we begin planting in the spring, or use it to jump-start compost breakdown for other plant materials. Not only does the passage of plant material through the bodies of our livestock decompose it for swift integration into the soil, but it converts some elements into forms more accessible to plant roots. In addition, the manure of ruminants supports soil life with a superdose of the microbes of plant decay. Manure is a part of the nutrient harvest every bit as much as the meat, milk, and eggs that grace our tables, and it is a vital part of the feedback system that fuels and ensures the health of the homestead.

Grass management is, at the most basic level, manure management. The work that goes into making sure our animals are moving smoothly over the pastures and grazing evenly on each area, simultaneously guarantees that manure will be laid down consistently to fertilize the entire pasture, rather than being concentrated in the hot spots—water hole, mineral block, shade tree, barn—that develop and intensify in a conventionally grazed pasture. The results are quickly evident in the early months of a shift to rotational grazing: not only are we no longer seeing some pasture plants scalped while others are left to grow woody and coarse, but as the earth increases in fertility and moisture retention as a result of the generous spread of manure our animals have provided, seedlings germinate and mature plants spread to form a thick, unbroken cover of plant life that holds and cools the soil, retains moisture, stores carbon, and provides home for a rich, healthy, diverse living community. With manure delivering nutrients for every need, pasture plants grow taller, denser, with more leaf matter, providing more food for the grazing animals that are achieving this transformation.

In the Garden

Elsewhere on the farm the cycle of sunlight harvest requires, if possible, even more intense management. We don't automatically think of gardening as forage planning—after all, we consider grass a pest in the garden—but here, too, in fact, the farm's natural, solar-energy driven fertility subsists, as it does in the pasture, in the rhythm of managed plant cover and even manure application, in the care taken to foster and promote soil microbial life. In the garden, however, the return of nutrients to the soil is not automatic, as it is in the pasture, because, in general, from here the harvest is removed to other places before being consumed. Many, perhaps even most, nutrients leave the garden where they were assembled to feed humans in the house, or animals

Figure 10.29. One way to collect the solar harvest.

in the barn, coop, or pasture. This constant removal of crop cover, root mass, and necessary elements from the garden soil requires that systems be developed and maintained for renewing fertility in the garden soil as well.

How to provide this renewal is one of the management tasks that keep grass farming interesting. While green manure crops—crops grown for direct return to the soil—make up one source, farm animals also fill an important role. Where green manures like buckwheat and oats may be sown for biomass—large scale additions of organic matter to the soil—and nitrogen is fixed in the root nodules of leguminous green manures like clover, beans, and field peas, the manure and composted bedding of farm animals will provide concentrated applications of nitrogen and tons of decomposed plant material. Fortunately, the seasonal flow of animals over the grass-based homestead offers many opportunities for the attentive farmer to direct manure where it will be most valuable.

By far the least human labor is required when we manage farm animals so that they, and not we, spread the fertilizer in the garden. Our own methods, in their present manifestation, can serve as examples only of the many possible arrangements of this kind; your own homestead will suggest others.

Pigs

Seasonally, we run pigs over the garden. They are particularly useful where we are planning to break ground for a new cultivated area: a panel of polynetting or two strands of polytwine serve to keep the pigs enclosed on areas we want turned over for new beds, and feeding shell corn on the bare ground encourages an already willing rooting over. Depending on how long the pigs are paddocked in one place, quite a bit of manure may be applied directly to the garden in this way. John Seymour recommended a full year of pig plowing to prepare a moderate-size garden bed, giving the animals time to turn over soil, eat the roots of perennial weeds, and dig out rocks and stumps. In established gardens we may use the pigs as gleaning crew and postharvest cleanup; they can be relied on to find and eat pretty much anything the human harvesters miss, or turn it into the soil to break down. Where there is much crop residue (squash vines, cornstalks, bean haulms, and so on), we divide up the beds with polywire, allowing access only to as much as the pigs can eat in a day, since otherwise they dig and trample more than they eat. Timing is important; some forethought in garden planning helps us spread the pig harvest over as much of the year as possible, instead of having periodic flushes too large to be utilized completely.

Chickens

We have never discovered that happy (and possibly mythological?) synergy with the barnyard hen

Figure 10.30. Oats and field peas: ground cover, green manure, and harvestable forage.

whereby she—with access to the kitchen gardens—hunts, scratches, and forages in the paths only, and leaves the beds alone. On the contrary, every single time a chicken gets into our garden she rips out yards of new lettuce, pulls up the germinating corn, or pecks holes in every tomato within her reach. We do feed our chickens from the garden, but we prefer to manage the harvest ourselves! Our garden is now completely surrounded with tight horse wire, and if any hen persists in flying over it, we clip her wings; this way, from early spring to late fall our garden is a chicken-free zone. But when the fall crops are harvested, and the winter spinach, lettuce, beets, and carrots are under tunnels, our chickens take up their valuable role in the garden fertilization cycle. Low tunnels constructed especially for them, and chicken tractors covered with 6-mil plastic sheeting, provide warm, well-lit housing for our laying birds while giving them an area of unfrozen soil in which to scratch. These can be moved several times during the winter to facilitate manure spread while providing a clean floor for the chickens and avoiding the hard pack of bedding generated when chickens remain too long in one place. When the soil under the tunnel becomes muddy, we add carbonaceous bedding, which the chickens till into the top layer, adding their own hot manure to initiate rapid decomposition. Straw can pack down in a thick layer and be difficult to incorporate into the soil next spring—although so far we have not had much problem, just moving any heavy accumulations onto the garden paths for mulch, or trucking them to the compost bin in a wheelbarrow—so this fall we bagged dry leaves for a lighter, more tillable bedding.

Ruminants

Ruminants also play an ongoing part in the long-term, deep fertilization of our gardens. Early on we experimented with direct manure application, by winter feeding round bales on our larger gardens. Although this method did add a great deal of organic matter and nitrogen to the soil without any extra labor on our part, it also resulted in serious soil compaction, and poorly decomposed waste hay made cultivation almost impossible next spring, while the slow breakdown of such a large mass of plant material initially bound up nitrogen and starved out spring plantings. More careful timing might have improved our results; waiting for the ground to freeze hard, and moving the round bale ring several times over the winter, would probably mitigate the problems. And in any case, twelve years later that area of the garden still shows the positive results of its large dose of fertilizer.

Today our gardens are much larger than they were then, large enough (about an acre) that we need more organic matter than just the pigs and hens can supply, and we have not given up using cattle for direct application of manure. More recently, we have paddocked the lactating cows in the garden for a few weeks while the ground is frozen, feeding hay in a different area each day so that the whole garden gets its share of waste hay and manure. This winter we intend to try setting aside one garden area in four on which to feed the cattle at whatever point in the winter it becomes necessary to supplement with hay. Holding the animals on these small (1/10-acre) plots will mean a certain amount of compaction, but this will be minimized on frozen ground, while the heavy applications of manure and waste hay that result will, we hope, more than offset this disadvantage. Then in early spring we'll frost-sow field peas and oats without prior cultivation, letting frost heave and animal hooves ensure good soil-to-seed contact and removing the cows before the cover germinates. This will be followed over the gardening season by a succession of soil-building green manure crops—buckwheat in hot weather, and tillage radishes or oats and peas as a fall crop, probably—for an entire year of nutrient building. The manure will give a jump-start to the spring planting, and by delaying the need for cultivation we hope to give the heavy application of waste hay and other organic matter time to break down, before a late summer tilling and the planting of a winter cover crop. With an entire fallow year, and the addition of quantities of organic material, including hundreds of pounds of ruminant manure, we hope to lay down a fund of fertility for some time to come, boosting these select areas of the garden to a much higher level of health and productivity.

Green Manures and Crop Rotations

And while we're on the subject of manure and the solar harvest, we should touch on the topic of green manures and cover crops, and other ways we can manage the soil for fertility in and out of the garden.

Legumes, nitrogen-fixing plants like clover, vetch, and beans, have a role in the pasture as well as the garden, a role our rotational grazing practices assist, whether we notice them or not. These are not sown species, but natives and volunteers, plants that *want* to be where they are. We find that the percentage of legumes in our pastures—mostly tall plants like medium red clover and some vetch on the slopes, with Dutch white clover seeming to prefer the cooler valley floor—seems to increase over time, a reflection, we think, of the improved conditions for their growth. Periodic grazing and rest keeps the legumes in our pastures from being disadvantaged by constant heavy and selective grazing, while the healthy root dieback that results adds nitrogen, as well as carbon, to the soil.

In the garden we are always looking for ways to increase our use of green manures and cover crops. Already large for the human food plants, our garden is expanded to grow pig fodder and forages, but the increase in size is not as great as you might think, since lots of the things we feed pigs fit into our rotations between human food crops. So far, for example, we've been successful putting the late-summer beans and turnips (primarily pig fodder crops) in the spaces left when we harvest potatoes in July, alternating a row of beans with two of turnips: the beans in the rows where the potatoes were, the

Figure 10.31. Hens in a tractor groom garden beds and paths.

turnips in between, where the mulch under the potatoes has broken down. This keeps more constant cover on the garden so the soil has less time to heat up—vaporizing nitrogen and carbon and killing soil biota—and cuts down on wind and water erosion. In addition, the beans, being legumes, fix nitrogen in the soil, while the turnips, taprooted vegetables, penetrate hardpan, carrying water and carbon deeper into the soil and increasing the overall productivity of the farm, present and future. If nothing happens to set our plantings back, both the turnips and beans have time to mature before first frost. If the beans—pintos, generally—are fully mature we may thresh them for people food and give the pigs just the haulms. The turnips will be topped and bagged and stored in the root cellar or dry cave until other fall crops with shorter storage lives—things like cabbage, squash, and pumpkins—are used up. After all these products of the garden are fed out, the ash and minerals they contain will be returned to the point of origin as composted manure and bedding.

Oats and field peas are winter cover crops and spring green manures, but if they get away from us before we can till them in, they can be cut and dried before the grain is mature enough to shatter out, then put up loose in the hayloft. Pigs love them, and so do hens—any oat-and-pea straw we drop will be thoroughly scratched through by the chickens. If the pigs are penned, their bedding will be returned to the garden as compost, completing the cycle of sunlight, grazing, and increased soil carbon.

We grow heritage field corns that are sweet enough for the table: Golden Bantam and Country Gentleman, principally—Country Gentleman produces better for us, but we plant both varieties for the sake of backup. What we don't harvest for ourselves—most of the crop, since we plant about $\frac{3}{10}$ of an acre—we may cut green, starting in late summer, and feed as a whole plant. The pigs eat ears and leaves and quite a bit of the stalk; what they don't eat they chew up and add to their bedding. Such as is left in the garden will mature and dry on the stalk, when we can either harvest the whole plant, or just the ears. Dry corn has a long storage life, but we don't have a corncrib so we have to feed it out early or let the hens and mice take a lot. Or we can just let the pigs self-harvest, saving us the trouble of hauling bedding and manure back to the garden.

The $\frac{3}{10}$ of an acre of corn is, granted, far more than we would use for the house, but it doesn't really add to the size of the garden since it precedes potatoes in our rotation, and we grow a lot of potatoes. Corn is a heavy feeder, but if we plant legumes with the corn some fertility is added back for the potato crop, and the ground remains cool and protected under cover. If the year is good, the winter

Planning the Stored Winter Foods Schedule

A little thought helps us plan a schedule for utilizing farm-raised forages, based on the answers to certain questions:

How is this crop stored?
How long does it keep?
When is it at its best, or when does it provide the best nutrition for the animal?
If it is harvested by the animal, what is the window of opportunity (before beets freeze in the ground, bean plants turn to mush, peas shatter from their haulms)?
What is its season of availability? (dairy, leafy crops, fruits, acorns, leaf mast)

On our farm, order of use is roughly as follows:

Short-term storage: Green oats and field peas cut for hay, and sunflower seeds, are fed soon after harvest because we are unable to store them where they are protected from pest species. Green corn and cornstalks, squash and pumpkins, and cabbage crops are fed in the late summer and fall, to be used up before the weather is settled cold.

Root-cellar storage: We begin feeding turnips, sugar beets, and other root crops (excluding mangelwurzels) when the short-term storage vegetables are used up, or after hard frost.

Long-term storage: Grass hay, bean haulms, and dried beans are least subject to decay and may be fed out as needed, all year.

Delay: Some fodder is improved by storage. Mangelwurzels are supposed to be stored until January; fed earlier, they may cause scours.

Culls: Fruits and vegetables not of a quality for storage, or that are decaying in storage, are fed out as we sort them: squashes, potatoes, and alliums, primarily.

squash patch outproduces household needs by a large margin; most of the small, damaged fruits will be fed to the pigs and chickens before deep cold sets in, and, later, cull squashes as we sort them over the winter. A tenth of an acre of mangelwurzels—taprooted tillage plants—has produced as much as 1½ tons of roots for us, and that is a lot of pig food: we feed around 30 pounds, with hay and some dairy waste, as a daily ration for the sow, with, of course, the manure harvest as part of the benefit.

Animal Bedding, Composting

But with the succulent crops we grow specifically for pig fodder—roots like turnips, beets, and mangelwurzels—another tactic is necessary. Winter self-harvest in our climate isn't always practical since these forages, being mostly water by volume, quite naturally freeze in really cold weather, after which not even a crocodile could get a bite out of them. Self-harvest by the animals becomes in this case impossible, so we harvest root crops in the fall, store them in the root cellar, and feed them out over the winter to confined animals. Ruminants also, under certain conditions, may be confined in pen or barnyard. During really cold winter conditions, when animal well-being is necessarily compromised in the field, or periods of prolonged wet weather, when animals in the pasture will cause pugging and trailing, circumstances dictate the temporary confinement of our animals—when, in addition to bringing them their dinners, it becomes our job to manage and capture their wastes. Confinement puts the minerals and nitrogen so urgently needed in the garden in the barn, where we can sequester them. Deep bedding—a thick layer of carbonaceous material in pen or stall—is not only the best way to keep the animals clean and comfortable, it acts as a sponge for soaking up and binding volatile nitrogen in the form of urine and manure. Furthermore, this nitrogen provides the "heat" to break down straw, hay, leaves, shredded paper, or whatever we are using for bedding, turning the animal pen into a small recycling station for compostable materials.

Bedding Materials—Compostable Carbon

Which form of absorbent carbon you choose to put under your animals is a decision based on several factors. First, what's available? At what cost, in money and labor? Second, how absorbent is it? Will the animals eat it, and if so, is it safe for them to do so? How much will they eat? How hard will it be to move, either dry or wet? And does it meet the needs of the species in question—if chickens, as litter to absorb moisture and to scratch in; for water birds, as nesting and flooring material; deep, absorbent bedding for pigs and ruminants, and so on? And to these questions add two more: How readily will it decompose? What sort of compost will it make when it breaks down?

On the Sow's Ear we've used a good many types of bedding, with varying degrees of success. Straw is great under most animals, but we have to look out for very small creatures, like piglets and baby birds, getting tangled or tripped up. It makes great compost, too, being naturally aerated (grain stems are hollow) and mostly free of weed and grass seeds. Hay is often the natural choice for bedding in a stall or barn, where we spread the rejected hay from the bottom of the manger before we refill it with fresh hay; hay mixed with ruminant manure and urine makes an excellent compost, if not as weed-free as straw compost. We also like the idea of using shredded paper under our animals, and our research indicates that most newsprint and office paper is safe for livestock bedding (even if they eat some) and decomposes readily. Too bad collecting it would require a trip to town. Sawdust is good for bedding; we have been able to harvest sawdust-based litter from a local stable and have had very good results with applying it directly to our gardens. On the other hand, with wood chips, a material to which we have ready access in bulk, we have not had good results when it is used as bedding, finding that it quickly becomes saturated, after which the chips pack down into an impervious layer that caps and holds moisture where it is not wanted. In the compost pile, wood chips take much longer to break down than do more particulate bedding materials, and, applied too early to the garden, wood chip bedding has the negative result of robbing the soil of carbon during its lengthy decomposition process.

Figure 10.32. Manure makes a fertile garden.

Grass farming means managing nutrients the farm over. While in the generality of cases it is most economical for our livestock to spread their own manure as they do in the pasture, no single approach meets every situation. Resilience is a function of diversity of method as well as species: direct application of manure to pasture is complemented, not threatened, by periodic nutrient capture in deep bedding of temporarily confined animals. In the symphony of grass management, deep bedding is scherzo to the rotational grazing theme.

CHAPTER ELEVEN

The Independent Farmstead

Doomsday predictions can no longer be met with irony or disdain. . . . The pace of consumption, waste, and environmental change has so stretched the planet's capacity that our contemporary lifestyle, unsustainable as it is, can only precipitate catastrophes . . .

—*Pope Francis,* On Care of Our Common Home

How are we to know if we are cut out for the homestead; if, once we undertake building it, we will be glad to come home there? What kind of a place will it be, and what kind of people will we become, living with it? We don't pretend to know what persuasion of person is cut out for homesteading from grass, but we can describe some of the effects it has had in our own lives, and some of the benefits—and costs—we have experienced. These may be enough to scare off all but the most determined.

Figure 11.1. Planting potatoes for humans and pigs.

Living with plants and animals puts a great many things outside our control. They get born, in awkward places and at awkward times. They go east when we want them to go west. They get sick. They die. Seeds have to be planted in favorable weather if they are to sprout; fruits ripen and must be harvested, or rot where they grew. Instead of tightening our control over natural systems we have to submit to them, assist them, adjust our expectations; in doing so we find ourselves part of a larger whole that shares its influence with us, pouring out food, education, beauty—far out of proportion to its price.

Farm Time

The alarm goes off early when you milk at five. In high summer it is already light out and the birds are awake; in winter only the stars and coyotes keep us company as we walk down to bring up the cows. The sameness of milking chores lets our minds rouse more slowly than do our bodies, and by the time we are awake to the day's requirements, some of them are already behind us. Other people rise for other chores, and all of us being awake, we prepare and eat breakfast together, with time to discuss the day's demands and plan its pleasures. Farm chores punctuate our whole day—weeding in the morning before the day grows hot, noon feedings for the bottle lambs, a trip to the garden to pick beans for dinner, and afternoon milking and paddock shifts. Evening finds us together again after settling the animals for the night, and we share another meal: fresh food, carefully prepared. Now, when many things take us out before the sun has come up or after it has set, the night is our friend, no longer filled with lurking danger as when our lives only happened in well-lit places. We go to bed early, and sleep hard.

Animal Chores

Timing. Milking takes us about forty minutes, one cow, one person. Walking back and forth from the barn and cleaning buckets makes half that time. Two cows don't take twice as long, since the walking, bucket carrying, washing, and so forth aren't doubled: add maybe twenty minutes for each additional cow. If two people milk two cows, overall time is shorter than with a single milker and cow, since moving animals and paddock fences, and carrying buckets and milk cans, are shared labor. Moving paddock for two cows takes about five minutes, unless you are making a long move or setting up a cell for several days, and then who knows. Of course, if the paddock is on the south forty, there's back and forth time, too. Six sheep take longer to move because they dodge around a lot and don't want to be crowded.

We feed hogs twice a day, sometimes thrice. Depending on the waste nutrients available, this can be a very fast chore, or it can require some thought. Cutting up roots for small pigs takes a while, and if the piglets are very small we may even cook the roots on the woodstove first. A bucket of skim milk only has to be poured into the trough. If the hayloft is over the pigpen, throwing down forage takes only a minute, so that's where we store green-cut oats and peas, and bean haulms, too. Sometimes there are supplements to dole out: garlic is fed whole, once a week, and so is kelp—about ¼ cup per pig. It generally takes longer to plan what to feed them than to do it.

In summer we feed the poultry twice a day, morning and evening, with one more trip around sunset to shut them up for the night: ten or fifteen minutes, two or three times a day. Afternoon poultry chores take longest because that's when we gather eggs. We feed the cats in the barn, to encourage rat patrol, and we'd feed the dog there, too, but the poultry will fight him for it.

In winter the chores take longer than in summer, but the workload is less overall because the fields and gardens are dormant. Where temperatures drop below freezing for extended periods, water is the single biggest problem. If there is no frost-free water source, such as a constantly running spring tank or ram pump, filling tanks and draining hoses will sometimes require extended time out in the fresh, cold air, something we appreciate most when we are coming back up to the house when it's over. The meager milk in the March bucket would sometimes hardly repay the labor of the milker, but we don't milk in late winter for the milk alone—each day a ruminant is milked ensures all the following days, for the entire lactation, because once the cow is dried off, that's it for milk until she freshens again. Moving fence across frozen pasture can pose some

Figure 11.2. Winter chores.

challenges, with fence posts hard to drive into the frozen soil. It pleases us that the needs of the garden—manure- and urine-enhanced bedding for the compost bins—require us to confine the youngest livestock for some part of the winter, because every animal in the barn is one less we have to move fences for; the flip side, of course, is that we have to haul their hay.

The Garden

We try to maintain the solar harvest as many months of the year as possible. In the garden, this means planning and planting for winter vegetables in the low and high tunnels, which actually increases the work load more in July, August, and September than it does in the wintertime, when most weeds are dormant and our job is harvesting what is needed on a daily basis, and trying to get it before the voles and mice do (they love warm winter tunnels and the free food they find there). But an intensive solar harvest increases garden chores in other ways, especially the growing of those forage and fodder crops that supplement the grass diets of nonruminants like pigs and poultry. Jobs like tilling, planting, and weeding for the large root crops, extra corn, and big plantings of legumes eat up days in spring and midsummer that might otherwise be available for big projects like fence building or spring development. Looked at another way, though, any job that increases ground cover, adds carbon and nitrogen, and cools the soil in the garden actually saves labor, since these things decrease the time we will have to spend weeding and soil building.

We never have the manpower we'd like for all the garden jobs that need to be done, from February—when maple trees are tapped and the fruit trees (a plant chore, if not a garden one), vines, and berry canes need to be pruned—through long months of preparing beds, planting, weeding, and harvesting, until in late November the garlic is planted and the tunnels are covered. To keep the gardens as we would like them kept could easily mean four to fourteen hours of work a day, eight months of the year, for a single person, with some help still required for bigger jobs. Some of us are living for the day we can pour out our souls in garden keeping, and walk visitors through with the pride of artists exhibiting their magnum opus. Until then, however, the gardens will be large, fruitful, sometimes pest ridden and weedy, never producing all we know they are capable of, their fertility never as balanced as it could be, but—we give thanks to God—ever fountains of excellent food, fresh, delicious, the basis of vibrant health.

We have to devote a great deal of time to this life; every year that passes, it seems there is more to do, so that even as we learn to apply our labor more efficiently, we remain as busy as ever. Keeping a dairy animal, building grass with intensive grazing, managing the flow of nutrients from soil to grass to ruminant, milk to table and pigs and poultry, slaughter and harvest and preservation, are not things to be accomplished in spare moments, or with the husks of time left over from a busy day; they can't be tacked on to an already full life but have to be imperatives, that which gets done even when nothing else does. The homestead, in fact, has to be first, in the way our personal relationships have to come before any other consideration: like marriage, parenting. Love means always taking care of the beloved.

People

Homesteading eliminates some choices. Chores are real, coming around daily or even more often, and not to be put off or avoided—because something's life depends on our doing them. They impose limitations on our lives, but lend them form at the same time. As the coming in and going out of the tide imposes a rhythm on the lives of coast dwellers, and common prayer orders monastic life, daily chores break our day up into regular periods, drawing us back, over and over again, to an awareness of what we are doing, why we are here. Daily, even twice-daily milking chores and paddock moves, carrying water and food to confined animals, letting them out to graze or shutting them away from predators at night, ground us in our ecosystem even as

Figure 11.3. It's important to keep everyone informed.

our farm tasks tie our bodies to this place as home. Focusing our attention on pasture plants, animals, insects, we find ourselves drawn into their community, intimately concerned with their welfare. Building soil creates physical ties between us and the earth. If commitment, necessity, obligation are things you would rather avoid, you're not going to like farming.

Neither is this a good life for the determinedly solitary. Building and living on the independent farmstead is not to be undertaken alone; here, if anywhere, we need a partner—at least one or, better, several. A single person will want to seek like-minded companions who share her peculiar fixation with plants, animals, soil, and food. Couples need the buy-in of family members, not just as passive, if willing, observers but as companions and substitutes, as a second set of hands, and someone to take over a milking when the flu sends us to bed. The quickest death we can imagine to a milking career is to try doing it alone. Dairying is not the occupation for a loner.

And this someone can't be lukewarm on the farming life, dragged along by our enthusiasm but all the time with one eye on the clock, calculating how soon the job will be over. We can think of nothing that palls sooner than expecting help from someone who hates being asked. We both wanted this life, even before we knew what we were wishing for; our children, who have grown up on the farm, love it, even when there is a chore they'd just as soon avoid. Farming chores—especially chores related to dairying—are as regular as sunup and sundown, and more inevitable than any demands of human relationship. Undertaking them with someone who shares our sense of their beauty and function can create intimacy and cohesion, but forcing cooperation on the unwilling would soon undermine even the strongest determination.

Our children are integral to the daily working of the homestead. Our belief in the value, even imperative value, of small-scale, low-impact household sustainability, and in the beauty and utility of natural food production, means that teaching our children to husband the land—or say, rather, learning to farm alongside them and sharing the lessons the land teaches us—are, to us, imperatives. In this country where we are encouraged to think that everything is possible for us if we just want it badly enough, we have had to conclude that our lives, at least, don't contain the kind of time and resources it would take to make good on that promise. Building a childhood in which are packed all the exposures, lessons, team experiences, and tutorials that might really make it possible for a person to follow any life course he thought desirable would cost a packet of money; we think, too, that it would not leave much room for childhood. We couldn't afford to make it happen; we don't like the idea in any case. But we don't see our lives, or theirs, as being impoverished by the limitations imposed by real life; we believe enough in what we are doing that we are satisfied to spend time with our family in work, and meals, and play, passing on what we know, working out problems together, testing the value of our assumptions according to the corrections those assumptions are often dealt when reality flinches.

Communication and oversight are critical in farm life, especially where children are concerned. Sharing chores with the whole family is not just a way of imparting education and a sense of responsibility—it's a survival technique. Without the working cooperation of as many people as possible, basic maintenance on the homestead would be more than a part-time job. But too often have we discovered that some animal or other was being fed more or less than we had directed, or more or less often than we intended. We may give what we think are clear,

Figure 11.4. A young farmer.

lucid directions, and go our way believing that what we have said should be done, *will* be done; nevertheless, even if we have the child repeat our instructions back to us, we won't know until we see the results what interpretation may have been put upon our words.

Sometimes a report has reached us without getting sufficient attention, the "yes, yes" of the distracted parent who fails to appreciate the significance of what is being said. Early on, we remember, a child given responsibility for feeding hens and collecting eggs brought us the information that "the hens aren't eating all the table scraps." We pictured maybe some citrus peels at the bottom of the trough and thought nothing more of it. A week later the same observation was made, and again we passed it off; until the third, or fourth, or fifth week, we happened to go down to the henhouse ourselves, where we found a mound of putrefying garbage rummaged by rats, drawing flies, and smelling like a dumpster. The chickens *really* weren't eating all the table scraps; we were giving them far too much, and it was time to get a pig. Listen to your help, we learned; communication breakdowns can happen at both ends of the command chain.

Failure

Nature is a complicated being, and setting out to work with her is an ongoing exercise in humility, because we can't fully comprehend her, we can only pay attention and hope to swim with the current. If we succeed, for a moment, in inflicting our will upon her, it is only for a moment; the inevitable results of our violence will soon restore the balance. But Nature's is a good balance, the only balance, in fact—all other pressures, of waste and consumption and manipulation and poison, being just flutters against the universal equilibrium.

We keep farming because, despite an inescapable sense of our own lack of information, we are even more strongly certain that the natural world makes sense, that it tends to fullness and fruition, that order is more powerful than disorder, and that death is only one manifestation of vibrant and vigorous life. Chickens *should* scratch in the dust, and eat bugs, and lay eggs, and so they do; pigs love swill and mud and lying in the shade, and these things make them healthy and good natured. Milk is for drinking and for cheese, it isn't a well of lurking putridity only kept at bay by ultrapasteurization and near-freezing temperatures. Hard work satisfies, and problems give us some mental exercise;

being right is only temporary, and being wrong just means trying again.

Theodore Roosevelt once said, "In any moment of decision, the best thing you can do is the right thing, the next best thing is the wrong thing, and the worst thing you can do is nothing." As children in the classroom we used to be afraid, when the teacher called upon us, that we wouldn't have the right answer. Now, as farmers, we know we don't have it, but we no longer have the option of doing nothing, and right or wrong, we have to make choices. So—we don't know yet how to farm "the right way," we may never know, but there is an order in the natural world that, so long as we move forward, tends to nudge us in the right direction. We are going to fail, not once, but many times; believing in the rationality of Nature even more than we believe in our failures gives us the confidence to keep farming anyway.

Independence, Interdependence

Homesteading highlights our need for one another in different ways. Never in our lives before we committed ourselves to grass-based food production had we conceived of or experienced the dependence all living things have—consciously or unconsciously—on one another, and on circumstances and events completely beyond our control.

Take the weather as an instance. One of the abiding shifts we have experienced as farmers is our attitude toward meteorological patterns. Rainy spell and dry, heat wave and cold snap, are no longer things we observe from an easy chair in a climate-controlled house, or which inconvenience us for the few moments between front door and driver's seat; we're out in them, and not only in them, but intimately affected by them. When it rains, we get wet; the cows have to be moved to a new paddock regardless of precipitation. In hot weather we drink more water and wash more laundry, soaking our clothes with sweat but getting the chores done regardless. Long dry spells may keep us awake at night, wondering how long the forage is going to hold out, and if the weather will break in time for the cool-season grasses to make some growth before winter sets in; prolonged rain means we have to be more aware of impact as we drive the lactating cows across the pasture and up to the barn for milking. We marvel at the change in our own concerns and thought patterns as a result of our partnership with this ecology; we marvel, and are grateful. Our lives are immeasurably richer by it.

The behavior of animals is really important to us. A calf with ringworm in February will dominate dinner conversation until the bare places start to fill in again; if one of the sheep is snorting a good deal, we'll start watching for signs of bot flies. A rooster who does his job too aggressively is bound for the soup pot, and a duck with a swollen foot will have us applying massage and poultices. And not only domestic animals, but wild, mammal, bird, insect, mollusk: the howl of coyotes in our home valley means a loaded shotgun where we can grab it quickly at night; the dry scream of a hawk reminds us to keep young poultry in a tractor or creep until they are too large to be carried off. The life cycle of locusts has a lot to do with whether we plant new orchard trees this year, or put it off until next, and the depredations of slugs in the lettuce beds can even induce us to dedicate a bottle of beer as slug bait in the garden.

And our friends, our community, are more important in the daily business of our lives than we have ever experienced them to be before. The example, aid, and comradeship of other grass-based farmers in the Eastern Ohio Grazing Council, and at events like the North Central Ohio Dairy Grazing Conference and Ohio Ecological Food and Farm Association Conference, have demonstrated our dependence on the help, advice, and friendship of other people. Butchering four hogs in a weekend, we are glad for all the people who show up to share the labor, time-honored jokes, good food. An overnight trip, illness, or injury, for the owner of dairy animals, requires finding someone who can milk, will milk, and is available to milk in our stead. Farmers need friends; those tough frontiersmen who did it all themselves probably died young.

We find that, through our farming, we are more closely related to our community in other ways. Putting more time into farming and less into cash-producing activities means a lower-volume domestic cash flow, but it doesn't mean we are less able to help disadvantaged friends and neighbors. And it doesn't prevent our supporting local charities; it just broadens the forms of our contributions. We feel even more blessed to be able to raise a hog for the freezer of a local widow with young children than when writing a check for foreign aid; raising food for the monastery includes us, in a small way, in their many spiritual and charitable works for the local community. Less time for organized community activities in town is balanced by time with people who come to the farm to spend a day in the country, learn something about where their food comes from, or volunteer time for heavy jobs at the monastery. Groups of university students come for a few hours, or days, to enjoy family time, nature time, time for physical labor. Farming, we find, increases, rather than decreases, our involvement with the people around us.

We see the world differently now, and are moved, excited, even, by the changes in ourselves. We grew up in a mechanized culture, one that, in our lifetimes, has only become increasingly mechanized, digitized, prefabricated, processed, artificially enhanced, and genetically modified. Nothing in our educations, and little in our experience, fostered the idea that human beings are members of an interdependent ecology shared with every other living thing on the planet; voices that preached our partial responsibility for that ecology found no material in our minds from which to build on that principle. Security, and insecurity, in minds cut off from knowledge of that interdependence become constructs of dollars and cents, technology and political event. Today we find life far richer, far more exciting, as grass farmers, experiencing repeatedly the evidence that we human beings do belong on this planet, that nature conspires to our delight and comfort, and that we may ourselves play a humble part in assisting that conspiracy.

Seasonal Order . . . and Disorder

Homesteads are not tidy places. Maybe this is the simple life, but it requires a great many tools and materials, far more than we needed in the city, although far less of it extraneous or merely ornamental or convenient. Heavy winter clothes, stout boots; many large pots and pans; mason jars by the hundred, for milk and ferments and dried things and canning; more tools, many more. Keeping them all in their appointed places is more difficult because they are really in use, so they move around. Things don't stay where you put them: the hammer you need for fencing turns up in the barn where someone is using it to put shelves in the dairy (we'd better go buy another hammer, or maybe two more);

the bottle lamb you put in its pen after feeding turns up in the garden where he has followed the ten-year-old, looking to have his head scratched. Things pile up: lumber for the next building project, seeds for the spring planting, sacks for storing the fall root crops, buckets for every possible purpose. And food is untidy: no longer does it come washed and chopped and ready to cook, if not precooked; now we have to deal with dirt and bugs and peels and seeds and shucks and the ragged outer leaves of things. And there's mud. And manure.

Weather is never tidy; sometimes it's downright messy. Mud season, when winter is passing off but spring is not yet here, all the ruts in the lane are running with snowmelt, the barnyard is a morass where you can lose a boot, and the compost bins are ripe with all the winter's manure thawing at once. Late summer in a dry year, all the boots coming in dusty and grit sifting through the window screens and settling on everything, the perennial weeds in the pasture tall and brown, snatching at clothing with cocklebur and tickseed, all the grass tow-colored, and hay making leaving us with eyes and noses and collars and hair full of black dust and chaff. Late autumn is buckeye-brown with drifting leaves and damp earth, and whoever brings in the firewood drops bits of bark and wood chip on the floor.

But then, of course, there's also spring green-up. The soil is still damp and black, and the new seedlings glow against it. Whole banks are purple with violets, picked out in yellow and white; dandelions speckle the short, soft pasture grass like drops of sunlight. In May and June, at least, flowerbeds are exploding with iris, poppies, and peonies, and all the orchard trees shower petals of white and pink. The vegetable garden fills with all different shades and textures of green, glowing with the yellow of sunflower, the reds and oranges of tomato, magenta of beets and rhubarb. Roadsides are Delft-patterned with Queen Anne's lace and chicory. Later, after the discouraging browns and seres of late summer, ditch weeds take on color: tall ironweed with its umbels of true purple, spikes of goldenrod, and great mauve heads of joe-pye weed. The maple at the foot of the hill turns rosy-peach, at first just one branch reaching out to the northeast like a signal to the other trees in the hollow to be ready soon, while sumac turns bright red along the wood's edge; then all the trees, red and yellow of oak, yellow of sycamore and elm and walnut, red and peach and pink of maple. And in winter, the gray and white and black of snow sifting down into the hollow; the blue and white and gold of sunshine on snowdrift.

Figure 11.5. Building the summer kitchen.

Going out in bad weather is not the same when you have to do it because if you don't something will go unfed, or without water, or maybe die. The focus is different, and we find that the weather ceases to dominate our experience and becomes just mood music, except when it takes on titanic proportions and blasts things or breaks things or washes them away, and then we are thrilled with its power. Working before sunup, or after sundown, we take note of the cycles of the moon, and where the planets and constellations are in the different seasons; we know, on dark winter mornings, where to find Venus, Polaris, and Cassiopeia; the last trip to the barn on a moonlit winter night may end with us sledding until long past bedtime. If picking up hay in July makes us sweat, the ride in the back of the truck as we bring it up to be stacked in the hayloft cools us off again. Cold just means more clothing before we go out in it. Rain is good for the pasture, and wet clothes make dry ones feel like a luxury. Hours on our knees in the garden setting out seedlings gives us lots of time to think, or talk, or work out harmonies to whatever we are singing. No: homesteading isn't tidy; but it's ravishingly beautiful.

Figure 11.6. Butchering involves community. Courtesy of the Franciscan Sisters T.O.R. of Penance of the Sorrowful Mother.

Farm Education

Frankly, we find that our educations have not prepared us for this life. School taught us to expect that every question has a right answer, but we find that, in farming, some questions have no right answers, at least in the sense of correct solutions, and many wrong answers are only wrong some of the time. As students, we knew that if we were right we got an A, but in farming there are degrees of rightness, or rather, no right answer stands alone, and often the only reward for doing it right is another job following. We had to learn lessons in farm mathematics: on the homestead, one plus one almost never equals two, but instead three, or thirteen, if the addends are male and female; if one is a carnivore, maybe it equals just one, but fatter than before. Farm life is not divided up into neat segments of work and free time, a bell ringing to announce the transition; hard labor is often punctuated by moments of intense relaxation, while in some seasons there may be no free time at all. Divisions of labor are by necessity blurred; autonomy ceases to have any meaning.

As homesteaders we are indeed undertaking a life of tasks for which we have little training and less knowledge, but instead of spelling failure this deficit often works in our favor, keeping us open to possibilities that conventional training would overlook, or discard untested. Starting out with lousy land, we were forced to find intensive improvement techniques. Our first milking Jersey, Isabel, gave lots of milk but was a constitutional nightmare of stillbirths and "down" events, so we learned dedication early, to solve problems and roll with the punches. Living on one small income we can't usually fix things with money and have had to rely instead on endurance, creativity, and the effects of time. Not having a mentor, we sought information and advice from many people we would not otherwise have met, made friends, garnered many viewpoints.

Learning to farm, we find, is like learning to ride a bike: not in the sense that once you've learned you never forget—oh, no—the similarity, instead, is in the likelihood of bumped knees. Farming, like biking, is an art made of many smaller skills, and one can't help thinking the learning process would be easier if we could concentrate on just one skill at a time. Balance first, maybe, and then steering; the steady rhythm of pedaling before mounting, accelerating, slowing down. But it can't be done that way, and if we hope ever to learn to ride we have to take it on all at once and do our best, and not mind when we land in a ditch, but get up and mount and try again. Maybe we want to specialize in poultry; that leads to a search for healthy, organic, farm-raised feeds, which leads us to dairy, which requires that we get a pig to utilize surplus. It also means cheese making, and forage management, and sustainable stock water systems; before we know it we're up to our Adam's apple in a practice with many parts, each as indispensable to the whole as the steering and pedaling and balancing are to riding a bike. Our early wobbles may be a little embarrassing—grown-ups prefer to be adequate in every situation—but the goal is so enticing we just wobble on. In time, even, we realize that all the other bikes on the path are wobbling, too; that, with all that Nature has to throw at us in the way of variation, nobody is getting a smooth ride.

This bumpy beginning is one of the reasons we think most people are going to be better off if they don't start out thinking they are going to make money as farmers. The unicyclist in the circus gets paid—albeit not very well—because he has practiced long and hard and knows how to do all kinds of tricks, but while the toddler on his two-wheeler may get a dime and a trip to the candy store from a loving parent as the reward for his efforts, if he hopes ever to be a circus performer, he had better start out riding for the love of speed and grace. The farmer who has learned to supply his family's food needs from the proper management of sunlight and rainfall may be able sell his surplus to a small, appreciative, and discerning public, but the neophyte homesteader will be better off if he directs his initial labors primarily toward achieving an ecological balance. If from the beginning his homesteading efforts must earn him a living, sooner or later he will almost certainly have to choose between financial or ecological failure.

Farming stretches our limits in a way few other things can and forces us to embrace contrasts—even contradictions. Work and delay, order and turmoil characterize the same pursuits. Animals live and reproduce, die and decay. Periods of silence and solitude give way without warning to events requiring sudden activity and enforced cooperation. We have to be completely present in the moment, while needing in every undertaking to anticipate the next hour, week, season, year. Spontaneity is built in to a life occupied with living systems, where weather may postpone a job, or make its completion suddenly imperative, and a fence break or water outage can mean redirecting all the farm's energy for the day. But farming also means inescapable routine, wasting no daylight, and still being able, when circumstances require it, to get out of bed at midnight to help a calving cow, or sweep snow off a high tunnel. We have come to like this routine of unvarying repetition, overlaid with frequent and often erratic improvisation.

Homestead Assessments

The homesteader learns a new set of standards, reflective of values unconnected with monetary exchange. We ask a different set of questions about any new animal, product, activity, or enterprise: Do we need it? Is there time for it? Will it make the farm more resilient? Can we afford it, and not merely in a monetary sense—will it add to the sum fertility and productivity of the farm? If it is a new animal, the regenerative farmer will ask other questions, many of them:

- What are the characteristics of this organism?
  - How big is it?
  - What does it eat?
  - How must it be confined/housed?
  - How resilient is it in the environment?
  - How does it interact with its environment?
    - Does it graze, root, scratch?
    - Improve pasture?
    - Turn the garden?
    - Clear brush?
- Where does it fit into the farm ecosystem?
  - Which of its needs can it supply for itself?
    - Self-harvesting food?
    - Parasite control?
    - Reproduction?
  - In what ways does this organism interact with others in the farm ecosystem?
    - Does it control pests, parasites, predators of other species?
    - Does it feed other animals with milk, eggs, wastes, or offal?
    - Can it serve guard or herd duty?
    - Will it endanger other species on the farm?
  - How do these relationships apply in reverse; that is, which of its needs may be supplied by other farm species, either wild or domestic?
    - Chickens improve pasture for ruminants;
    - Grazing reduces cover for predator species;
    - Surplus dairy feeds chickens, pigs, and garden?
  - Which needs must be supplied by the farmer?
    - Shearing, docking, trimming, dehorning, castration;
    - Fodder management, all year-round;
    - Planning and timing of breeding and birthing;
    - Care for sick or injured animals?
  - Which, if any, must come from off the farm?
    - Feed, at least temporarily?
    - Genetics/breeding?
    - Services like shearing, shoeing, hoof trimming?
  - What liabilities does it bring to the farm?
    - Is it dangerous, either to other species, animal or plant, or to human beings?
    - Is it a restricted animal or plant (gooseberries in a pine forest, heritage pigs in Michigan)?
    - Especially, what wastes are produced, and can they be turned into values?
  - Does/can this organism reproduce in kind?
    - Unaided, or with assistance?

Then, only after we have considered the species' compatibility with the farm ecology do we inquire:

- What and how many benefits/products will it produce in excess of what is needed or desired for farm use?
- Is the cash or trade value of the product greater than the nutritive/ecological value of the product if it remains on the farm?
  - For example, should we sell surplus milk, or corn, or hay, or would it be better to feed it to the livestock?

According to the answers to these questions, new animal and plant species will add security to our ecosystem, or they won't find a place here.

Dinner

And naturally, our diets have changed! And not only our diets but all our eating habits as well. Chore time and farm projects provide the underlying architecture of our day, with the secondary result that we come together for meals. Cooking from the farm means meal preparations require more planning in any case, favoring regular mealtimes and planned menus over the casual pass through the kitchen and hastily assembled sandwich. While we have more people cooking, there is less impromptu food preparation. And our diet? Fresh, whole, and couldn't be more local. Spring, summer, and fall we have lots of vegetables right from the garden, of course, but even in winter our table is filled with produce from the high and low tunnels, and potatoes, roots, winter squash, and alliums from the root cellar and dry cave. Dairy products, so abundant on our farm, make part of every meal; most days we go through a couple of gallons of milk for drinking alone, and we can't keep cheese on the shelf, we eat it so fast. And of course beef, pork, lamb, and poultry are all abundant on our farm, and we don't apologize for eating lots of anything the production of which is such a benefit to the land, the ecosystem, and the community. We still buy food: grains, mostly, but coffee, tea, spices, and some luxuries—tortilla chips come to mind! We're not ascetics, and we're not afraid of a little self-indulgence; we just like good food, and farming.

Figure 11.7. The sustainable farmstead is one that keeps finding new ways to extend the solar harvest.

Tools

We don't buy them before we need them, but we spend money on tools. The right tool can make the difference between a good job and a bad job, a good day and a bad day, a long-term fix or doing it over again soon. Fortunately, we like secondhand clothing—and furniture—and vehicles, so we can allot some money for the important things. Anyway, the cows don't care how our jeans fit or how much rust is on the truck; neither do our fellow homesteaders. We don't borrow money, so we don't have to impress the banker, either.

Tools for the independent farmstead:

- Fence: two or, better, three reels; twine, tape, or netting; ground rod and charger; a power source.

- Basic hand tools for the shop and garden; you'll have your own list. Be generous, but avoid fancy single-use tools.
- Shovels, sledges, and spud bars. Wheelbarrows. A hand-operated winch; maybe a chain saw, too.
- Buckets, barrels, hoses, hose fittings.
- Good boots. Weather-appropriate outdoor work clothing. Really.
- Books: build a good homesteading library, and keep adding to it.
- A vehicle for hauling: lumber, firewood, dirt, manure. Something you can get dirty.

There will be lots of others; some that are necessary to us will be extraneous to you, and some that you can't do without won't be important on our farm, but overall the lists will be similar.

Belonging

Stretching out on the grass in the pasture feels different when we are a vital, cooperative part of the ecosystem. We labor with the grasses, the bugs in them, the predators and scavengers, the grazing animals and their impact, the rain that recharges the soil, the sunlight from which comes all the energy, and somehow the land welcomes us as it never did when we were just strangers looking for a moment with Nature. Day and night, winter and summer are our clock and calendar. Snow isn't just something to be shoveled, or even surf for our sleds; it's water for grazing livestock, precipitation recharging the aquifer, sandbags steadying the high tunnel against winter storm winds. Rain is time: extending limits, as for pasture regrowth, or recharging stock water reservoirs; or establishing limits, flooding a hayfield, drowning new plantings. Sunlight is life.

We love the way stewardship of the earth requires the vigorous use of our bodies. Running 5 miles a day never felt as good as the long swing of booted legs over a rough grass pasture, the lift and thrust going uphill, the driving heel as we go down. Fresh milk, meat, and vegetables are the best we ever ate; we seldom get sick and recover quickly. We've developed muscle appropriate to our jobs. It's no trick to carry two 40-pound bales, split wood, or lift a calf into the truck—and there's no upper-body workout to beat hand-milking. Heavy exertion and large animals mean we get hurt in new ways, but heal faster. Do you love dirt? Black dirt under your fingernails, brown dirt on the knees of your jeans; mud on your boots, dust in your hair, grime under your collar? It's all part of the homesteader's life. And tides: of work, of weather, of food and drink; of contemplation and rest.

Sunlight, soil, water, grass. The animals that eat them. Food. Security. Resilience. Local, natural food production needs to be freed from the taint of sentimentalism or elitism; it needs to be moved front and center for everyone who draws life from this planet, the white-collar Jack in his office building, the laborer downing a well-earned beer, the SUV-driving soccer mom and the Burkina Faso woman carrying water home to her children. Our enemy is not the other guy, it is our own consumption, the dispraise of food, water, shelter, and companionship in favor of more, bigger, faster, and shinier. No one can tell us exactly how much longer the soil, air, and water will hold out before they can no longer satisfy even our legitimate needs; it might not help us to know.

Every year in which we harvest sunlight with ruminants we see the soil under our feet deepen, see it grow in fertility, see its drought resistance increase. Learning how the earth conspires to feed us, and feeling our own participation in the process, makes us less subject to the ups and downs of

Figure 11.8. Spring pasture.

political or economic climate, even of emotional and spiritual variation. Leisure, and pleasure, aren't things we save up money to buy; our well-being is not a function of our Fitbit readout, credit rating, or the amount of money we have in the bank. The daily enjoyment of working together enhances the camaraderie of an evening beer or glass of homemade wine. What is happiness, if it is not working with friends at something worthwhile, sharing good food and sleeping well, with the confidence that the next day will bring more work, more friends, good food, well-earned rest?

How badly do we want this? Enough to give up time: my time, your time, time at the gym, time in front of a screen? Is it worth sweating for, worth getting our toes stepped on—hard? A mere taste for the outdoors and a poetic appreciation for the theory of self-sufficiency isn't going to get us very far when it's time to plant the garden in the spring, working from early morning until late at night, or turn gallons of milk into cheese, or fill stock tanks and drain hoses in negative temperatures; it won't fuel us for three days in cold weather while we stand out in the shop cutting up beef for the freezer. Building a homestead, we are going to find out if we love trees more than television, cows better than a rust-free car, a good pair of boots over whatever clothes are in fashion.

But inevitably, even as it demands regimentation, this life rewards self-discipline. If we work hard, we sleep well; good food fuels our work instead of padding our waistlines. If we get off the farm for social and community events less often, more people come here to spend time with us and our animals, to share our incomparable food. They come for vacations, for quiet, for peace. Many come for a small experience of food production from the ground up; some come for longer, to learn how to do it themselves.

And we stay here, tied to the land by our labor, satisfied by the land in our labor, fed by the land from our labor. The good things are really good, and the bad things aren't just imaginary. If success doesn't mean God's blessing, neither does failure mean His displeasure—and disaster doesn't mean His rejection. Death happens often. Coming to know death we are made more comfortable in the thought of our own, seeing it as part of a continuation that somehow includes even the deceased. Life without air and sun and the pattern of leaves against a blue sky would not be so hard to part with, in any case. Failure is giving up; success is still being here when the seasons change.

We find it is a good life.

Acknowledgments

Our heartfelt and abiding thanks to the founders of the Eastern Ohio Grazing Council, who first determined the need for sharing and disseminating information and experience in intensive grazing as it is practiced in the Ohio River Valley; an extrafervent thank you to Beth Krupzak, Kevin Swope, and especially Clint Finney of Spring Valley Stock Farms, for entertaining our questions, exploring possibilities with us, and making room for us in a community where we are still relative newcomers. Your help, kindness, and especially your fellowship have been a support to us since the building of our very first paddocks.

And thank you, too, to the Franciscan Sisters T.O.R. of Penance of the Sorrowful Mother, for making us part of your family and letting us play in your yard.

Resources

Supplies

Fencing

Kencove Farm Fence Supplies: electrical fence materials, made in the United States. We've always had good results working with Kencove. www.kencove.com

Animal Watering Systems

Jobe floats: low pressure livestock valves, widely available.

Hudson valves: another option for low pressure stock watering systems, widely available, try your local farm store.

Step-in pig waterer: the Portable Step-in Pig Nipple, contact lucasdougherty27@gmail.com. Pig water that will go anywhere your pig can.

Bell-Matic poultry waterer: we find these useful on relatively level land. Widely available online, or try the local farm store.

Poultry nipples: especially useful where slope may compromise the stability of an open water reservoir. Available online or at the farm store.

Dairy

Detect-Her heat detection tail paint: useful for helping determine when a cow is in standing heat. Look online.

Dr. Naylor products: mastitis indicator pH cards, udder balm. We have to order the pH cards online; Dr. Naylor udder balm is readily available at the local farm store.

Dairy filters: we use Schwartz Filter-Clean disposable paper filters. Make sure whatever you get is for gravity-flow systems. Probably not at the farm store; you'll have to look online.

Estrotect™ heat detection patches: these are our best tools, after direct observation, for determining when one of our cows is in standing heat. You can order them online from a farm supplier, or directly. www.estrotect.com/estrotect-heat-detector

Portable electric milking machine: we don't use one of these, but we hear from reliable sources that they are a good assist for dealing with small teats, arthritis, or vacation help. Adam and Faith Schlabach, Misty Morning Farm, Virginia. www.mistymorningfarmva.com/products/

Water

Ram pumps, sling pumps, nose pumps: Rife Hydraulic Engine Mfg. Co., made in the United States. www.riferam.com

Garden

Mangelwurzel seed: Berlin Seeds, Johnny's Selected Seeds

Support and Information

Organizations

Eastern Ohio Grazing Council: the Eastern Ohio Grazing Council (EOGC) works in cooperation with Carroll, Columbiana, Harrison, Jefferson, Stark, and Tuscarawas County Soil and Water Conservation District (SWCD) and Natural Resources Conservation Service (NRCS). This organization has been a huge help in our self-education. We hope your own county has, or will soon have, something similar. www.carrollswcd.org/eastern-ohio-grazing-council.html

The Livestock Conservancy: for information on heritage or endangered livestock and poultry breeds. www.livestockconservancy.org

The Sustainable Poultry Network: for information on sustainably raised standard-bred poultry. www.spnusa.com.

Books

Asher, David. *The Art of Natural Cheesemaking: Using Traditional, Non-Industrial Methods and Raw Ingredients to Make the World's Best Cheeses.* White River Junction, Vt.: Chelsea Green Publishing, 2015.

Bonsall, Will. *Will Bonsall's Essential Guide to Radical, Self-Reliant Gardening: Innovative Techniques for Growing Vegetables, Grains, and Perennial Food Crops with Minimal Fossil Fuel and Animal Inputs.* White River Junction, Vt.: Chelsea Green Publishing, 2015.

Coleman, Eliot. *Four-Season Harvest: Organic Vegetables from Your Home Garden All Year Long, 2nd Edition.* White River Junction, Vt.: Chelsea Green Publishing, 1999.

Coleman, Eliot. *The Winter Harvest Handbook: Year-Round Vegetable Production Using Deep-Organic Techniques and Unheated Greenhouses.* White River Junction, Vt.: Chelsea Green Publishing, 2009.

The Core Historical Literature of Agriculture. Albert R. Mann Library, Cornell University. chla.library.cornell.edu. A free archive of over two thousand public domain agriculture texts published between the early seventeenth century and the late twentieth century, and an irreplaceable resource on traditional farming feeds and methods.

Kimball, Herrick. *Anyone Can Build a Tub-Style Mechanical Chicken Plucker: Complete Instructions for the Kimball Whizbang.* Moravia, N.Y.: Whizbang Books, 2003.

Salatin, Joel. *Pastured Poultry Profit$: Net $25,000 in 6 Months on 20 Acres.* Swoope, Va.: Polyface, 1993.

Salatin, Joel. *Salad Bar Beef.* Swoope, Va.: Polyface, 1996.

Van Loon, Dirk. *Small-Scale Pig Raising.* Brattleboro, Vt.: Echo Point Books & Media, 2014.

Magazines

Farming Magazine. P.O.Box 85, Mt. Hope, OH 44600

Graze. P.O. Box 48, Belleville, WI 53508, 608-455-3311, www.grazeonline.com

The Stockman Grass Farmer. P.O. Box 2300, Ridgeland, MS 39158-9911, 601-853-1861, www.stockmangrassfarmer.com

Index

About the Authors

Franciscan Sisters T.O.R. of Penance of the Sorrowful Mother

Shawn and Beth Dougherty have been farming together for over thirty years, the last twenty in eastern Ohio on their home farm, the Sow's Ear, where they and their children raise grass, dairy and beef cows, sheep, pigs, and poultry. They identify intensive grass management as the point of union between good stewardship and good food; their ongoing goal is to rediscover the methods and means by which a small parcel of land carefully husbanded, with the application of ruminants, pigs, and poultry can be made to grow in fertility and resilience while feeding the animals and humans living on it.

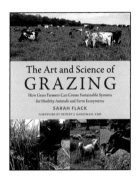

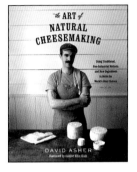

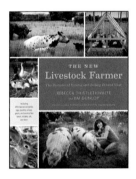

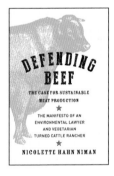

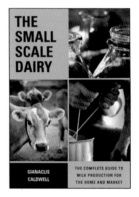

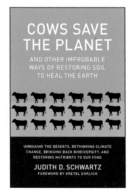